Ergonomic Solutions for the Process Industries

Ergonomic Solutions for the Process Industries

Dennis A. Attwood
Joseph M. Deeb
Mary E. Danz-Reece

AMSTERDAM • BOSTON • HEIDELBERG • LONDON
NEW YORK • OXFORD • PARIS • SAN DIEGO
SAN FRANCISCO • SINGAPORE • SYDNEY • TOKYO

Gulf Professional Publishing is an imprint of Elsevier

Gulf Professional Publishing is an imprint of Elsevier
200 Wheeler Road, Burlington, MA 01803, USA
Linacre House, Jordan Hill, Oxford OX2 8DP, UK

Recognizing the importance of preserving what has been written, Elsevier prints its books on acid-free paper whenever possible.

Library of Congress Cataloging-in-Publication Data
Application submitted.

British Library Cataloguing-in-Publication Data
A catalogue record for this book is available from the British Library.

ISBN: 0-7506-7704-X

For information on all Gulf Professional Publishing publications visit our website at www.gulfpp.com

03 04 05 06 07 08 10 9 8 7 6 5 4 3 2 1

Printed in the United States of America

Contents

Preface

Let's begin by understanding what the title means to users of this book. First, the distinction between *ergonomics* and *human factors* has been debated in every major national technical human factors and ergonomics society for decades. Our solution is simple—we make no distinction between human factors and ergonomics and we use the terms interchangeably throughout the book.

Second, the term *process industries* is intended to include not only the integrated oil and petrochemical industry, in which we work, but also industries where process technology is the kernel that controls the production of the product. The product may be electricity from nuclear or fossil fuel power plants, treated waste from municipal facilities, clean water from desalination operations, or manufactured products from virtually any continuous process.

Third, the term *solutions* was chosen intentionally. In the more than 60 years since the start of the Second World War, when ergonomics began, academic research in this area has mushroomed. Many great scholars, including Sanders and McCormick (1993), Grandjean (1988), Welford (1968), and Chapanis (1959), have written books on the general subject of human factors. In addition, books have been written on specific topics in the field, such as Konz (1979), van Cott and Kinkade (1972), and Attwood (1996). Each provides students and practitioners with the basic research that they can use to set hypotheses and develop tools. Our objective in this book is not to provide more academic information to ergonomics specialists but to focus this book on the nonspecialist users of human factors, to use the theory created by academics to develop simple tools and procedures that the nonspecialist can use to apply ergonomics inside the plant gates. To do so, it is necessary to provide some theory, but only enough to explain and justify the application.

With this in mind, it's important to understand our position on human factors and ergonomics. We believe that an educated, experienced plant practitioner can implement any of the human factors tools and processes contained within these pages. This does not imply that we believe human factors and ergonomics are mere "common sense." We believe that, at the plant level, operators and supervisors who are dedicated to making a difference, have the right tools, and have been trained to use them properly can identify the issues, set priorities, collect and analyze the data, develop interventions, and measure their effects. The specialists have their place, but their place is not performing the routine human factors duties that require local knowledge of the people and plant. In our opinion, the human factors/ergonomics specialist is a resource that should be used to develop programs, train the practitioners, and provide the detailed knowledge required to mobilize the plant staff.

So, we invite you to use the knowledge and the tools that are contained in this book. We hope that we have adequately explained how and why to use the years of human factors/ergonomics knowledge that is at the heart of this book in a deliberate, systematic way.

REFERENCES

Attwood, D. A. (1996) *The Office Relocation Sourcebook.* New York: John Wiley & Sons.

Chapanis, A. (1959) *Research Techniques in Human Engineering.* Baltimore: The Johns Hopkins Press.

Grandjean, E. (1988) *Fitting the Task to the Man.* London: Taylor and Francis.

Konz, S. (1979) *Work Design.* Columbus, OH: Grid Publishing Inc.

Sanders, M. S., and McCormick, E. J. (1993) *Human Factors in Engineering and Design.* Seventh Ed. New York: McGraw-Hill.

Van Cott, H. P., and Kinkade, R. G. (1972) *Human Engineering Guide to Equipment Design.* Washington, DC: American Institutes for Research.

Welford, A. T. (1968) *Fundamentals of Skill.* London: Methuen and Company Ltd.

Acknowledgments

Many people have contributed to this project. It is our pleasure to acknowledge, in alphabetical order, the following individuals who reviewed the manuscript and provided us valuable recommendations for revisions: John Alderman, Risk Reliability and Safety Engineering; John Bloodworth, ExxonMobil Chemical Americas; Larry Csengery, Shell Global Solutions; Dave Fennell, Imperial Oil Resources; Bernd Froehlich, ExxonMobil Chemical Central Europe; John Gelland, ExxonMobil Refining and Supply; Mike Henderek, ExxonMobil Corporate Safety Health and Environment; Dave Johnson, ExxonMobil Refining and Supply; Jere Noerager, Human Factors Consultant; Debby Rice, ExxonMobil International Medicine and Occupational Health; Tammy Smolar, ExxonMobil Biomedical Sciences, Inc.; Eric Swensen, ExxonMobil Biomedical Sciences, Inc.; Dan Taft, ExxonMobil Development Company; Evan Thayer, ExxonMobil Biomedical Sciences, Inc.; Theo van der Smeede, ExxonMobil Chemical Belgium.

We also extend our sincere thanks to the management of ExxonMobil Biomedical Sciences, Inc. (EMBSI) and ExxonMobil Research and Engineering (EMRE) for giving us permission to use much of the information contained in this book. We especially thank Patty Sparrell, EMBSI manager, and Al Lopez, EMRE vice president (retired), for their efforts on our behalf.

We dedicate this book to our families for their support and understanding during the its preparation.

Dennis A. Attwood thanks Pamela, Gordon, Cathy, and Sean
Joseph M. Deeb thanks Carol, Alexander, and Lauren
Mary E. Danz-Reece thanks Timothy, Eugene, Vivianne, and Eric

Disclaimer

The information in this publication represents the authors' own views about the application of human factors analysis in the process industry. The information consists of general guidance of facts, concepts, principles, and other information for developing and implementing ergonomics or human factors. The information in this book is furnished without warranty of any kind from the authors or the publisher.

The book is not intended to provide specific guidance for the operations of any company, facility, unit, process, business, system, or equipment; and neither the authors nor publisher assumes any liability for its use or misuse in any particular circumstances. Readers should make their own determination of the suitability or completeness of any material or procedure for a specific purpose and adopt such safety, health, and other precautions as may be necessary.

The information provided in this publication is no substitute for internal management systems and operating facilities standards that may include company-specific, facility-specific, or unit-specific operating and maintenance procedures, checklists, equipment descriptions, or safety practices. Users are advised that the information included here should not be applied if it contradicts government regulations governing their sites.

1

Introduction

1.1 INTRODUCTION

At 03:00 hours of his fifth consecutive night shift, a process control operator in a chemical plant received a "Group 1" alarm on the visual display monitor of the plant's distributed control system (DCS). He was not alert at this time of the morning. He recalled that, during the shift change meeting, a colleague reported that this alarm had gone off several times during the previous shift. Each time the problem was traced to a faulty transducer on a fin-fan. So, the operator acknowledged the alarm without checking further. If he had checked, he would have found that this time, the "group" alarm was notifying him that power to the DCS had been lost and the entire system was now on battery. In 4 hours, when the batteries discharged, the DCS failed and the control valves on the furnaces went fully open. The ensuing temperature increase burst the tubes in the furnace and started a major fire. Without power to the DCS, the control valves could not be closed. The damage was $30 million. Fortunately, nobody was hurt.

Human factors/ergonomics is defined as the systematic process of designing for human use through the application of our knowledge of human beings to the equipment they use, the environments in which they operate, the tasks they perform, and the management systems that guide the safe and efficient operations of refineries, chemical plants, upstream operations, and distribution terminals. Neglecting any of the elements, depicted in Figure 1-1, could lead to the failure of the entire system, not just the physical plant control system but the much broader system and structure under which it operates. In the hypothetical example, the design of the alarm system and the shift schedule did not consider the limitations and capabilities of the human operator, and the system failed. System failure could take many forms, including injury or the loss of property,

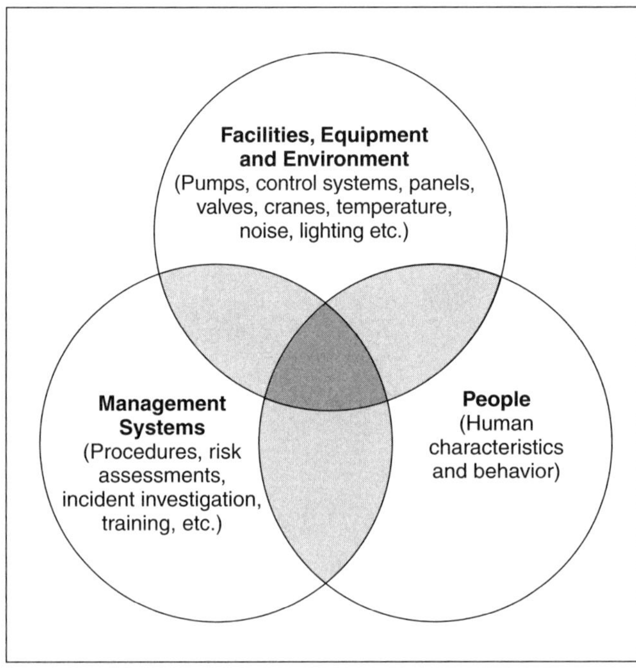

Figure 1-1 The human factors process.

hardware, or information. The key word in the definition is *systematic*. Clearly, human factors and ergonomics have been considered to some extent in the design of process plants for many years. But, the implementation of this technology has not been systematic in most instances. Human capabilities and limitations have typically been considered in new applications because the old design did not work properly or a user was injured because the equipment was too difficult to reach or operate. This trial-and-error approach to human factors can also cause the work system to fail and lead to retrofits that can be very expensive. The systematic application of ergonomics can also be termed a *right-the-first-time* approach.

The principles of human factors can be applied in any operation where humans interact with their working environment. To understand how humans interact with the systems they operate it may be useful to briefly review the classic model of the human/system interface shown in Figure 1-2.

A system, shown in the bottom box of the Figure 1-2, may take many forms. It may be as complicated as the process control console of a

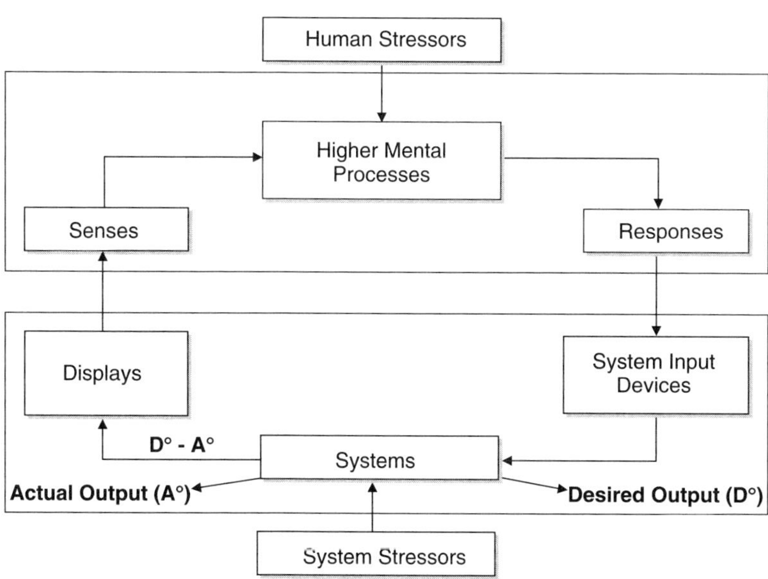

Figure 1-2 The human/system interface.

chemical plant or as simple as a toggle switch. Each system has external inputs, performance criteria, and external outputs. The state of the system can be displayed to the operator in many ways; for example, by using visual displays such as gauges, dials, lights, switch position, or printed text; using auditory displays such as horns, bells, or the human voice; or even using tactile displays such as vibrating pagers. The information received by the human's senses is processed and compared with stored experiences and a response is initiated. The ability of the human operator to process the information may be impaired by factors such as fatigue, over-the-counter medications, or stress. The operator's response may involve speaking, touching a keypad, or turning a valve. This, in turn, affects the system. The system changes its state and the changes again are displayed to the human. In continuous operations, such as driving a fork-lift, this entire process may be repeated many times a second.

This book is about applying human factors/ergonomics technology in the process industry. In this book, *human factors* and *ergonomics* are synonymous and the terms used interchangeably. The book deals with the analysis of the ergonomic issues in current plants and the development and implementation of proposed solutions. Issues range over the entire spectrum of ergonomics and include facility and equipment design, the working environment, the design of procedures and permitting systems, process control buildings and facilities, and safety management systems. The book also deals with the design of new facilities and equipment.

The book is targeted to on-site safety and health professionals, safety engineers, and project personnel from both process companies and consulting design organizations. It can best be described as a how-to book, a guide to getting it right the first time, providing useful tools to simplify the application of ergonomics for those without formal training in the technology. For this reason, the book is also intended to be used in universities as an introductory text for engineers specializing in process design and operation.

Ergonomic legislation worldwide requires companies to reevaluate the design of their facilities and equipment and implement systems that reduce the number and severity of injuries to plant personnel. This book provides companies with the information that makes it easier and cheaper to implement effective programs.

The standard for presenting each ergonomics topic is to

- Provide limited theory to outline the human/system interface issues.
- Present the "tools" developed to address these issues.
- Demonstrate how the tools are used.
- Provide a case study to show the reader how to apply them.

At the end of each chapter, we provide review questions so you can test your knowledge of the information in that chapter.

The following paragraphs outline the contents of the book.

1.2 CHAPTER REVIEW

1.2.1 Chapter 2. Personal Factors

This chapter introduces the capabilities and limitations of workers that affect their performance in the process workplace. Topics include

- Sensory capabilities, mainly vision and hearing, and how they vary with age.
- Cognitive capabilities, including attention, perception, memory, and decision making.
- Physical capabilities, including body size, muscular strength, and endurance and how they vary among different major groups (nationality, gender, age).

1.2.2 Chapter 3. Physical Factors

The chapter highlights two major physical activity factors, manual handling and cumulative trauma disorders.

1.2.2.1 Manual Handling

The reader is introduced to the notion that the body is limited in the amount of force it can apply and continually exceeding that force can be injurious. *Manual handling* is defined in terms of the types of tasks performed in operations and maintenance, from the basics of lifting through to the more complex tasks involving unusual or dynamic body positioning. Finally, methods of assessing the safety of manual handling tasks is described with examples and computer-based tools.

1.2.2.2 Cumulative Trauma Disorders

This section focuses on the effects of repeated soft tissue injury on the ability of workers to perform over long periods of time. The section identifies those critical risk factors that determine how people get hurt and provides methods to evaluate a job for risk factors and modify the job to reduce the potential for injury.

The tools used to assess and modify tasks with the potential for acute or cumulative injury are presented. A case study shows how these tools can be used to improve the work system.

1.2.3 Chapter 4. Environmental Factors

This chapter discusses the effects of four major environmental variables on performance: lighting, noise, vibration, and temperature (hot and cold). It reviews the limitations of the human operator to work where the environmental stressors are at high levels. It also provides recommendations for mitigating against the effects of environmental stressors on performance.

1.2.4 Chapter 5. Equipment Design

The chapter is divided into three major sections: controls, displays, and field control panels.

1.2.5 Chapter 6. Workplace Design

This chapter has three objectives:

1. To identify the principles involved in the design, installation, operation, and maintenance of workplaces. A list of the literature that supports each of these principles is provided.
2. To review the techniques in place to analyze work situations that determine workplace and workstation design.
3. To provide guidance (models) on the evaluation and redesign of existing and new (grassroots) workstations. A case study is presented on the use of these models in the design of a control room.

1.2.6 Chapter 7. Job Factors

This chapter discusses the nonequipment factors in a job that can affect performance. The following paragraphs summarize the topics included in this chapter.

1.2.6.1 Work Schedules: Fatigue and Rotating Shifts

This section examines the effects of work schedules on the performance of human operators. It provides guidance on the alleviation of fatigue through the use of coping strategies implemented by the shift worker, the company, and the family.

1.2.6.2 Stress

A certain amount of stress is required by the body to maximize performance. Too much or too little stress can affect performance and health. This chapter talks about sources and causes of stress and the strategies designed to help cope with it.

1.2.6.3 Job Analysis

This section describes the use of task analyses and the methodology used to identify and rank critical tasks.

1.2.6.4 Team Processes

This section describes the growing use of team-based processes to improve safety performance and the methods that can be used to create high-performing teams in process plants. The analyses used to create high performing teams include

1. The cognitive problem-solving style (Kirton Adaptive-Innovative Survey).
2. Drexler-Sibbet High-Performance Team Model.
3. ACUMEN.
4. SYMLOG (the systematic multilevel observation of groups).

1.2.6.5 Behavior-Based Safety

Behavior refers to the acts or actions by individuals that can be observed by others. It is what a person does or says not what he or she thinks, feels, or believes. Behavior-based safety (BBS) programs are based on the reinforcement of safe acts or actions and the elimination of at-risk acts or actions. This chapter reviews the principles on which behavior-based safety programs are founded, the techniques used to identify at-risk behavior, and the strategies used to create safe behavior. It also reviews some of the most popular commercially available behavior-based programs and provides a gap analysis that can be used at a site to determine whether it would benefit from the implementation of a BBS program.

1.2.7 Chapter 8. Information Processing

1.2.7.1 Human Error Theory and Methodology

The reader is provided with limited theory on the principles of human error and the contribution of error to major accidents in the process industry. The theory is used to derive principles for recognizing and dealing with the causes of error in process operation.

1.2.7.2 Plant Signs and Labels

Plant signs and labels help plant personnel identify equipment and caution operators on its use. So, signs and labels serve three critical purposes in plant operations. First, they ensure that the references to equipment provided in procedures and on process and instrumentation drawings (P&IDs) are the same as on the equipment. Second, labels and signs provide critical learning information for new operators unfamiliar with the process plant. Third, they provide prompts and cues for experienced operators on critical operating information. Consequently, it is critical that the labels and signs are accurate, complete, consistent, meaningful, and legible.

1.2.7.3 Procedures

Procedures are a core part of every process operation. They ensure jobs are performed in a consistent manner. Procedures are also a major aid for training new workers. This section provides information on how to determine whether a procedure is required and how to develop and analyze procedures from a human factors viewpoint.

1.2.7.4 Training

Training is a major factor in preparing a new process employee for his or her job and maintaining the skill level of the process plant. The objectives of every training program are to maximize skills and knowledge while minimizing the time required to learn them. This section examines the human factors requirements of a training program and illustrates how to determine whether the training program is effective.

1.2.7.5 Mental Workload

Humans operate best when they are neither too busy nor too bored. Today, the efficiencies that each process company is trying to achieve means that process operators are likely more busy than bored and, in some cases, too busy.

This section provides the reader the tools necessary to assess operator workload and identify those activities in an operator's job description causing overload and provides methods for balancing the "boring" and "busy" cycles that may occur in a single job.

1.2.8 Chapter 9. The Use of Human Factors in Project Planning, Design, and Execution

Each phase in the design and execution of process facilities requires consideration of the human/system interface. This chapter is about the inclusion of human factors in local and major projects. It provides a model that shows how to manage human factors in projects; when to apply human factors tools during the planning, design, and implementation of the project; and the tools available. The distinction between the way major capital projects are conducted and the way local projects are conducted profoundly influences the way human factors is considered in each type of project and the design of the tools used. It is important for the reader to remember that not all "projects" are created equal.

Finally, the chapter provides the reader a description of the tools developed to assist process and mechanical engineers to analyze the requirements of their particular project and specify the best design.

1.3 PROPOSED MODEL FOR THE SYSTEMATIC IMPLEMENTATION OF ERGONOMICS/HUMAN FACTORS

This section introduces a model that guides the systematic process of designing for human use. The purpose of the model is to provide a systematic and deliberate method for solving human factors issues in the most cost-effective way. It describes how

a. Issues are identified and priorities set.

b. Analyses are conducted.

c. Solutions are implemented.

d. Results are measured and follow-up actions taken.

e. Systems are affected

The model is based on the flowchart in Figure 1-3.

1.3.1 Develop or Adopt Standards

Standards determine the performance boundaries for the process plant. They set minimum safety performance, specify job requirements, and

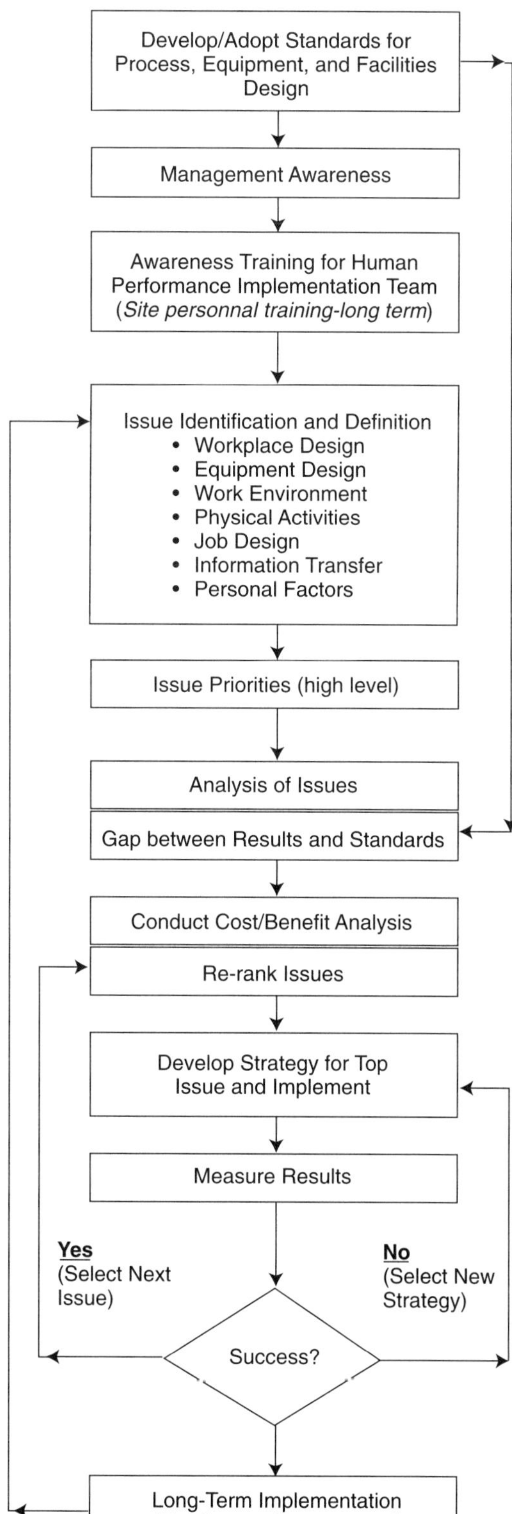

Figure 1-3 Model for human performance assessment and intervention.

ensure that equipment and facilities meet minimum engineering design requirements. They provide a benchmark for measurement. They are used to set priorities for issues. They help guide the selection of strategies and measure the effectiveness of the implementation. For this reason, the model begins with the development or adoption of standards and ends with performance measured against standards.

Standards may be developed by companies (in-house standards), industry associations, international organizations, or legislation. For example, in-house standards or guidelines may be developed for the design of plant equipment and facilities. Standards may also be developed for the design of offices. Style guides may be created to guide the design of procedures to ensure that each is formatted the same way. Industry technical organizations, such as the American Petroleum Institute (API) or the Canadian Chemical Producers Society (CCPS), develop guidelines for member companies to adapt to their operations. Industry standards serve several purposes. For one thing, it is more efficient for member companies to pay for part of the development of a set of guidelines than for each company to develop its own. Moreover, industry standards are generally consensus documents. They contain the best information from each of the contributing companies. So, the standards should be as good or better than those developed by any one member company. Finally, they provide consistency for regulators. Regulators who are comfortable with the standards tend to spend less time auditing the companies that adopt them.

Most regulatory agencies have developed minimum performance and design standards to which the process industries are required to adhere. Examples include

- OSHA Regulation (Standards-29CFR) Process safety management of highly hazardous chemicals; 1910.119.
- California Code of Regulations, Title 8, Section 5110, Repetitive Motion Injuries, Article 106—Ergonomics.
- Washington State, WAC 296-62-051 Ergonomics.
- Contra Costa County, California, Ordinance 98-48, Section 450-8.016 (B) Requirements to develop a written human factors (HF) program.

- European Union Council Directive 90/270/EEC "Work with Display Screen Equipment," May 29, 1990.
- British Columbia, Canada, OHS Regulation Part 4, General Conditions, Ergonomics, 1998.

International bodies have developed standards that cut across country boundaries. Examples include

- ISO/DIS6385: 2002—Ergonomic principles in the design of work systems.
- International Standards Organization (ISO) Ergonomic requirements for office work with visual display terminals, 1992.
- ISO 2631-1 (1997) Mechanical vibration and shock—Evaluation of human exposure to whole-body vibration.

1.3.2 Management Awareness Sessions

We all know that no program succeeds without full management support. And, management will not support something that it does not understand. With this is mind, it is important that management understand the science of human factors, how it can benefit the plant in both the short- and long-term, and how it will be implemented.

In preparation for management awareness sessions, information about the status of human factors in the plant should be collected.

Existing worksite data should be reviewed and summarized. Available data could include

- Absenteeism.
- Workers Compensation costs.
- Occupational illness statistics.
- Injury statistics.
- Turnover.
- Previously identified human factors issues.

- Results from incident investigations.
- Results from risk assessments.

Stakeholders should be interviewed to determine

- What they think the issues are.
- What they want the program to achieve.
- Any resistance to the initiative and why.
- What they believe is their role in the program.
- How they feel they can participate.

Management sessions should be short, graphic, and to the point. They should comprise at least the following information:

1. Review of worksite data (from above).
2. Stakeholder opinions (from above).
3. Description of the program.
    - Proposed program activities, such as training, issue development (observations, incident analysis, risk assessment), issue priorities, analyses, strategy development.
    - Program schedule that maps activities against a timeline.
    - Standards of performance used.
    - How performance is measured and stewarded.
4. What benefits are expected.
5. What costs are anticipated, both level of effort from worksite personnel and costs for equipment and outside support.
6. Long-term benefits.
7. Roles of management and worksite personnel.

1.3.3 Educate Site Personnel

All site personnel should be educated about human factors (HF). The length of the training and the amount of theory and practice the training contains depend on the roles and responsibilities assigned the attendees.

Technical staff, for example, require a different human factors focus than worksite operations personnel. Safety committee members need additional training time and to become more proficient with ergonomic methods than general worksite staff. Safety specialists, in addition, may have the responsibility to conduct training and awareness sessions with all worksite personnel, to make them aware of human factors/ergonomics and the program being launched for the site. With this is mind, a detailed training awareness session should be tailored to the roles and responsibilities of the worksite personnel. Table 1-1 provides a "training requirements model," which assigns training courses to key positions by job category. These courses may have to be developed by an outside HF specialist for use by internal trainers. We strongly recommend that each site develop a matrix similar to Table 1-1.

The outcomes of HF training programs include

1. *Awareness.* Participants are aware of the scope of human factors and the capabilities and limitations of workers.
2. *Knowledge and skills.* Participants gain sufficient knowledge and hands-on skills to use the tools and techniques to identify and analyze human factors issues.
3. *Sharing.* Training sessions provide an opportunity for the participants to discuss worksite human factors issues.
4. *Worksite issues identified.* Training can be an opportunity to identify worksite issues. If the training is graphic and contains generic examples, most attendees can make the mental leap from the training examples to their own worksite issues. If their issues can be captured and given priority, they have a good start to developing an issue identification list. Based on our experience, the strategy of teaching by examples and case studies has been successful in encouraging worksite personnel to accept human factors.

1.3.4 Identify Issues

The key to an effective human factors program is the identification of potential worksite issues. A number of processes have been developed to identify human factors:

TABLE 1-1

Recommended Human Factors Training Modules by Key Positions

Key Positions	*HF Awareness (3 hr)*	*Behavior-Based Safety (4 hr)*	*HF in Design (8 hr)*	*Human Factors for Site HF Teams (8 hr)*	*Human Factors for Site Workers (7 × 1-hr modules)*	*e-Training Module on Human Factors (2 hr)*	*Musculo-skeletal Injury Prevention (4 hr)*	*Office Ergonomics Awareness for Office Workers (1 hr)*	*Office Ergonomics for Practitioners (4 hr)*	*HF in Risk Assessment (4 hr)*	*Ergonomics for Drivers (3 × 1-hr modules)*	*HF Refresher Training (½-day every 3–5 years)*
Sr. Managers	✓											
Operations/ Plant Managers		✓		✓			✓					
Technical Managers		✓		✓								
Design Engineers		✓	✓	✓						✓		
Process Control/ Application Engineers			✓	✓								
Project Engineers/ Managers	✓	✓	✓							✓		

Instrument Technicians	✓		✓								
Safety/Health Advisors		✓	✓	✓		✓	✓	✓	✓	✓	
Risk Assessment Specialists		✓	✓	✓						✓	
Designated Change Authorities	✓									✓	
Team Leaders/ Supervisors (1st & 2nd Line)		✓		✓		✓	✓				✓
HF Site Implementation Team		✓		✓		✓	✓				✓
Operations/ Mechanical Employees		✓			✓	✓					
Office Employees/ Managers	✓							✓			
Office Ergonomics Practitioners									✓		

- Team training, as already described.
- Observation of behavior.
- Musculoskeletal risk assessments.
- Task analyses.
- Office ergonomics assessments.
- Incident analyses.
- Risk analyses, such as job safety analyses or last minute risk assessments.
- Employee interviews.
- Site assessment by HF specialists.
- Site safety inspections by safety specialists.

Each issue should be captured in a simple database that categorizes it by plant location or human factors/ergonomics focus. Each issue should also be identified by date of submission and source.

It is important that the processes used to identify human factors issues record every issue proposed and that each participant feels free to propose any issue that he or she wants. The success of this program depends on encouraging site personnel to identify as many issues as possible, no matter how insignificant they may initially appear. You will find that the process of proposing issues can be valuable. In addition, the process of setting priorities screens the issues and ensures that only the most important ones are considered.

1.3.5 Setting Priorities

No company can afford to spend money and commit resources on issues that have little benefit to the operation. So, it is essential to set priorities on the list of issues generated and continue to work only on those that have an impact on plant metrics, such as safety, health, or productivity.

Before continuing, it is important to note that every issue has an owner. So, every issue is important to someone in the plant. People who contribute to the identification process expect that their issue(s) will be resolved. So, they must be told whether their issues are going forward or not. And, if they are not going forward, why not?

There are many ways to conduct an initial screening of potential issues. Each method typically revolves around cost and benefit or risk. We find that worksite personnel can easily categorize issues into the consequence/probability risk matrix shown in Table 1-2.

Column headings specify the probability that an issue will occur. The high level, for example, could be one or more occurrences per year. The low level would then be less than one occurrence per year. Row headings specify the consequences if an issue occurs. Consequences can be expressed in several ways, as shown in Table 1-2. The consequence categories can include safety and health, profit and loss, the effect on the public, and the effect on the environment.

The highest priority issues are those having a high probability of occurring and a high consequence when they occur. A simple process for identifying the most important issues is explained on page 21.

TABLE 1-2

A Simplified Risk Matrix for Ranking Human Factors Issues

		Consequence Categories				*Probability*	
		Safety	*Financial*	*Public Disruption*	*Environmental*	*High >1 time per year*	*Low <1 time per year*
Consequence	HIGH	LTI MA RWC	>$100k	>1-DAY	Fine/ Notification		
Consequence	LOW	FA	<$100k	<1-DAY	No fine or Notification		

Where:

LTI = Lost time incident

MA = Medical aid incident

RWC = Restricted work case incident

FA = First aid incident

1.3.6 Analyze the Issues and Assess the Gap against Standards

In this step, the top priority issues identified are analyzed. The results are compared against the performance standards established in the first step of the model.

The type of analysis selected depends on the issue topic. For example, if the issue is musculoskeletal (e.g., lifting, reach, posture), the analytical tools could include a task analysis followed by one of these:

1. National Institute of Occupational Safety and Health (NIOSH) lifting analysis or equations.
2. University of Michigan, three-dimensional analysis tool or checklists.
3. Lumbar motion monitor (Marras et al., 1993).
4. Cumulative trauma disorder (CTD) analyses such as rapid upper limb assessment (RULA) (McAtamney and Corlett, 1993).

Each analysis provides a result that can be compared against previously established standards.

If, for example, the issue is the inadequate design of a visual display, standards have been established to cover all aspects of display design, including character size (IPC, 1974), contrast (Sanders and McCormick, 1993) and lighting (Grandjean, 1988). Various analytical techniques are identified and explained in the chapters to follow.

If an analysis indicates that the issue is in compliance with accepted standards, then it is noted, and the "owner" is notified. If the analysis indicates that the issue is not in compliance with accepted standards, it enters the next phase of screening and cost/benefit analysis.

1.3.7 Conduct Cost-Benefit Analysis

Every solution to an issue has a cost: capital, personnel time, schedule disruption, public criticism. And every solution has a benefit: standards compliance, profit, risk reduction, safety improvement, public endorsement, productivity improvement. This step estimates the costs and the benefits associated with solutions to each issue. We find that, on a high level,

TABLE 1-3

Simplified Cost/Benefit Matrix

	Cost	
Benefit	*High*	*Low*
High		
Low		

worksite personnel can make the estimates for costs and benefits using the simple 2 × 2, *cost/benefit matrix* shown in Table 1-3. The process for selecting the most cost-effective solutions is explained in the following box.

The "DOT" Process for Prioritizing Issues

In sections 1.3.5 and 1.3.7 of this chapter, the systematic model required that a list of many issues be reduced to a small set of high priority issues by using either a simple 2 × 2 risk decision matrix, or a 2 × 2 cost/benefit decision matrix. The procedure explained below, is a simple way to identify the top priority issues. We term this procedure the "DOT" process.

The table on page 22 shows a 2 × 2 matrix that is identical to Table 1-2 in the text. The set of columns is labeled PROBABILITY and the set of rows CONSEQUENCE. (If "Cost/benefit" was being determined, the columns would be labeled COST and the rows would be labeled BENEFIT as in Table 1-3).

One of the columns is labeled HIGH, the other LOW. Similarly, one of the rows is labeled HIGH, the other one LOW.

Let us assume that the objective of the process is to identify those issues that have the *highest risk*, i.e., the highest probability of occurring and the highest consequence when they occurred; in other words the issues that fall into the cell "HIGH Probability" and "HIGH Consequence." We provide each participant in the procedure with a set of dots. (For this example, we'll use GREEN, but the color doesn't

Continued

Risk Matrix for Ranking Human Factors Issues

		Consequence Categories				*Probability*	
		Safety	*Financial*	*Public Disruption*	*Environmental*	*High >1 time per year*	*Low <1 time per year*
Consequence	HIGH	LTI MA RWC	>$100k	>1-DAY	Fine/ Notification		
	LOW	FA	<$100k	<1-DAY	No fine or Notification		

Where:

LTI = Lost time incident
MA = Medical aid incident
RWC = Restricted work case incident
FA = First aid incident

matter.) The number of GREEN dots given to each participant is normally about 1/3 of the number of issues being evaluated. Participants are instructed as follows:

1. Familiarize yourself with each issue
2. Identify issues that are duplicates or very similar and combine them into one issue. (you don't want to be splitting votes between similar issues)
3. Decide "Which issues, if resolved, would have the highest *probability* of occurrence?"
4. Place one GREEN dot on each issue that he or she feels has the highest probability of occurrence

At the end of this process, some issues will be covered by many dots, some by not so many, and some by no dots at all.

The second step is to supply each participant with another set of dots having a different color from the first set. In this example, our second set will be RED. The number of dots provided to each participant is 1/3 of the number of issues *that have at least one GREEN dot on them from step 1.* Their task is as follows:

5. Identify those issues that, if they occurred, would have the highest consequence, where consequence might include injury, loss, public disruption, etc. as explained in the text.

At the end of step 5, some issues will be covered with many dots of each color; some with only GREEN dots and some will have no dots at all. The final step is to prioritize the issues.

In most instances, the top issues are obvious by looking at the number of GREEN and RED dots covering them. But, if the top issues are not clearly defined, participants can vote with a set of ten dots numbered from "1" to "10" as follows.

6. Each member is instructed to rank the issues that are covered with at least one GREEN **and** one RED dot from highest priority (#10) to the lowest priority (#1). The participant does this by placing a numbered dot on each of issues he or she chooses.
7. The numbered dots on each issue are arithmetically totaled and the issue with the highest total is the one that enters the next step in the model in Section 1-2-4.

The above process can be used for both "Risk Assessment" and "Cost/Benefit" decisions.

1.3.8 Set Priorities on Issues

As part of the process for assigning costs and benefits, the highest priority issues are also identified.

1.3.9 Develop and Implement Strategy for Top Issue

We find that there is no one optimal intervention strategy for any issue. Alternate strategies may be available to accomplish the same result or several may be applied concurrently to improve the solution.

Human factors intervention strategies can be grouped by category as follows:

1. *Engineering/interface design.* Changes or additions are made to the design of the physical workspace, equipment, or environment to improve worker performance.

2. *Training.* To improve the knowledge and skill of the worker. Training could be given both on and off the job. On-the-job training could also include "work hardening," a process whereby the worker gradually becomes adapted to the physical and mental stresses of the task.

3. *Job or task design.* The task conditions are modified. Modifications might include
   - Changing the activity sequence in a task.
   - Improving procedures or permitting systems.
   - Reassigning task steps to different work posts.
   - Modifying work or shift schedules.

4. *Selection.* In some limited circumstances, a rational intervention strategy might consist of choosing people with capabilities that match the job. We would not ask a physician to perform an engineering analysis or ask a color-deficient person to install color-coded cables. "Selection" must be performed responsibly and within the laws that many jurisdictions have enacted to prevent discrimination.

5. *Behavior. Behavior* is defined as "what people do or say" (Geller, 1996). Behavior is observable and measurable. It can be affected by interface design, job design and training, or strategies specifically designed to change behavior. These can include
   - Creating positive consequences for desired behavior.
   - Removing negative consequences for desired behavior (make it easier).
   - Influencing risk perception among workers.
   - Creating negative consequences for undesired behaviors.
   - Providing coaching and corrective feedback for undesired behaviors.

In the chapters that follow, detailed examples of potential interventions are proposed for each of these categories.

Burke (1992) developed an "intervention discovery" guide for musculoskeletal issues. The guide consists of a series of 38 worksheets that walk the user from the initial issue through to a potential solution. Similar "smart" guides could be developed for other types of issues.

1.3.10 Measure Results

If you cannot measure the effects of an intervention, you cannot know whether performance has improved. We find that the effects of any intervention can be measured. The measures may be quantitative, such as total recordable injury rates (TRIR), or qualitative, such as results of an employee survey both before and after the intervention strategy was implemented. In the chapters to follow, example performance measures are developed for use in this process.

The measurement selected should tell us whether the intervention not only improved performance but by how much. If performance improves to the levels required by established standards, the intervention was a success and the next issue in the priority queue can be selected for action. If, however, performance does not improve to a satisfactory level, the strategy may have to be modified and reimplemented.

1.3.11 Improve Management Systems

Major process mishaps that occurred in the 1980s spawned an initiative in the1990s to develop and implement management systems whose purpose is to improve the safety of the process industry (CCPS, 1993). Many of the base management systems deal with and are influenced by the human factors topics that are covered here. For human factors to be successful over the long term, the results from implementing strategies must be used to continuously improve each management system by integrating human factors into the systems.

REVIEW QUESTIONS

Test your understanding of the material in this chapter.

1. Identify an ergonomic initiative within your company, then list the major standards that determine your initiative's success. Categorize the standards by group:
   Internal standards.
   Industry standards.
   Technical association standards.
   International standards.

2. Explain how your safety management systems will change as a result of success with your initiative.

3. Identify the risk management process that you will use to determine which ergonomic issues that you identify will go forward for analysis.

4. Briefly discuss how you would measure the effectiveness of a human factors training program.

5. Standards are used for the following reasons (True or False):
   a. To conduct a gap analysis of the results and issues analysis.
   b. To obtain funding for an ergonomics program.
   c. To provide a basis for measuring the results of strategy implementation.
   d. To ensure that the cost/benefit analyses are valid.

6. Which of the following items are desired outcomes of a human factors training program?
   a. Behavioral observations.
   b. Identification of worksite issues.
   c. Complete task analyses.
   d. Awareness of the capabilities and limitations of workers.

7. Which should be conducted first at your site, management awareness or site personnel training?

8. Discuss briefly how you would prioritize the potential human factors issues identified at your site.

9. Discuss briefly how you would conduct a cost/benefit analysis for the issues identified in a process control system analysis.

10. What, in your opinion, would be a valid intervention strategy for a valve that was found to be too high to reach and operate?

REFERENCES

Burke, M. (1992) *Applied Ergonomics Handbook.* Boca Raton, FL: Lewis.

CCPS. (1993) *Guidelines for Auditing Process Safety Management Systems.* New York: Center for Chemical Process Safety of the American Institute of Chemical Engineers.

Geller, E. S. (1996) *Working Safe.* Radnor PA: Chilton Book Company.

Grandjean, E. (1988) *Fitting the Task to the Man.* London: Taylor and Francis.

IPC. (1974) *Applied Ergonomics Handbook.* Guildford, Surrey: IPC Science and Technology Press.

Marras, W.S., Lavender, S.A., Leurgans, S.E., Rajulu, S.L., Allread, W.G., and Ferguson, S.A. (1993) "The role of dynamic three-dimensional trunk motion in occupationally-related low back disorders." *Spine*, 18(5), 617–628.

McAtamney, L., and Corlett, N. (1993) "RULA: A survey method for the investigation of work-related upper limb disorders." *Applied Ergonomics*, 24(2), 91–99.

Pheasant, S. (1996) *Bodyspace: Anthropometry Ergonomics and the Design of Work.* Second Ed. Philadelphia, Taylor and Francis.

Sanders, M. S., and McCormick, E. J. (1993) *Human Factors in Engineering and Design.* Seventh Ed. New York: McGraw-Hill.

2

Personal Factors

2.1 INTRODUCTION

In the human/system interface (HSI) model (Figure 1-2), the operator continuously receives information from the environment and processes it. There may or may not be follow-up action. If a response is required, then the appropriate information is transmitted back to the system in the form of manual manipulation of controls (e.g., valves) and buttons, lifting bales of rubber, climbing ladders and stairs, and so forth. This, in turn, changes the status of the system. The operator receives new information from the displays about the status of the system and, again, may or may not take action to bring the system to the desired level or maintain its status. Furthermore, the information the operator receives does not comc only from displays of machines. As an example, when using a hand tool such as a saw, the external status of the line to be cut is perceived by the operator and continuous corrections are made to end up with a straight cut. Thus, the word *display* has a more general meaning than just a temperature or pressure gauge.

This HSI model functions as a closed-loop system. This means that, as long as we interact with something (i.e., machine, tools, other humans, or even animals), we continuously receive information, process information, look for status changes, receive new information, and so on. The first set of limitations that can affect the efficient working of this HSI closed-loop system is the sensory, perceptual, and cognitive capabilities of the operator. The second set of limitations is the physical capabilities that allow the body to move in the desired way. This chapter discusses both sets of limitations.

At the end of this chapter is a case study related to examining human force capabilities when interacting with valves at different positions. In addition, a set of review questions are included to help check your understanding of the material covered in this chapter.

2.2 SENSORY AND COGNITIVE CAPABILITIES

Among the human sensory mechanisms are the five senses (vision, hearing, touch, smell, and taste). These senses deal with stimuli external to the body and are referred to as the *exteroceptors*. In addition, we have senses that deal with the position of our body in space and assist us in jumping, climbing stairs and ladders, reaching, and the like. These are referred to as *proprioceptors* and are embedded in the muscles, tendons, joints, and inner ear. This section covers the exteroceptors, specifically vision and hearing, for two reasons:

1. The majority of information we need to perform tasks is received through the visual and hearing senses, such as written procedures, warning labels and signs, control position, dials, warning lights, and alarms.
2. The majority of human factors research has been conducted in the visual and auditory senses.

2.2.1 Visual Sense

Vision has been the most studied and researched sense. In addition, given current technologies, vision is probably the most overloaded sense at work. Our visual system receives energy from the environment around us and converts the energy into something meaningful for us to understand. Our visual system works on two levels: physiological, includes the eyes and brain, and psychological, involves both the immediate visual sensation (energy received) and our interpretation of it based on our knowledge and experience (the perception process).

In general, we receive visual information either directly (direct observation of a coworker similar to the observation process of a behavior-based safety program) or indirectly (seeing things through a computer, television, radar, or the like).

The most important part of the eye is the retina, which covers the back of the eye. It converts energy from the environment (i.e., light) into electrical energy and passes it to the brain through optic nerves. The retina is made up mostly of cones and rods (so-called because of their shape). Cones function at high illumination levels and are responsible for daylight vision. Cones are also important in visual acuity (ability of the eye to see details) and differentiate extremely well among colors. Rods, on the other hand, respond to low illumination levels, such as at night. Rods also differentiate between shades of black and white.

Because of the characteristics of the cones and rods, the eyes function well over a wide range of illumination levels. However, several factors determine visual capabilities and effectiveness, and have important implications in designing visual displays. These factors are discussed next.

2.2.1.1 Accommodation of the Eye

Accommodation refers to changing the shape of the lens in the eye to properly focus the image on the retina. In other words, accommodation is the ability of the eye to focus on objects at different distances. For example, the ability to read signs on the side of the road is an example of far vision. The ability of a control room operator to monitor displays is an example of near vision.

In normal accommodation, the lens of the eye flattens to focus a far object on the retina and bulges to focus a near object. The eyes may not accommodate an object for two reasons:

- *Nearsightedness.* Individuals cannot focus on far objects. The lens tends to remain in a bulged condition and the light from the far object is focused in front of the retina. Nearsightedness is also referred to as *myopia.*
- *Farsightedness.* Individuals cannot focus on near objects. The lens tends to remain in a flattened condition and the light from the near object is focused in the back of the retina.

With aging, there is a reduction in the ability of the eyes to accommodate. This is related to the gradual loss of the elasticity of the lens. This reduction is measured by the closest distance (near point) at which the eyes can focus. The table below lists average near point distances against age.

Age	*Near Point in cm (in.)*
10	7 (2.8)
15	8 (3.2)
30	11 (4.4)
45	25 (10)
50	50 (20)
60	85 (34)
70	100 (40)

2.2.1.2 Visual Field

Visual field is defined as "what can be seen when head and eyes are kept fixed (motionless)." We divide the visual field roughly into three areas (Grandjean, 1988):

1. *Optimal field* of view with an angle of view of 1° where objects are sharp and seen clearly with details.
2. *Middle field* with a visual angle ranging between 1° and 40°, where objects are not sharp; however, movements and strong contrasts are noticed.

3. *Outer field* with a visual angle ranging between 40° and 70° where objects are not sharp or clear. Objects must move to be noticeable.

It is important that the objects we need to see clearly are located in the optimal field area. This subject is discussed in detail in Chapter 5, "Equipment Design."

2.2.1.3 Adaptation Process

Adaptation is the process or ability of the eyes to adjust to different light levels. We need time to adapt to darkness or brightness before we can discriminate details. For example, when we first move from bright sunlight to a dark movie theater, we have difficulty finding our way around. After a while, we begin to see distinct objects and faces. This process is called *dark adaptation*. Dark adaptation is the process of transitioning from cone to rod vision. The process can take from 30 minutes to an hour depending on the initial illumination levels. Conversely, moving from darkness to bright sunlight, we experience high glare and find it difficult to see details. The process of transitioning from dark to light is called *light adaptation*. Light adaptation is the process of transitioning from rod to cone vision. Light adaptation is completed relatively fast, a minute or two. The process of adaptation has practical work implications:

1. Allows sufficient time for the eyes to adapt when performing tasks at low levels of lighting.
2. Avoids excessive changes in illumination levels, since the constant need for adaptation results in visual fatigue.
3. Avoids excessive differences in brightness in all important surfaces within the visual filed.

2.2.1.4 Color Vision

Color, in the workplace, is used to help the operator to distinguish objects in the visual field. For example, the green-colored areas on a temperature display indicate acceptable levels, yellow means caution, and red means unacceptably high levels. It is important to point out here that the eyes do not have the same sensitivity to color under different lighting conditions.

When it is bright, the eyes are more sensitive to red and less sensitive to blue. The reverse occurs when it is dark. Therefore, we need to consider the lighting conditions under which the colored-coded temperature display is viewed. More details on lighting are provided in Chapter 4.

People with normal color vision are capable of distinguishing among hundreds of different colors. Complete lack of color vision (color blindness) is extremely rare.

Color deficiency, such as the inability to discriminate between red and green is more common. Usually, these colors are seen as poorly saturated yellows and browns. Overall, about 8% of the male population and less than 1% of the female population have this form of color deficiency. Therefore, color should always be used as a redundant (additional) method of coding, and not the primary method.

2.2.1.5 Visual Acuity

Acuity refers to sharpness of vision. Visual acuity is the ability to see fine details; for example, detecting differences in the position of two controls, recognizing the presence of an object in the visual field, inspecting a tank for cracks, or even reading the small details on a bottle of a prescription drug. A small distinction between visibility and visual acuity is warranted at this point. The ability the see the pressure gauge on a field display panel refers to visibility, whereas the ability to identify the numbers on the gauge refers to visual acuity.

Visual acuity depends largely on the individual's visual capabilities, such as accommodation of the eyes, visual field, the process of adaptation, color vision, and age (discussed in Section 2.2.1.6) and situational factors. Situational factors include:

- *Size of the object.* As the size of the object increases, so does our ability to see the object (Figure 2-1). However, this depends on the viewing distance. The further away we are from the object, the less is our visual acuity to see details. For example, we can see many more ground details when sitting in an airplane at an altitude of 2000 feet than at 35,000 feet. Similar logic can be applied in the work environment. If we need to design a sign in the plant or a visual display in the control room, we need to consider what will be the greatest viewing distance (Figure 2-2). For example, a process control room operator needs to read information from monitors and displays located at different distances from his or her eyes. Yet, the sizes of characters on most screens are designed for a viewing distance of about 71 cm (28 in.). This may have effects on increased eye strain and visual fatigue as well as the possibility of misreading the information, leading to human error. Formulas are available to calculate the recommended height of letters or figures (Grandjean, 1988; Sanders and McCormick, 1993; Konz, 1990). For favorable viewing conditions (good lighting, good contrast, noncritical display),

$$H = D/200$$

where H is the character height in meters and D is the viewing distance in meters. For unfavorable viewing conditions (low light levels, reduced contrast, critical displays),

$$H = D/120$$

For example, the size of letters to be used for a sign that summarizes evacuation instructions and must be clearly visible from 10 meters (30 feet) away under dim lighting conditions is

$$H = D/120 = 10/120 = 0.08\,\text{m or } 8.0\,\text{cm (3.15 in.) high}$$

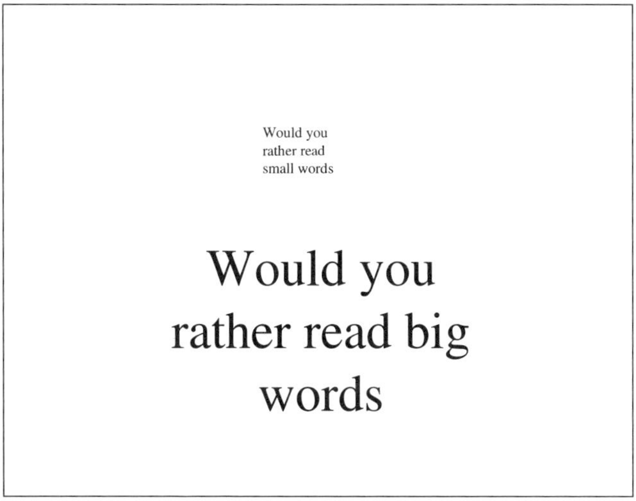

Would you
rather read
small words

Would you
rather read big
words

Figure 2-1 Size of object and visual acuity.

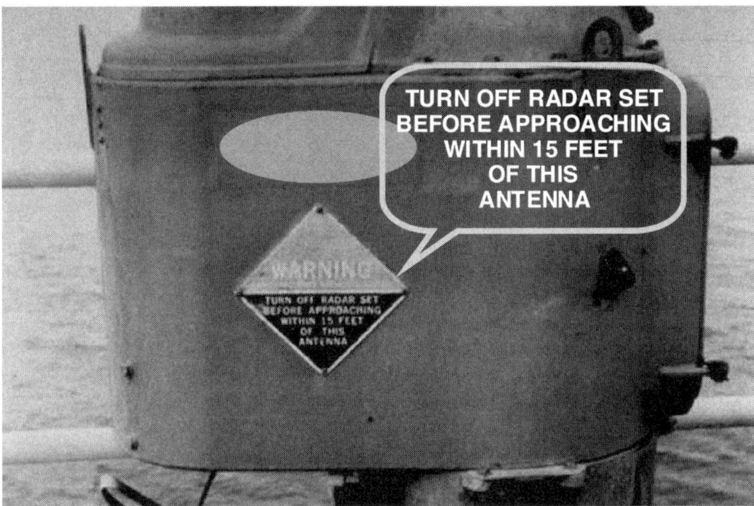

Figure 2-2 Poor design of a warning sign; the text is too small to be read from 15 feet away.

- *Contrast.* This refers to the difference in brightness between a visual target and its background. Generally, as the contrast increases, the ability to see details (visual acuity) increases. For example, detecting dark (black) characters on white background (90% contrast) is much easier than detecting dark characters on relatively darker background (Figure 2-3).
- *Illumination.* Up to a point, visual acuity and contrast sensitivity (the ability of the eye to detect the smallest difference, contrast, between a lighter and a darker spatial area) increase, or improve, with increasing illumination. However, high, intense levels of illumination are not always warranted, as they may contribute high levels of glare, weaken information clarity, and lead to temporary blindness. Therefore, the amount of illumination necessary to perform visual tasks is task dependent. Illumination is discussed in detail in Chapter 4.
- *Exposure time.* Visual acuity increases with increased exposure time. The longer we view and focus on an object, the more details we see.
- *Observer or target movement.* Our visual acuity decreases as the observer or target move. This is so because movement affects expo-

Dark letters on a dark background are hard to read

Dark letters on a light background are easy to read

Figure 2-3 Example of poor and good contrast.

sure time. For example, consider the difficulty in reading the details of an object on the side of the road while driving a car at 100 km per hour (65 miles per hour).

2.2.1.6 Age

The fact that our visual abilities and visual performance decline with age should not be a point of dispute or debate. For example, our visual acuity is reduced by about 30% between the ages of 20 and 60 years. Faster depreciation of vision generally takes place after age 40, and we wonder why we cannot read the fine print on the bottle of a prescription drug.

This decrease in visual acuity is related to the following changes in the structure of the eye:

- Yellowing of the lens, where the amount of light reaching the retina is significantly reduced.
- Decrease in the accommodation power of the eye around the age of 40, which affects depth perception and distance vision. This is caused by deterioration in the muscles that contract and change the lens shape. Depth perception is our ability to judge our distance from objects and the distance between objects in three-dimensional space.
- Retina and nervous system changes around the age of 60, affecting the size of the visual field and sensitivity to low quantities of light.

If the workforce is older, special considerations must be made to accommodatc workers who are more likely to experience visual difficulties. Some of these considerations are:

- Higher levels of lighting and larger targets such as characters.
- More viewing time to absorb the information.
- Higher level of contrast between the target and its background, such as black characters on white background.
- Targets set at the same distance from the eyes, since speed of accommodation decreases with age. For example, the displays viewed by a control room process operator should be at the same distance from the eyes.

2.2.2 Auditory Sense

Hearing is second only to vision in providing information through which we can understand and adjust to our surroundings. For example, a field operator and a control room operator communicate through a radio to change a state of a system such as opening or closing specific valves. Other areas where hearing is key is in the design of warning signals on machines, a buzzer on a fire alarm, speech communication, and the like.

The two defining characteristics of any auditory stimulus are frequency and intensity. The corresponding subjective attribute (what people hear) for frequency is tone and, for intensity, is loudness. This area is discussed in detail in Chapter 4 in the section on noise. In many situations, the designer has to choose between using visual and auditory displays or even a combination of the two. The decision on when to use an auditory or visual presentation of information depends on:

- The nature of the message. An alarm that calls for immediate action should be auditory. The message here is short and simple and deals with events in time. Moreover, a person need not look at an alarm to hear it.
- Characteristics of people involved, such as young versus older people (refer to the noise section in Chapter 4).

- Conditions under which information must be received. For instance, if the visual system is overburdened, then a buzzer makes sense to alert operators of equipment failure without having to look at a visual display.

In deciding on the type and role of auditory displays, a number of factors must be considered:

- Signal characteristics, which include sound frequency, intensity, duration, and rate of sound occurrence like the on/off emergency signal. Signal characteristics are discussed in Chapter 4.
- Individual characteristics, which include the hearing ability and experience of the person, age, and length of exposure to signal.
- Environmental characteristics, which include the type and amount of background (masking) noise.

2.2.3 Cognitive Capabilities

In the previous section, we discussed how people receive information from the environment through their visual and auditory senses. This section reviews how we process and interpret this information and the associated limiting factors.

The senses (eyes, ears, etc.) receive signals (data) from the environment and pass them to the brain through nerves. Such data are referred to as *sensory information*. The brain acts on the sensory information through complex processing mechanisms to give it meaning. This is referred to as *information processing*. Information processing influences people's behavior, and people's behavior at work is a consequence of information processing.

Information processing involves several different processes: attention, perception, memory, and decision making. These processes are interdependent and serve different functions in the flow and interpretation of information. As we will see and repeat later, receiving and interpreting information requires knowledge, experience, mental alertness, skills, and the ability to formulate new ideas. More discussion on information processing in terms of procedure writing and evaluation, vigilance, and human error can be found in Chapter 8.

2.2.3.1 Attention

Consider the following two examples to illustrate the process of attention:

1. When driving a car, we want to ensure that we are watching the road, steering the car, maintaining velocity by looking at the speedometer and adjusting the pressure on the gas pedal, listening to the radio, and perhaps talking to a passenger and drinking coffee.
2. A control room process operator scans and monitors a number of displays (each display presenting different information, temperature and pressure of different units), talks to coworkers, answers the phone, and communicates through the radio with an operator in the field.

It is clear that we are continuously bombarded with information from the environment around us. However, because we have limited capacity to process the information we receive, we must select out the important messages and ignore all other information. Attention is the process of information reduction. It allows us to select important information, and it controls which information is processed further. Occasionally, we do not "pay attention" and miss important information that may be the cause of task errors and accidents. For example, while driving the car we decided to change the radio station (most important activity at that moment). Now the attention is diverted from the road and, suddenly, a bumper-to-bumper incident takes place. The limiting factors here are:

1. We behave like a single-channel information processing device. We make active decisions about one thing at a time, even though we think we are attending to all things going around us, at all times.
2. We process information at a constant rate because of our limited channel capacity in terms of speed and quantity.

For example, we behave roughly like the automatic feeder of a photocopy machine. If we put in a stack of 100 pages to copy, a single feeder (single channel) takes in a fixed number of pages (process at constant rate) per minute or hour. Four types of situations define the direction of attention (Goldstein and Dorfman, 1978; Broadbent, 1982; Brouwer et al., 1991; Parasuraman and Nestor, 1991; Wickens, 1992):

- Selective attention is a conscious decision to attend voluntarily to important (relative to the individual) information. For example, the control room process operator scans the displays to detect unexpected but important events, such as excessive temperature or pressure.
- Focused attention allows us to filter out unwanted information and attend to one source of important information. For example, the driver changing the radio station is focused on the radio as the one source of important information at that moment.
- Divided attention occurs where we need to attend to several things at once. For example, the control room operator monitors displays and talks to coworkers while eating.
- Sustained attention, also known as *vigilance*, is where we must monitor a situation for a long period of time. For example, operators scan the radar in an airport control tower.

2.2.3.2 Perception

After the signals from the environment enter the brain through the eyes or ears, for example, they are interpreted to produce a recognizable pattern of sight or sound. This process, called *perception*, is entirely subjective in nature because it is greatly influenced by the individual's knowledge, experience, expectations, feelings, and wishes (Anderson, 1995). A few facts about perception follow:

- Perception is a complex, active process that can occur with or without external input. Internal input could be a stomach ache, eye pain, or the like.
- It overrides what we know to be true.
- It takes place without awareness (subconsciously).
- It need not correspond with reality.
- Our behavior is determined by perceived rather than actual information.

The main interest for an ergonomist is to design the actual outcome of a job to fit in with the operator's perception of the expected outcome.

Two issues can be briefly discussed here: adaptation and direction of movement.

Adaptation is the process of adapting to our environment, which affects our expectations in making a decision and response. For example, if a control room process operator, over a period of time, experiences a high number of false alarms, the expectation is that the next alarm is also false. Therefore, over time, the alarms no longer arouse the operator, and the next true alarm will not be responded to accordingly. Therefore, we should keep the occurrence of alarms to a minimum and ensure that attention is drawn to priority conditions so that arousal level stays high.

The direction of movement or expected relationship between controls and displays needs to meet population stereotypes. Population stereotypes are what most of the population will do on most occasions. For example, in North America, we push a light switch up to turn on the lights. In other parts of the world, the switch is pushed down (see Chapter 5 for more detail).

2.2.3.3 Memory

Memory is the mechanism of storing and retrieving information. The information stored in the memory is based on our learning and past experiences. For example, driving a car, operating machines, using tools, monitoring a process in the control room—all depend on our capacity to learn and store information. Three stages of memory can be identified. Each stage has different levels of information processing, storage capacity, representation of information, and time span for retention. A simple model of memory is in Figure 2-4.

This simple model shows that the stimuli are captured and perceived by the senses (eye, ears, etc.) and held briefly in the sensory memory storage. The filtering process, through attention, passes on the information considered relatively important to the individual to the short-term memory storage. Further rehearsal (i.e., repeating, rereading, reviewing material) of the stimuli allows information to pass on and be stored in long-term memory (Bailey, 1989; Wickens, 1992; Anderson, 1995; Best, 1995). Each of the stages is briefly presented.

Sensory memory (SM) is relatively automatic in nature, meaning we need not attend to the information for it to be maintained at this stage.

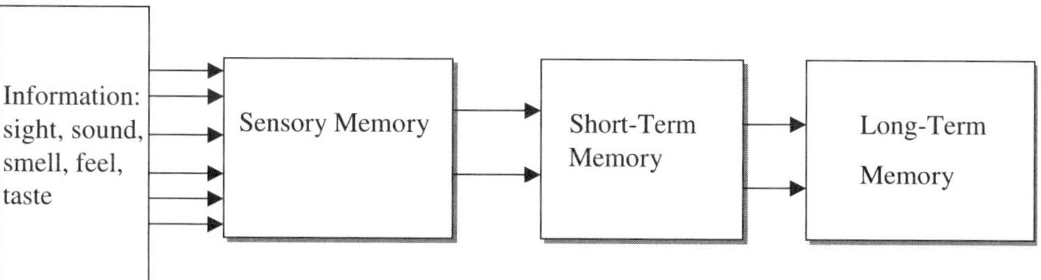

Figure 2-4 Stages of memory storage.

This memory is specific to the sensory system (visual, auditory, etc.) being stimulated. For example, visual information is held in the visual sensory storage (iconic memory), and auditory information is held in the auditory sensory storage (echoic memory). Iconic memory storage lasts for about 1 second, while echoic memory storage can last for 3–5 seconds before the information is lost or decayed. For example, a phone number for an advertised product is flashed briefly on the television screen. As we start writing down the first digits we forget the last ones simply because it was lost from the visual sensory storage. Or, someone is talking to us and we do not hear a single word the person has said. The reasons some of the information is lost in the sensory memory are: (1) decay, where information fades very rapidly; and (2) because material considered important and attended to (paying attention) is passed on to short-term memory, the rest is lost.

In addition, the sensory memory stage serves as a limiting factor in a task sequence. For example, a control room process operator has the task of scanning many pages of alarm messages on a display throughout the shift. While the operator thinks he may have scanned all the important messages, one or more messages did not register and may not have been passed on into the long-term memory for proper action. This is because the messages were not attended to properly and the memory trace was lost.

Short-term memory (STM), also referred to as *working memory*, is the stage where we maintain information we consider important and pay direct attention to. It is also the stage where we try to organize and encode information into the long-term memory through the process of rehearsal. An example of encoding is Morse code, where either of two codes consisting of dots and dashes or long and short sounds is used for transmitting messages by audible or visual signals. The characteristics of STM are:

- *Short duration and decay.* STM storage lasts for a few seconds. If we pay attention and use the process of rehearsal, then we strengthen the information in the STM and avoid decay; for example, looking up a phone number and then holding it in the STM through rehearsal (repeating the number over and over again) until we have dialed it. Sometimes, we think of a few items we want to do as we drive to the office. If we do not constantly rehearse the items or write them down, we get to the office and cannot remember a thing. The information in this case is lost or decayed simply because of other interfering information and competing demands on the STM. This is why some people use a small audiotape recorder to record things, write phone numbers on the palm of their hand, or write a sequence of valves, to open or close, on their hands or gloves.
- *Distraction, disruption, and interference.* STM faces interference, distraction, and disruption all the time, depending on the competing demands; for example, being interrupted in the middle of a conversation by a phone call and not knowing where we were in the conversation.
- *Limited capacity.* The storage capacity in the STM is limited to seven plus or minus two (five to nine) independent chunks or single pieces of information (Miller, 1956). These chunks could be five to nine digits, words, steps in a procedure, or the like. For example, if you were given two 10-digit telephone numbers to memorize (which exceeds the five to nine rule), you may make mistakes when attempting to remember both of them. However, if you group or chunk the information using your knowledge and association already learned, you can do a much better job remembering. For example, would you rather have to remember these two numbers:
  8008474869 and 2255252752
  or 800-847-4869 and 225-525-2752
  or 800-VISIT-NY and CALL-ALASKA

  Some of the guidelines to improve STM limits are:

Minimize the load in terms of time and number; for example, a touch tone phone is faster to enter numbers and, therefore, the memory time is less than when operating a rotary phone.

Use chunking as just discussed: Use three to four letters or numbers per chunk (Bailey, 1989). For example, 800-847-4869.

Use letters instead of numbers where possible and meaningful; for example, CALL-ALASKA.

Keep letters and numbers separate (Preczewski and Fisher, 1990); for example, car license plate B68RN2 is harder to remember than 628 BRN or BRN 628.

Create meanings for characters; for example, the number 628 BRN can have the following meaning: 6 + 2 = 8, and BRN can be an abbreviation for the color word *brown*.

The transfer of information from STM to long-term memory (LTM) relies on the attention and rehearsal processes. The rehearsal process must involve *not* just rote repetition but the use of associations (information analyzed, compared, and related to other information) between the new information and information already in LTM. This makes retrieving information later on much easier; for example, storing the number 225-252-2752 versus CALL-ALASKA. LTM contains all the learning, knowledge, and experience we obtain during our lifetime. Retrieving information from LTM presents challenges for some of us. We often have difficulty recalling a 25-step procedure or meeting times, names, or even more serious events such as birthday and anniversary dates. However, we do better in recognition (i.e., faces, checklists) than recall. Some of the characteristics of recall and recognition are:

Recall	*Recognition*
Limited	Almost unlimited
About 2000 words	About 100,000 words
Identify about 8 colors	Discriminate between 500 colors

The factors that affect our ability to remember are:

- Initial learning: how well we analyze, compare, and relate new information to that already learned determines how much and how well we learn. Factors that affect this process are motivation and the desire to learn, prior learning, anxiety or stress—too much anxiety and stress can affect learning, and the method of learning.

- Importance and meaningfulness of the material we are trying to learn.
- Interference, which is best illustrated with an example. An individual who learned to drive a car on the right-hand side of the road is in a different country, learning to drive on the left-hand side. After a while, she is back driving on the right-hand side. Learning the new left-hand side driving interferes with her previous learning and affects her ability to drive safely on the right-hand side for a while, until relearning take place.
- Age: memory ability declines with age and can affect day-to-day activities, as most of us Baby Boomers experience.

2.2.3.4 Decision Making

Decision making is the process of evaluating alternatives and selecting an action. We tend to search for information related to the problem at hand, estimate the probabilities of different alternatives, and attach meanings and values to anticipated outcomes. Therefore, decisions are a choice among courses of actions. People who must make too many decisions too quickly have to trade off speed of decision making against accuracy of decision outcome. In addition, the time to make an accurate decision is related to the amount of uncertainty in the decision. Naturally, the more uncertainty we have, the longer it takes us to search for the information, estimate probabilities of different alternatives, and attach values to outcomes. Possibilities to improve human decision making include:

- *Design or redesign support systems.* Designing the support system appropriately improves decision making and therefore performance. For example, if we eliminate the need to make manual mathematical calculations by the control room process operator and build that into the process control system, we enhance the speed and accuracy of the decision-making process and eliminate any possibility for human error due to miscalculation.

- *Decision tools.* Decision tools or aids can eliminate a number of the steps the operator has to take to reach a decision; for example, using an expert system to calculate and predict future outcomes based on a large database of previous events. For instance, if there is a continuous rise in temperature or pressure of a boiler, the expert system can predict what may happen, based on the data from previous events, and present the operator with specific decisions to make. The decisions may be to keep the process going or shut the system down. To arrive to such a decision without the aid of an expert system requires the operator to gather information from different displays and sources; consult procedures, supervisors, and coworkers; and so forth. This increases the mental workload (discussed in Chapter 8) and creates an environment for human error.
- *Training.* Training proves especially useful when a system is in an abnormal state. Abnormal situations do not happen frequently, but when they happen, operators may not be ready simply because they have not been regularly trained on such events. Therefore, placing a simulator of the process in the control room center provides the process operators with training, especially in abnormal states. This helps develop the process of decision making and action taking during both routine and abnormal situations.

2.2.4 Summary of Information Processing

The following is a summary of points to remember on information processing.

- Inattention:
  Minimize conflicting demands.
  Ensure redundant checks in critical situations.
- Forgetting when distracted: minimize distraction.
- Inability to retain enough information:
  Chunk information to ensure no more than 7 ± 2 concepts are addressed at once.
  Develop logical associations between the information and the action.

- Inability to retain information for prolonged periods: design systems to retain information until it is no longer needed by the operator.
- System demands exceed information processing ability: design calculations into the process control system.
- Inability to recall information from long-term memory:
  Replace "recall" with "recognition" through the use of cues and reminders, such as checklists, pull-down menus, help screens, set points.
  Conduct refresher training (using procedures).
  Design systems to eliminate the need for procedures or the use of long-term memory.
- Performing activities out of sequence or too slow: design a logical relationship between one activity and the next, such as building sequence into the task.
- Confusion due to information presentation:
  Ensure that plant equipment operates in ways that the operator expects and understands.
  Make the design compatible with operator's expectations.

2.3 PHYSICAL CAPABILITIES

This section of the human system interface (HSI) model, discussed in Section 2.1, deals with converting a decision to an action. This is where the operator communicates with the system using bodily actions such as pulling a lever, pressing a button, or speaking over the radio to communicate with others in the plant. For the body, or parts of it, to move, reach, and push a button, or turn a valve requires the involvement of the musculoskeletal (muscles, bones, and connective tissues) system, which provides the primary components for muscular activity. Musculoskeletal movement is also referred to as *motor performance*.

When an operator is called on to perform a physical activity like turning a valve, the amount of work required must be within his or her physical capabilities. The aspects of anthropometry (operator's relevant body dimensions) and biomechanics (mobility of bones, joints, and muscles) define these capabilities. Muscles, in particular, play a key role as limiting factors in many physical tasks and activities. This section considers the factors that limit motor performance to ensure that the operator can reach and apply the necessary force to activate a control. It focuses mainly on two areas of interest: muscular strength and endurance and anthropometry.

2.3.1 Muscular Strength and Endurance

Muscles have limits to their strength and their ability to maintain that strength and resist fatigue (also referred to as *muscular endurance*). *Strength* is defined as the maximum force one can exert voluntarily. It is measured in kilograms (kg) or pounds (lb). The work that muscles may have to do is classified into two types:

- Static work, also referred to as *static effort*, is characterized by long-term muscular contraction without body motion occurring; for example, when a posture must be held for long periods like sitting, standing in one place, bending, and twisting to perform a maintenance job.
- Dynamic work, also referred to as *dynamic effort*, is characterized by body motion accompanying muscular tension; for example, repeated muscular contractions with rest periods in between, such as walking.

Most physical activities have both static and dynamic work components. Take, for example, carrying jars filled with products from a sampling point in the plant back to the laboratory. The upper body, especially the arms, shoulders, and back muscles, is under static effort to support the weight and maintain (stabilize) the body in an upright position. However, the lower body, especially the legs, is under dynamic effort to allow the body to move.

Strength is strictly defined as "the maximal force muscles can exert isometrically in a single, voluntary effort" (Kroemer, 1970; Chaffin, 1975). However, strength under dynamic conditions can also be measured (Sanders and McCormick, 1993). Information about isometric muscular strength capability is needed to specify, for example, the minimum valve resistances and the forces required to safely carry bags of chemicals. Most tasks or activities that operators perform require the integrated exertion of many muscle groups. For example, lifting a bag of chemicals off the floor requires squatting, extending arms, grabbing the bag with both hands, lifting the bag with the arms and shoulders, lifting the body using the legs, and the like. In this complex situation, the maximum force that can be exerted is determined by the weakest link in the different muscle groups involved.

2.3.1.1 Factors Affecting Strength

There is a wide range of muscular strength capabilities among people, from heavy weight lifters to sedentary TV watchers. In addition, many variables play the role of limiting factors and can influence muscular strength:

- *Age.* Strength increases rapidly in the teens, reaches its peak in the middle to late twenties, levels off, and starts decreasing, slightly in the forties and faster in the fifties and sixties. Most tasks and jobs are designed around the young worker. But, the strength required may exceed the capacity of older groups. This is truly an issue today, since the current workforce contains many older workers.
- *Gender.* In general, women have less-developed muscles and their overall mean strength is about 67% that of males (Hettinger, 1961; Roebuck, Kroemer, and Thomson, 1975; Lauback, 1976). However, we must not generalize this figure (Wilmore, 1975; Redgrove, 1979; Pheasant, 1983; O'Brien, 1985; Chaffin and Andersson, 1991; Sanders and McCormick, 1993). The reasons are:
  1. This 67% is an average value from various groups of muscles. The range is 35 to 85% of male mean strength.
  2. For the lower extremities, the average strength of females is comparable with males. This may reflect the use of lower extremities by both males and females for daily regular activities. But, in general, men use their upper extremities more than women.
  3. Trained female athletes are stronger than many untrained men.

  In terms of designing manual tasks, such as lifting, the maximum weight that can be lifted is determined by the weakest member of the two genders involved.

- *Occupation.* Relates to the different levels of muscle use (exertion) by people in different occupations; for example, a process operator versus a desk clerk given that neither is a fitness fanatic.
- *Physical training.* Physical training, especially weight lifting, tends to increase muscular strength and endurance by as much as 30–50%.
- *Others.* Other factors that have been associated with influencing muscular strength are:
  1. Gloves. Wearing gloves, on average, results in a loss of about 20% in grip strength. Of course, this depends on the thickness of the gloves: the thicker the gloves, the less the strength.
  2. Body dimension, such as height and weight. Body weight is an important determinant of arm strength. Likewise, height correlates highly with pulling strength.
  3. Body position. For example, the strength of elbow flexion is at a maximum with the elbow at about 90° but drops off to about half this value at extreme angles.
  4. Diet. Inadequate and improper nutrition decrease body strength.
  5. Other personal factors: genetics, health, fatigue, drugs, motivation.

2.3.1.2 Endurance and Fatigue

As mentioned earlier in this chapter, *muscular endurance* refers to the ability of the muscle to continue to work. In the case of static work, the length of time (endurance) a force can be maintained depends on the proportion of the available strength being exerted. For example, maximum force can be maintained for a very short time (few seconds). However, the smaller the force exerted, the longer it can be maintained (Caldwell, 1963, 1964; Monod, 1985; Deeb, 1988). The applications of these data are important. If we expect an operator to apply and maintain static force to hold an object (i.e., a drill weighing 15 kg or 33 lb) for a period of time, the designer must consider endurance. In the case of dynamic work, *endurance* is defined in terms of force and frequency of repetition. However, the endurance time is significantly longer. For example, 25% maximum voluntary contraction (MVC) can be maintained for about 10 minutes under static contraction compared with over 4 hours under dynamic work.

Fatigue is a product of the work performed, whether dynamic or static. The degree and severity of fatigue constitutes the level of discomfort, pain, distraction, and reduced performance that may lead to accidents. It is, therefore, very important to design tasks by eliminating factors that induce fatigue.

2.3.2 Anthropometry: Body Size

Anthropometrists, measurers of the human body, have collected body size data for many years. For example, the measuring units of foot and hand have been derived from the dimensions of body parts. The term *anthropometry* is derived from two Greek words: antropo(s), or human, and metricos, or measurement. Anthropometry is used extensively by ergonomists to design tools, equipment, plants, manufacturing lines, clothes, shoes, and the like to ensure the proper fit to the person. Therefore, to achieve proper fit, it is important to have details on the dimensions of the appropriate body part. For example, the size of the hand is used to design the dimensions of controls such as switches and push-buttons, while details of arm reach are necessary to position controls at appropriate distances.

Two primary categories of anthropometric data interest ergonomists:

- *Structural anthropometry*, also referred to as *static anthropometry* or *static dimensions*. These are measurements with the body in a still or fixed position; for example, stature or height, weight, head circumference.
- *Functional anthropometry*, also referred to as *dynamic anthropometry* or *dynamic dimensions*. These are measured with the body engaged in various work postures, indicating the ranges of motion of individual body segments; for example, arm reach.

It is important to point out that static anthropometry data are often measured on unclothed (nude) individuals, mainly to ensure consistent results. Therefore, corrections must be made to account for increases in body size due to clothing, such as a process operator working outside during an Alaskan winter and another during a Texas summer. In addition, allowances must be made for wearing safety shoes and hard hats, which could add about 10–12 cm (4–5 in.) to the stature that must be considered in the design.

2.3.2.1 Sources of Body Size Variability

The various genetic, biological, and physiological differences between humans influence the way they vary with respect to body dimensions in terms of height, weight, shape, and the like. This can be noticed if you observe people in a shopping mall. Therefore, we need to be very careful using anthropometric data if they are to be of value (Pheasant, 1982a, 1982b).

The common practice in ergonomics is to specify anthropometric data in terms of percentiles. A percentile refers to a percentage of the population with a body dimension up to a certain size or smaller. For example, if a 95th percentile height (stature) is 170 cm (66.9 in.), it indicates that 95% of the population have heights up to 170 cm. Or, 95% of the population have heights of 170 cm or less and 5% are taller than 170 cm. If, on the other hand, the 5th percentile stature is 150 cm (59 in.), it indicates that 5% of the population are shorter than 150 cm and 95% are taller.

When a particular design (i.e., placing a valve hand wheel at a given height) is expected to be used by many people, we need to consider the following variables in influencing body size and adjust the design accordingly:

- *Gender.* Men are generally larger than women at most percentiles and body dimensions. The extent of the difference varies from one dimension to another. For example, hand dimensions of men are larger than women sizes—hand finger thickness of men is about 20% larger, while fingers are about 10% longer for men than for women (Garrett, 1971). Women, in general, exceed men in five dimensions: chest depth, hip breadth and circumference, thigh circumference, and skinfold thickness. O'Brien (1985) proposes a complete list of anthropometric differences between men and women.

- *Age.* Body dimensions generally increase from birth to early twenties, remain constant to around age 40, and decline afterward into old age as part of the normal aging process. For example, stature or height reaches full growth at around age 20 for males and 17 for females (Trotter and Gleser, 1951; Damon, Stoudt, and McFarland, 1971;

Roche and Davila, 1972; Stoudt, 1981). Decline in stature is more pronounced in women than in men. Therefore, it is important to define the user population early in the design cycle.

- *Nationality and culture.* Nationalities and cultures differ in body sizes. For instance, Asians tend to be somewhat shorter on average than Northern Americans, while certain cultures from southern Sudan (Africa) tend to be taller. Therefore, it is important for the designer to define and use the anthropometry data related to the user population nationality and culture. Table 2-1 (and Figure 2-5) presents an example of anthropometric values for three different nationalities: North American, Japanese, and Hong Kong.
- *Occupation.* Differences in body size dimensions among occupational groups is common and well documented. For example, manual workers, on the average, have larger body sizes than sedentary workers. Sanders (1977) found truck drivers to be taller and heavier

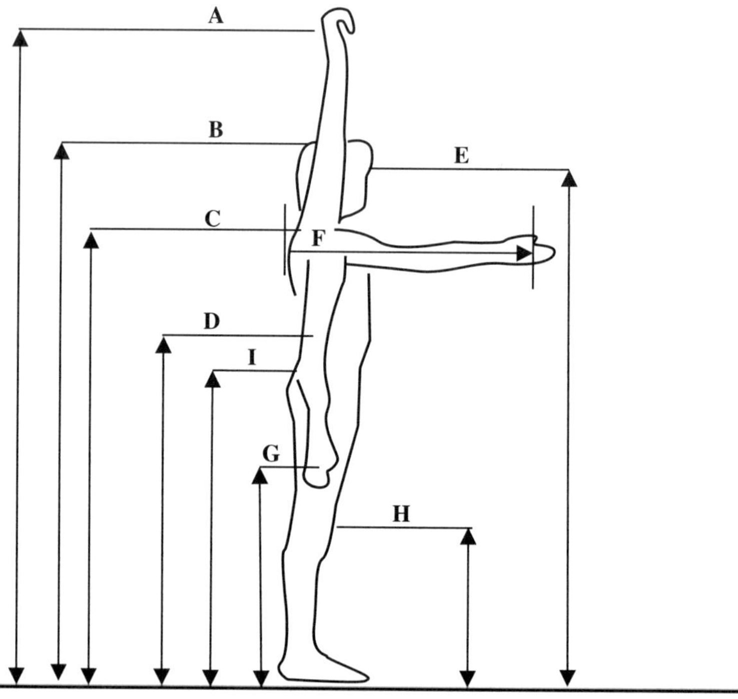

Figure 2-5 Anthropometry figure to guide Table 2-1.

TABLE 2-1

Selected Anthropometry Data for Different Nationalities in Centimeters (cm) and Inches (in.)

	North Americans				*Japanese*				*Hong Kong*			
	95th% Man		*5th% Woman*		*95th% Man*		*5th% Woman*		*95th% Man*		*5th% Woman*	
Anthropometric Dimensions	*(cm)*	*(in.)*	*(cm)*	*(in.)*	*(cm)*	*(in.)*	*(cm)*	*(in.)*	*(cm)*	*(in.)*	*(cm)*	*(in.)*
A. Vertical grip/reach	217.5	85.7	177.8	70.0	207.5	81.7	168.0	66.1	210.5	82.9	168.5	66.3
B. Stature/head height	184.4	72.6	149.5	58.9	175.0	68.9	145.0	57.1	177.5	69.9	145.5	57.3
C. Shoulder height	152.4	60.0	121.1	47.7	143.0	56.3	107.5	42.3	146.0	57.5	118.0	46.5
D. Elbow height	119.0	46.9	93.7	36.9	110.5	43.5	89.5	35.2	108.0	42.5	87.0	34.3
E. Eye height	172.7	68.0	138.2	54.4	163.5	64.4	135.0	53.1	164.0	64.6	133.0	52.4
F. Forward grip/reach	88.3	34.8	64.0	25.2	75.0	29.5	57.0	22.4	77.0	30.3	58.0	22.8
G. Knuckle height	80.5	31.7	64.3	25.3	80.5	31.7	65.0	25.6	81.5	32.1	65.0	25.6
H. Knee height	59.2	23.3	45.2	17.8	53.0	20.9	42.0	16.5	54.0	21.3	41.0	16.1
I. Waist height	110.5	43.5	86.4	34.0	89.5	35.2	70.0	27.6	92.0	36.2	71.5	28.1

Note: Add about 4 cm (1.5 in.) for shoes and 7.5–10 cm (3–4 in.) for hard hat.

than the general civilian population. This difference among occupations may be the result of

1. Diet.
2. Exercise.
3. Physical activities imposed by the job (i.e., manual handling).
4. Imposed selection, such as individuals need to be a certain height to be accepted in a particular job.
5. Self-selection, such as individuals with a given height choose a particular job for practical or sociological reasons.

   We must take great care of not using anthropometric data obtained from groups of one occupation, such as armed forces, to design the environment of another, such as office workers.

- *Historical trends.* The average size of people has been increasing over the years. For instance, the average adult height in Western Europe and the United States increased about 1 cm (0.4 in.) per decade (Sanders and McCormick, 1993). This is so, perhaps, because of better diet, medical care, hygiene, and living conditions. As designers, we need to consider present-day users as well as future generations to ensure proper systems design few decades down the road.

- *Body position.* Posture affects body size. For example, restraints such as seat belts, affect data applicability of forward reach.

- *Clothing.* As mentioned earlier, almost all anthropometric data are obtained from nude individuals. Therefore, the type (material) and amount of clothing add to body size and can also create restriction of movement such as affecting overhead and forward reach. Another example is the use of gloves where allowance must be made to accommodate different thicknesses of gloves.

2.3.2.2 Principles of Body Size Application

When determining the proper anthropometric data to be used in a design, the following must be observed: we need to carefully define the population or group we are designing for and ensure that the data are reasonably representative. For most purposes, a range of dimensions from the 5th to the 95th percentile is generally acceptable. The range can also increase, if possible, from the 2nd to the 98th percentile or even larger. The choice of design percentile is largely a matter of cost. Three general principles of body size application to specific design problems are accepted (Chapter 6 covers these principles in detail):

1. *Design for the average.* The average value is taken as the 50th percentile, meaning 50% of the population is above and 50% below this value.

2. *Design for the extreme.* In designing for the extreme, the ergonomists constantly apply the following two principles:
   - Design fit or clearance dimensions for the largest individual.
   - Design reach dimensions for the smallest individual.

   It is frequently the practice to use the 95th percentile male for the clearance or fit dimensions and the 5th percentile female for reach dimensions. Therefore, it is safe to say that a design that would accommodate individuals at one extreme would also accommodate virtually the entire population.

3. Design for the adjustable range. Designing for the adjustable range is generally the preferred method to accommodate individuals of varying sizes. This use of adjustability can be seen in car seats, office chairs, desk heights, bench heights in a maintenance shop, adjustable tables for manual materials handling jobs, and the like. However, this may not always be possible:
   - If we strictly use the range, then we must accommodate people from 3 feet tall to 9 feet tall and weighing from 23 kg (50 lb) to 227 kg (500 lb). This is why the 5th and 95th percentile values are traditionally advocated.
   - Adjustability may not have any practical value and the cost outweighs the benefit; for example, an eye wash station in a plant or a bathroom toilet height.

2.4 CASE STUDY

The force an operator is able to apply to a valve hand wheel depends on the location of the wheel relative to the operator, the orientation of the valve stem, the diameter of the wheel, and the quality of the interface between the wheel and the hands. The objective of this study was to create a database of human force capabilities, to apply a hand wheel to open or close a valve, that can be used by plant designers in the selection, placement, and orientation of valves. A summarized version of this study is presented here. For more details, consult Attwood et al. (2002).

2.4.1 Method

2.4.1.1 Participants

Sixty-six volunteer process operators participated in the study. Data from nine volunteers were rejected: Seven subjects were removed because of irregularities in their data; two women were removed to provide uniform gender data. Analyses were performed on 57 men.

Subjects were screened for health problems before participation. Subjects signed a consent form.

Warm-up and practice sessions were conducted. Each subject was provided with a small token of appreciation.

2.4.1.2 Equipment

Force data were collected on the "primus" device manufactured by the Baltimore Therapeutic Equipment Company, Hanover, Maryland (Figure 2-6).

2.4.1.3 Procedure

- Nine valve positions were identified to represent typical heights and stem angles.
- Valve positions were tested randomly within low, medium, and high heights.
- Operators completed three exertions at each valve position.
- Operators were instructed to assume standard position and exert maximum force on valve wheel in a counterclockwise direction, increasing force smoothly to maximum and holding for 6 seconds.

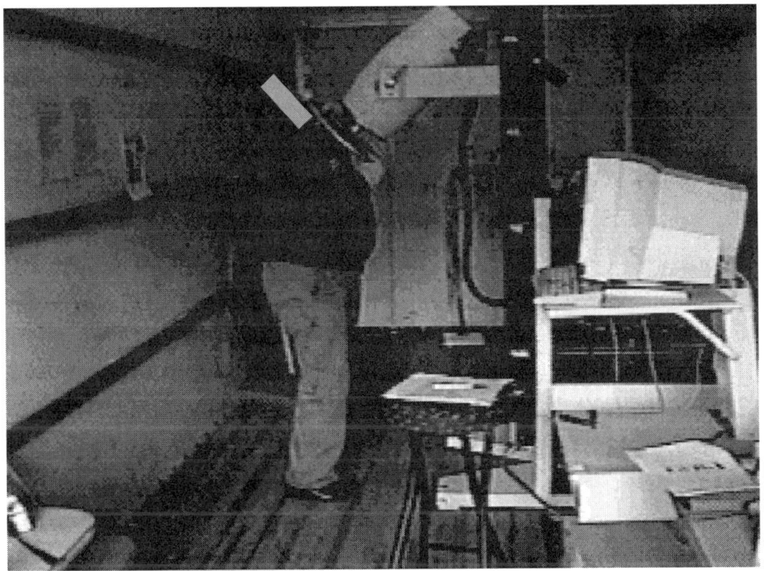

Figure 2-6 Subject participating in study.

- Operators were allowed 30 seconds rest between each repetition and 2 minutes rest between different valve positions.

2.4.2 Data Collected

- Age, gender, height, weight, and functional reach were compared with population norms.
- Average force, 3–5 seconds into repetition.
- Peak force per repetition.

2.4.2.1 Data Analyses

The maximum average force was identified from three repetitions per participant for each valve position. Also determined were the mean and standard deviation of force for each of the nine valve positions, as well as the 5th percentile of force for each valve position and the relationship between force and the stature of the operator.

2.4.3 Conclusion

The data suggested a relationship between stature and force capability. The valve locations that produced the highest turning forces differ from those typically used by designers. The 5th percentile force capability is much lower than recommended historically. Figure 2-7 presents the 5th percentile force (in lb) as a function of valve height and stem orientation.

The recommendations for existing operations are

- Identify critical manually operated valves that are operated frequently, in an emergency, and have time constraint for operation.
- Measure the valve force required to crack critical valves.
- Modify those that exceed operator force capability (for the existing location and orientation).

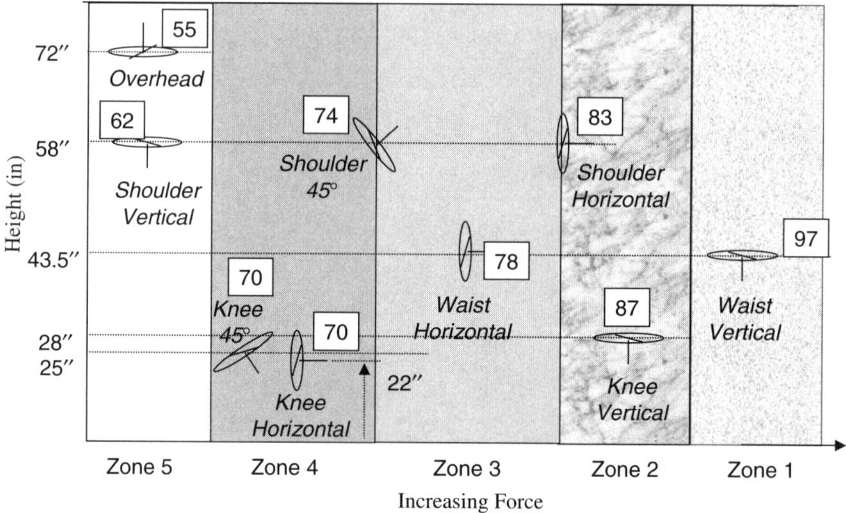

Figure 2-7 The 5th percentile force (in lb) as a function of valve height and stem orientation. *Note*: Means within the same column are not statistically different from each other but are different from those in different columns. Each column is defined as a "Zone."

REVIEW QUESTIONS

1. What is the function of the retina of the eye?
2. Define accommodation of the eye.

3. What is nearsightedness and farsightedness?
4. What are the areas of the visual field?
5. Describe the process of adaptation of the eye.
6. Why should we not use color as the primary method of coding?
7. What situational factors affect visual acuity?
8. Describe how attention is essential in processing information.
9. Why is perception a subjective and not an objective process?
10. Describe the three stages of memory: sensory, short-term, and long-term.
11. Describe what to do to improve the limited capacity of short-term memory.
12. What factors affect strength?
13. What is anthropometry?
14. Discuss the concept of adjustable design.
15. Why there is no such thing as an average human?
16. Give examples of the two categories of anthropometry.
17. List the sources of body size variability.
18. What does percentile mean?
19. Discuss the three principles of body size application.
20. Why design for the extremes? Give examples.

REFERENCES

Anderson, J. R. (1995) *Cognitive Psychology.* Fourth Ed. New York: W. H. Freeman.

Attwood, D. A., Nicolich, M. J., Doney, K. P., Smolar, T. J., and Swensen, E. E. (2002) "Valve Wheel Rim Force Capabilities of Process Operators." *Journal of Loss Prevention in the Process*, 15, pp. 233–239.

Bailey, R. W. (1989) *Human Performance Engineering.* Second Ed. Englewood Cliffs, NJ: Prentice-Hall.

Best, J. B. (1995) *Cognitive Psychology.* Fourth Ed. St. Paul, MN: West Publishing.

Broadbent, D. (1982) "Task Combination and Selective Intake of Information." *Acta Psychologica* 50, pp. 253–290.

Brouwer, W. H., Waterink, W., Van Wolffelaar, P. C., and Rothengatter, T. (1991) "Divided Attention in Experienced Young and Older Drivers: Lane Tracking and Visual Analysis in a Dynamic Driving Simulator." *Human Factors* 33, pp. 573–582.

Caldwell, L. S. (1963) "Relative Muscle Loading and Endurance." *Journal of Engineering Psychology* 2, pp. 155–161.

Caldwell, L. S. (1964) "Measurement of Static Muscle Endurance." *Journal of Engineering Psychology* 3, pp. 16–22.

Chaffin, D. (1975) "Ergonomics Guide for the Assessment of Human Strength." *American Industrial Hygiene Association Journal* 35, pp. 505–510.

Chaffin, D., and Andersson, G. (1991) *Occupational Biomechanics.* Second Ed. New York: Wiley.

Damon, A., Stoudt, H. W., and McFarland, R. A. (1971) *The Human Body in Equipment Design.* Cambridge, MA: Harvard University Press.

Deeb, J. M. (1988) "An Exponential Model of Isometric Muscular Fatigue as a Function of Age and Muscle Group." Dissertation, Department of Industrial Engineering, State University of New York at Buffalo.

Garrett, J. W. (1971) "The Adult Human Hand: Some Anthropometric and Biomechanical Considerations." *Human Factors* 13, pp. 117–131.

Goldstein, I., and Dorfman, P. (1978) "Speed Stress and Load Stress as Determinants of Performance in a Time-Sharing Task." *Human Factors* 20, pp. 603–610.

Grandjean, E. (1988) *Fitting the Task to the Man.* London: Taylor and Francis.

Hettinger, T. (1961) *Physiology of Strength.* Springfield, IL: Charles C Thomas.

Konz, S. (1990) *Work Design: Industrial Ergonomics.* Third Ed. Worthington, OH: Publishing Horizons.

Kroemer, K. H. E. (1970) "Human Strength: Terminology, Measurement, and Interpretation of Data." *Human Factors* 12, pp. 297–313.

Lauback, L. L. (1976) "Comparative Muscular Strength of Men and Women: A Review of the Literature." *Aviation, Space, and Environmental Medicine* 47, pp. 534–542.

Miller, G. (1956) "The Magical Number Seven, Plus or Minus Two: Some Limits on Our Capacity for Processing Information." *Psychological Review* 63, pp. 81–97.

Monod, H. (1985) "Contractility of Muscle during Prolonged Static and Repetitive Dynamic Activity." *Ergonomics* 28, pp. 81–89.

O'Brien, M. (1985) "Women in Sport." *Applied Ergonomics* 16, pp. 25–39.

Parasuraman, R., and Nestor, P. G. (1991) "Attention and Driving Skills in Aging and Alzheimer's Disease." *Human Factors* 33, pp. 539–557.

Pheasant, S. T. (1982a) "A Technique for Estimating Anthropometric Data from the Parameters of the Distributions of Stature." *Ergonomics* 25, pp. 981–992.

Pheasant, S. T. (1982b) "Anthropometric Estimates for British Civilian Adults." *Ergonomics* 25, pp. 993–1001.

Pheasant, S. T. (1983) "Sex Differences in Strength—Some Observations on Their Variability." *Applied Ergonomics* 14, pp. 205–211.

Preczewski, S. C., and Fisher, D. L. (1990) "The Selection of Alphanumeric Code Sequences." Proceedings of the Human Factors Society 34th Annual Meeting, Santa Monica, CA, pp. 224–228.

Redgrove, J. (1979) "Fitting the Job to the Women: A Critical Review." *Applied Ergonomics* 10, pp. 215–223.

Roche, A. F., and Davila, G. H. (1972) "Late Adolescent Growth in Stature." *Pediatrics* 50, pp. 874–880.

Roebuck, J., Kroemer, K., and Thomson, W. (1975) *Engineering Anthropometry Methods.* New York: Wiley-Interscience.

Sanders, M. S. (1977) "Anthropometric Survey of Truck and Bus Drivers: Anthropometry, Control Reach, and Control Force." Westlake Village, CA: Canyon Research Group. Cited in M. S. Sanders and E. J. McCormick, *Human Factors in Engineering and Design.* New York, McGraw-Hill, 1993.

Sanders, M. S., and McCormick, E. J. (1993) *Human Factors in Engineering and Design.* New York: McGraw-Hill.

Stoudt, H. W. (1981) "The Anthropometry of the Elderly." *Human Factors* 23, no.1, pp. 29–37.

Trotter, M., and Gleser, G. (1951) "The Effect of Aging upon Stature." *American Journal of Physical Anthropometry* 9, pp. 311–324.

Wickens, C. D. (1992) *Engineering Psychology and Human Performance.* Second Ed. New York: HarperCollins.

Wilmore, J. H. (1975) "Inferiority of Female Athletes: Myth or Reality?" *Journal of Sports Medicine* 3, pp. 1–6.

3

Physical Factors

The objective of this chapter is to provide an overview and methods on how to reduce work-related risk factors that contribute to musculoskeletal disorders (MSDs). A musculoskeletal disorder is an injury that affects the bones, muscles, or other related tissues of the body joints.

In petrochemical operations, the two primary types of physical work tasks on which to focus are manual handling tasks and repetitive tasks. Since numerous physical work tasks are done at a plant, it is important to identify the tasks that hold high priority; that is, those that should have reduced risk factors to reduce the potential for MSDs. A qualitative screening approach assists in selecting the physical work tasks that need further attention and possible risk reduction. The screening approach categorizes work tasks into high-, medium-, and low-priority groups based on relative risk.

In some cases, further assessment of a task is needed to focus in on the specific risks that need reduction and guide interventions to reduce the risks. This chapter includes an overview of the main tools for assessing risk factors in manual handling and repetitive tasks. The focus is on assessment tools for the practitioner to use to evaluate the tasks and less reliance is placed on self-reporting by the individuals doing the task, since people tend to be inaccurate at estimating weights and forces, especially in unfamiliar situations (Wiktorin et al., 1996).

This chapter also offers a framework for an ergonomics program so that risk identification and reduction can be sustained on an ongoing basis. The ergonomics program recommended in this chapter is a specific application of the general program model presented in Chapter 1.

3.1 MUSCULOSKELETAL DISORDERS

Musculoskeletal disorders affect the bones and muscles of the body and the tissues that form the body joints. There are two categories of disorders, based on the type of event that caused it: (1) conditions caused by an acute trauma, such as a slip or fall resulting in, for example, a strained back, bruised leg, or sprained ankle; and (2) conditions due to exposure to a repeated, or chronic, type of physical activity, resulting in, for example, soreness from inflamed tendons or ligaments. Conditions associated with repeated exposure to physical activity are called *cumulative trauma disorders* (CTDs). *Repetitive strain injury* (RSI) is another term for a cumulative trauma disorder specifically related to repetitive tasks. Examples of CTDs include

- Tendon disorders: tendinitis, tenosynovitis, DeQuervain's disease.
- Nerve disorders: carpal tunnel syndrome, ulnar or radial nerve damage.
- Neurovascular disorders: thoracic outlet syndrome, white finger (Raynaud's syndrome).

Some people also consider low back pain a type of CTD, if a single acute cause cannot be identified. CTDs can result from, be precipitated by, or be aggravated by intense, repeated, sustained, or insufficient recovery from exertion, motions of the body, vibration, or cold. CTDs generally develop over periods of weeks, months, and years. Physical activity risk factors related to CTDs include (Occupation Safety and Health Administration, 1995)

- *Application of force.* Higher forces translate into higher loads on the muscles, tendons, and joints, which can quickly lead to muscular fatigue.
- *Repetitive motion.* This is defined as performing the same motion every 30 seconds or less or where 50% of the work cycle involves similar upper extremity motion patterns.
- *Awkward posture.* An awkward posture requires more muscular force because muscles cannot work as effectively.
- *Contact stress.* Tools, objects, or equipment that create pressure against the body (usually the hands and arms) can inhibit nerve function and blood flow.
- *Overall muscular fatigue.* Insufficient recovery time between muscular contractions may lead to overall muscular fatigue.

The magnitude of risk associated with a specific quantity of exposure to these factors is not well defined, but there is consensus that exposure to high levels or combinations of these risk factors increases the risk of the CTDs. For example, force and repetition are recognized multiplicative factors; that is, if they are simultaneous risk factors, the interaction between them likely results in an increased risk of CTD compared to a single factor. Also, tasks with high repetition and high forces are 14 times as likely to be associated with a type of CTD in the wrist called *carpal tunnel syndrome* (CTS) than low-repetition, low-force tasks (Silverstein, Fine, and Armstrong, 1987).

Because there can be multiple confounded causes of an MSD, and particularly a CTD, it is often difficult to determine whether a condition is work or nonwork related. The ergonomist's goal is to identify the most critical risk factors that may be present in the work setting that could aggravate or contribute to the cause of an injury. The critical risk factors are those that the ergonomist believes are most important to address for the best possibility of reducing the risk of an injury.

To identify the critical risk factors, an ergonomist conceptually categorizes the physical work activities into two types, those that consist mainly of single exertions involving the entire body and those that are more repetitive and most likely involve more intensive use of the upper body or arms and hands. The physical work activities that involve whole-body exertions typically involve carrying or moving an object, so they are called *manual handling tasks*. The ergonomist uses different approaches for assessing the risk factors depending on whether the task is a manual handling task or a more repetitive task.

3.2 MANUAL HANDLING TASKS

Manual handling tasks are activities during which workers move objects from place to place by lifting, lowering, carrying, pushing, or pulling. In petrochemical plants, examples of manual handling activities include

- Lifting a sack of additive and holding the sack while pouring the contents into a mixer.
- Picking up boxes of packaged motor oil from a pallet.
- Pushing and pulling a loaded pallet jack.
- Tilting and rolling drums of chemicals in staging areas.
- Grasping and turning a handwheel to manually open and close a valve.
- Holding and pouring a large bottle of liquid additive into a blender.
- Picking up a hose, dragging it, and holding it in position to connect at a manifold station.
- Picking up and carrying plant spare parts or toolboxes.

3.2.1 Manual Handling Risk Factors

There are general guidelines that can be used to screen out the tasks where the force required could be a concern. A task that involves any of the following features may pose risk that could be reduced:

- Repetitive or prolonged kneeling or squatting.
- Significant sideways twisting of the body.
- Repetitive or prolonged reaching with hands above shoulder height.
- Performance in a confined area that restricts natural posture.
- More than 16–27 kg (35–60 lb) handled and cannot be held close to the chest or torso.
- More than 27 kg (60 lb) handled.
- More than 8 kg (17 lb) lifted from or below knee height or above shoulder height.
- Unbalanced carrying or lifting (e.g., the bulk of the load supported by one side of the body).
- More than 5 kg (10 lb) handled by only one hand (e.g., a heavy tool).
- Unstable, unbalanced, difficult, or awkward to handle loads being moved.
- Object of more than 75 cm (30 in.) in two dimensions carried or lifted.
- Object weighing more than 8 kg (17 lb) carried more than 9 m (30 ft).
- Combinations of physical tasks performed for more than 1 hour at a time.
- Physical tasks requiring heavy exertion performed more than once per minute for continuous periods of more than 15 minutes.

For manual handling tasks in petrochemical plants, the ergonomist's concern is usually about the maximum force required in a single or just a few exertions. The MSD risk of a manual materials handling task is determined by comparing the force actually required to execute the task to an estimate of the maximum force that would be safe for this person to exert in similar circumstances. For a lifting task, the actual weight that must be lifted is compared to a recommended weight limit for that situation. For pulling, the force required to pull an object can be measured with a force gauge and compared to the recommended maximum pulling forces that can be safely exerted under various circumstances (see Figure 3-1).

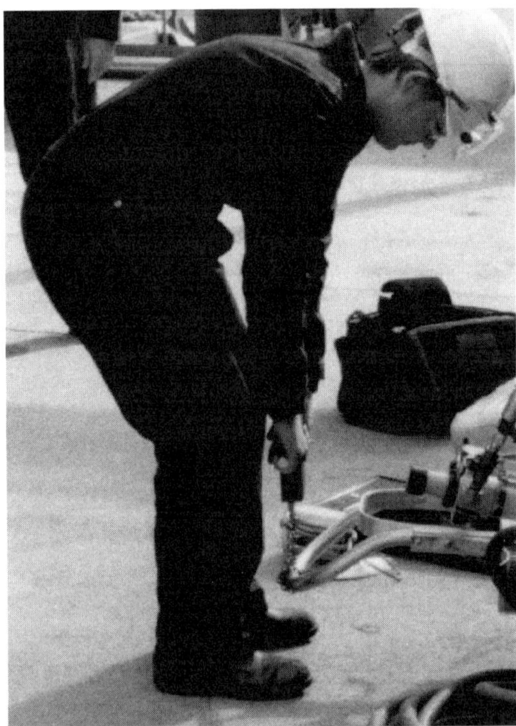

Figure 3-1 Using a force gauge to measure a pushing/pulling task.

Fewer highly repetitive lifting tasks are encountered in petrochemical operations than in other industrial operations, such as parcel package mailing and delivering. But, there may be occasions when the recommended weight limits or forces would be determined for multiple exertions (e.g., several lifts per hour, as in a warehousing operation unloading a pallet and stacking the boxes on shelves).

For lifting, aspects of the object, task, individual's technique, and the environment influence the amount of load that can be safely handled in various situations (as listed in Figure 3-2).

The size, weight, stability, and contents of the load influence the amount of weight that can be handled safely. So does the location of the load when it is picked up: The height of the load from the floor and horizontal distance of the load from the body are critical factors in assessing risk of a lifting situation. The number of times the person must lift a load during a time period is also considered. The technique the individual uses to move

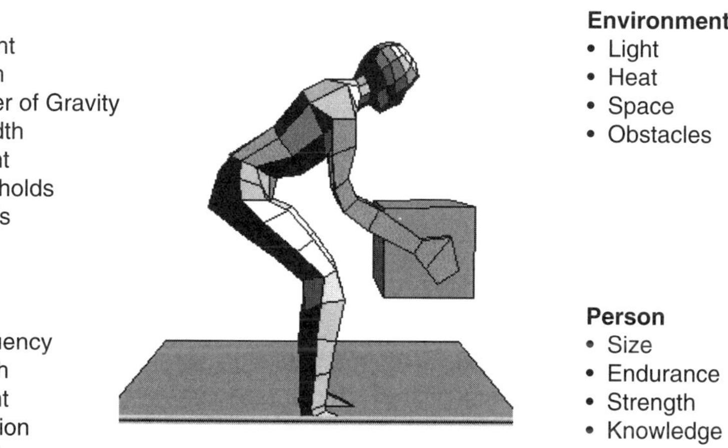

Figure shown was rendered using the University of Michigan 3D Static Strength Prediction Model (1995), with permission.

Figure 3-2 Risk factors that influence the weight of the load.

the load also influences the risk, as does individual factors such as the person's strength, size, and endurance. In addition, environmental factors such as quality of lighting in the work area and path and housekeeping factors such as obstructions and condition of the floor surface also influence how safely a load can be handled.

A safe lifting task, of course, is one in which the weight of the load does not exceed the weight that could be safely lifted. An optimal lifting task is one in which the object is compact, easy to grasp and handle, and requires no awkward movements to maneuver the load. The object is easily accessible from between waist and elbow height and oriented so that it can be approached from straight on without bending or stooping toward it. This configuration is the least stressful on the body for lifting and most favorable for safely lifting a load.

Situations that deviate from the optimum scenario contain risk factors. Examples of these situations include bulky and unstable objects, objects on the floor or overhead, and objects that are located so that twisting the torso or deep stooping is required for access.

General guidelines for manual handling tasks are

- Avoid static work as much as possible.
- Analyze and redesign jobs requiring the manual handling (lifting, pulling, and pushing) of heavy material and equipment, as necessary.
- Provide mechanical assistance for jobs in which manual forces exceed the limits of the operator.
- Provide mechanical assistance for jobs requiring frequent handling of heavy loads.
- Design manual handling jobs to avoid twisting as much as possible.

3.2.2 Methods for Evaluating Manual Handling Tasks

Three main tools are available to evaluate MSD risk of manual handling tasks: (1) checklists and postural observation, (2) National Institute of Occupational Safety and Health (NIOSH) lifting equation, and (3) mathematical models based on biomechanical principles. By design these tools tend to yield conservative results. Evaluations from any single tool should not be solely relied on, but should be considered as input into the prioritization and decision-making process. The training and skill required to use these tools depends on the type of tool. The checklists and postural observation tools described in this chapter can be effectively used by personnel who have minimal ergonomics training. Use of other tools, such as the NIOSH lifting equation, is more complex and requires more training for proficiency. For example, Dempsey et al. (1991) show that trained individuals can effectively apply the NIOSH lifting equation but untrained individuals make critical errors when using the equation. Use of mathematical and biomechanical models is not appropriate for the layperson and best left to the trained specialists. Training to use the models is available, however, and those that may have to use the model frequently get more out of the training than those who use it infrequently.

3.2.2.1 Postural Observation Checklists for Manual Handling Tasks

The objective of postural observation methods is to identify the tasks or portions of tasks most physically stressful. To use most postural observation methods, all that is needed is the worksheet and a pencil, but for most methods it is recommended that the task be videotaped so that it can be reviewed while scoring the worksheet. Two available methods are rapid entire body assessment (REBA), developed by Hignett and McAtamney (2000), and Ovako working posture analyzing system (OWAS), developed and validated by Karhu, Kansi, and Kuorinka (1977). Subsequently, OWAS was applied in industrial settings and the analyses published (Karhu, Harkonen, Sorvali, and Vepsalainen, 1981; Kivi and Mattila, 1991; Mattila, Karwowski, and Vilkki, 1993; Scott and Lambe, 1996). The OWAS method is intended to be applied to whole body analyses, such as for manual handling tasks, and therefore, it provides very little precision for upper limb assessment. The method was developed specifically so that it could be used successfully by untrained personnel and provide unambiguous answers. The procedure and forms for using the OWAS can be found in Karhu et al. (1981).

Attachment 1 is a worksheet that can be used to evaluate and set priorities for materials handling tasks to identify those of highest priority for reducing MSD risk. The worksheet provides a rating system for each aspect that affects the safety of a manual handling task, such as nature of the object, weight lifting, posture used, repetition, and other conditions of the task. The ergonomist fills out one worksheet per task and tallies the score for the entire task. Then, tasks are compared and tasks with the higher scores receive the higher priority for follow-up.

3.2.2.2 Calculation of Weight Limit for Two-Handed Lifting Tasks

NIOSH used a conservative approach to develop an equation to calculate the recommended weight limit (RWL) for simple two-handed lifting tasks (Waters et al., 1993; Waters, Putz-Anderson, and Garg, 1994). The equation is applicable to specific types of lifting tasks and requires training for effective application (Dempsey et al., 1991). Table 3-1 is a checklist that allows the user to determine when the NIOSH equation should *not* be used. If the task meets any of these conditions, then the equation cannot be used and other means of determining a recommended weight limit should be applied.

The NIOSH lifting equation is

$$\text{RWL (Recommended Weight Limit)} = \text{LC} \times \text{HM} \times \text{VM} \times \text{DM} \times \text{AM} \times \text{FM} \times \text{CM}$$

The variables are defined in Table 3-2, and Figure 3-3.

TABLE 3-1

Criteria for Determining When the NIOSH Lifting Equation Is *Not* Applicable

Lifting or lowering with one hand.
Lifting or lowering for over 8 hours.
Lifting or lowering while seated or kneeling.
Lifting or lowering in a restricted work space.
Lifting or lowering unstable objects.
Lifting or lowering while carrying, pushing, or pulling.
Lifting or lowering with wheelbarrows or shovels.
Lifting or lowering with high-speed motion (faster than about 30 in. per second).
Lifting or lowering with unreasonable foot or floor coupling (<0.4 coefficient of friction between the sole of the shoe and the floor).
Lifting or lowering in an unfavorable environment (i.e., the temperature significantly outside the 19–26°C [66–79°F] range, relative humidity outside the 35–50% range).

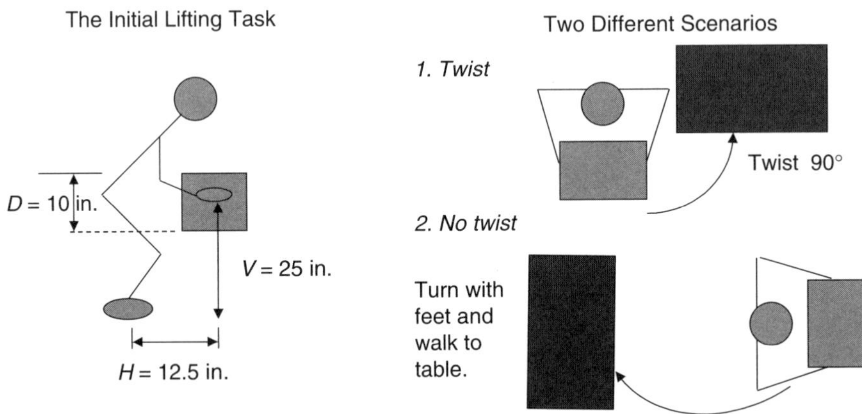

Figure 3-3 Demonstration of the use of the NIOSH (1991) lifting guidelines.

The multipliers and the task variables (which determine the values of the multipliers) listed in the equation are summarized in Table 3-2. Finally, Tables 3-3 and 3-4 outline the values for the frequency multiplier (FM) and the coupling multiplier (CM).

Figure 3-4 illustrates use of an electronic spreadsheet programmed with the NIOSH equation to determine the recommended weight limit (RWL)

TABLE 3-2

Summary of NIOSH Multipliers and Task Variables

Multiplier Name	*Multiplier Abbreviation*	*Metric Equations*	*U.S. Customary Equations*	*Task Variable*
Load constant	LC	23 kg	51 lb	None
Horizontal multiplier	HM	(25/H)	(10/H)	H
Vertical multiplier	VM	$1 - (0.003\|V - 75\|)$	$1 - (0.0075\|V - 30\|)$	V
Distance multiplier	DM	$0.82 + (4.5/D)$	$0.82 + (1.8/D)$	D
Asymmetric multiplier	AM	$1 - (0.0032 \times A)$	$1 - (0.0032 \times A)$	A
Frequency multiplier	FM	See Table 3-3	See Table 3-3	None
Coupling multiplier	CM	See Table 3-4	See Table 3-4	None

H = horizontal distance of hands from midpoint between the ankles. Measure at the origin and the destination of the lift (cm or in.).

V = vertical distance of the hands from the floor. Measure at the origin and destination of the lift (cm or in.).

D = vertical travel distance between the origin and the destination of the lift (cm or in.).

A = angle of asymmetry, angular displacement of the load from the sagittal plane. Measure at the origin and destination of the lift (degrees).

F = average frequency rate of lifting measure in lifts/min. Duration is specified <1 h, <2 h, or <8 h, assuming appropriate recovery allowances.

C = coupling factor from Table 3-4.

TABLE 3-3

Frequency Multipliers

	Work Duration					
	<1 h		*<2 h*		*<8 h*	
Frequency (lifts/min)	$V < 75$	$V \geq 75$	$V < 75$	$V \geq 75$	$V < 75$	$V \geq 75$
0.2	1.00	1.00	0.95	0.95	0.85	0.85
0.5	0.97	0.97	0.92	0.92	0.81	0.81
1	0.94	0.94	0.88	0.88	0.75	0.75
2	0.91	0.91	0.84	0.84	0.65	0.65
3	0.88	0.88	0.79	0.79	0.55	0.55
4	0.84	0.84	0.72	0.72	0.45	0.45
5	0.80	0.80	0.60	0.60	0.36	0.35
6	0.75	0.75	0.50	0.50	0.27	0.27
7	0.70	0.70	0.42	0.42	0.22	0.22
8	0.60	0.60	0.35	0.35	0.18	0.18
9	0.52	0.52	0.30	0.30	0.00	0.15
10	0.45	0.45	0.26	0.26	0.00	0.13
11	0.41	0.41	0.00	0.23	0.00	0.00
12	0.37	0.37	0.00	0.21	0.00	0.00
13	0.00	0.34	0.00	0.00	0.00	0.00
14	0.00	0.31	0.00	0.00	0.00	0.00
15	0.00	0.28	0.00	0.00	0.00	0.00
>15	0.00	0.00	0.00	0.00	0.00	0.00

TABLE 3-4

Coupling Multipliers

	Coupling Multipliers	
Couplings	*V < 75 cm (30 in.)*	*V > 75 cm (30 in.)*
Good	1.00	1.00
Fail	0.95	1.00
Poor	0.90	0.90

for a lifting task that involves picking up a box located on a shelf 25 inches high and moving it to a shelf 10 inches higher (35 inches from the floor). The calculations indicate that the RWL is about 40 lb, for this situation. Figure 3-5 illustrates the use of the NIOSH equation to determine the RWL for a lifting task identical to the one in Figure 3-4, except the new task involves a 45° torso twist. The RWL for the task with twisting is reduced by about 12.5% to 35 lb.

3.2.2.3 Biomechanical Models

A biomechanical model is a series of calculations that mimic the mechanics of the human body. The models can be used to predict the limits on force or strength exerted by a person while positioned in a specified posture. One widely used computerized model that was developed by the University of Michigan (1995; see Figure 3-6) is applicable to lifts that are infrequent and performed slowly.

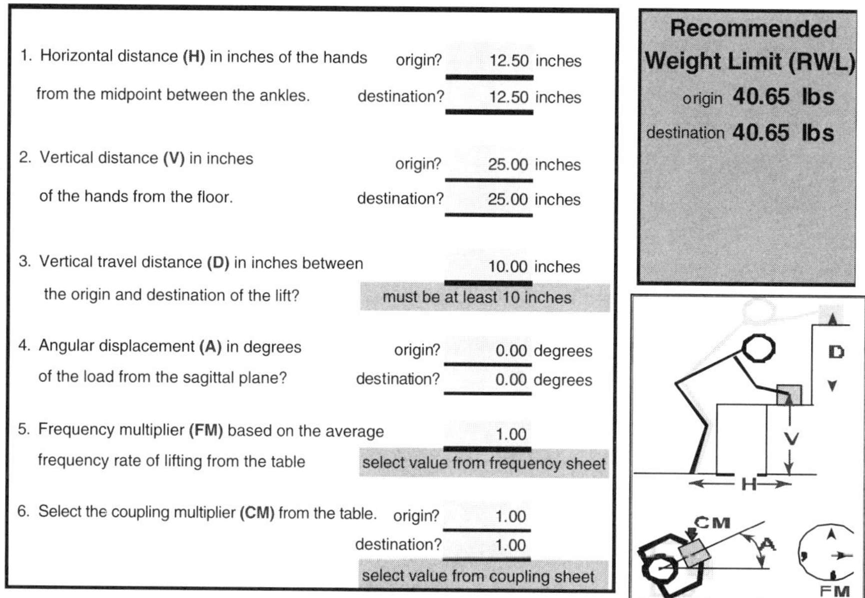

Figure 3-4 NIOSH equation and figure, no twisting.

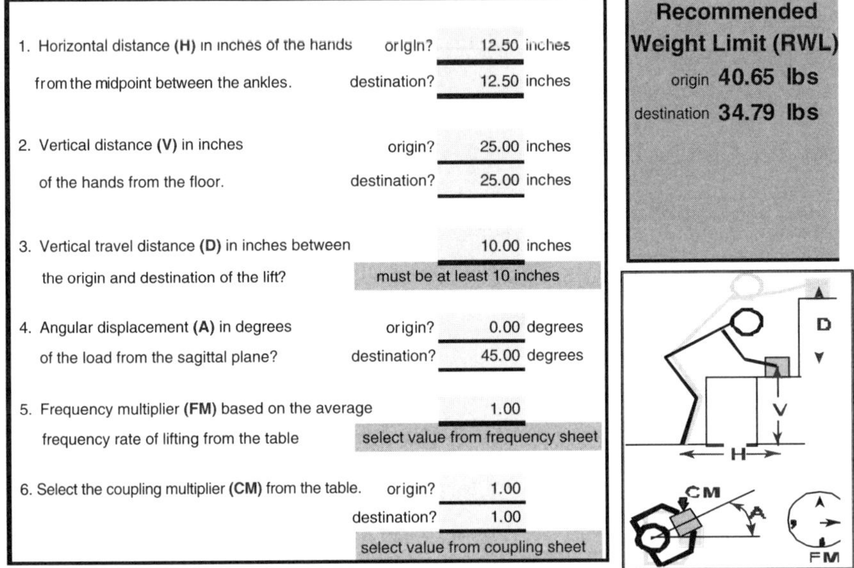

Figure 3-5 NIOSH equation and figure, twisting at the waist.

This model estimates the limit based on two criteria:

1. Estimated percentage of the population that would be capable of exerting strength required to do the task.
2. Estimated compression forces on the spinal disc between the fifth lumbar and first sacral vertebrae in the region of the low back.

The percentage capable and the estimated compression forces can be used as guidelines to assist the practitioner in assessing the MSD risk of a particular task.

For specific situations, the model estimates the percentage of the population capable and the compression force on the spine, based on the following inputs:

- Size of the person doing the task.
- Angle position of each body joint that determines the posture used to do the task.

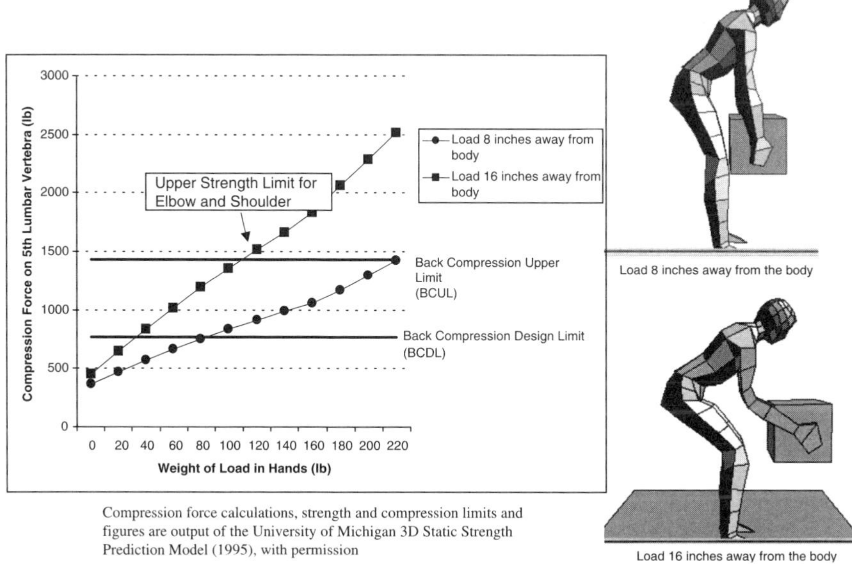

Figure 3-6 University of Michigan model: Plot of load (lb) versus compression force (lb) for 8″ horizontal and 18″ horizontal for best lifting posture.

- Weight of the object or the force being applied by the hands to the object being handled.
- Description of how the hands apply force to the object (e.g., lifting upward, pushing downward, pushing or pulling horizontally).

When the percentage capable is less than 99% for men and 75% for women, the force required by the task is categorized as the strength design limit (SDL), which is similar to an "action limit" (NIOSH, 1981). When less than 25% of men and less than 1% of women are capable, the task requires force at the strength upper limit (SUL), which is, in effect, a "maximum permissible limit." If the maximum permissible limit were exceeded, there may be an increased risk of MSD injury.

If the compression force exceeds 770 lb (3400 Newton), then the back compression design limit (BCDL), or "action limit," is exceeded. Compressive forces over 1430 lb (6400 Newton) exceed the back compression upper limit (BCUL), or "maximum permissible limit." The SDL and BCDL guidelines were developed to be "consistent with the biomechanical and psychophysical criteria used to develop the NIOSH lifting equation." Whenever the upper limits are exceeded, engineering controls or alterations in work practices are highly recommended. If the design limits are exceeded, decisions regarding workplace changes must be made with regard to

1. The percentage capable of safely performing the task considered.
2. The costs and benefits associated with the introduction of engineering controls.
3. Work practice alterations to accommodate the percentage of the population not capable of performing that task.

The NIOSH equation and the biomechanical models can be used for doing sensitivity (i.e., what if?) analyses on various situations to explore how dimensions of the work area can affect the ergonomic risk factors. Various task parameters, such as lifting height or weight of the load, can be increased or decreased and the equations used to recalculate the recommended weight limit, the percentage capable, and the spinal forces. The recalculated values can then be compared to the original case to determine how much risk reduction can be achieved with various modifications to the work setup.

Other tools also can be used to determine if the force requirement of a manual handling task exceeds recommended limits. Numerous tables of acceptable forces are listed in Ayoub, Mital, and Nicholson (1992) and published papers can be consulted for specific situations (e.g., Woldstad, McMulkin, and Bussi, 1995, for hand wheels). Suggested limits for pushing and pulling a cart, such as a flatbed, can be determined by referring to tabled values of maximum acceptable forces published by Snook and Ciriello (1991). The tables were developed by measuring the maximum acceptable forces that industrial workers chose to exert during pushing and pulling tasks. The maximum acceptable forces are indexed in tables according to height of the hands while pulling or pushing, number of times per minute the pushing or pulling is required, and gender. To determine if a pulling or pushing task requires excessive force, the actual force can be compared to the corresponding maximum acceptable force. To do this, the actual force required to execute the push or pull is measured with a force gauge. More extensive tables for pushing and pulling, as well as lifting, tasks can be found in Ayoub et al. (1992).

3.3 HAND-INTENSIVE REPETITIVE TASKS

For hand-intensive repetitive tasks, the emphasis of the assessment is on identifying which of the five CTD risk factors (force, repetition, awkward posture, contact stress, and muscular fatigue) are significant enough to warrant reduction. In petrochemical operations, some jobs consist of many different tasks, each possibly containing various combinations of the risk factors. Highly repetitive tasks that are repeated hundreds of times per hour are not as likely, compared to other industries such as electronics assembly and word processing.

3.3.1 Risk Factors

If the task involves any of the following features, then it may pose a risk of MSD. More commonly, more than one of the risk factors is present:

- Reaching frequently reaching behind the body or more than 30 cm (12 in.) away from the body.
- Repetitive or prolonged reaching with hands above shoulder height.
- Repetitive or prolonged work with the hands above elbow height.
- Repetitive or prolonged finger-pinch gripping or hand grasping.
- Repetitive or prolonged rotation of the forearm (e.g., motion while using a screwdriver).
- Repetitive or prolonged awkward wrist posture (see Figure 3-7).
- The same or similar hand, arm, or upper body motions repeated frequently.
- Force is exerted by the hands over prolonged periods of time.
- Tool handle or grasped object digs into palm or fingers.

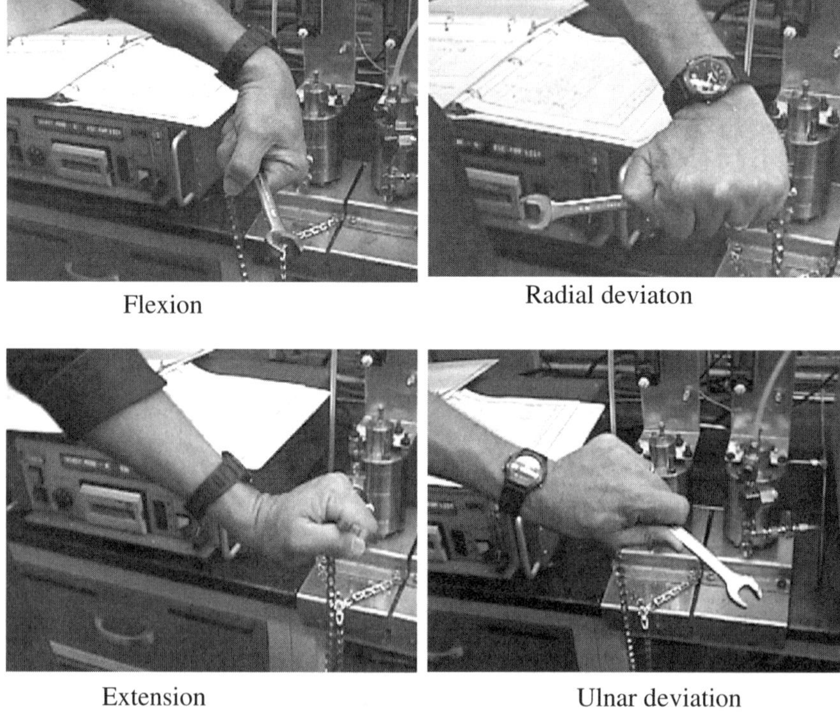

Figure 3-7 Examples of improper wrist posture.

- High force is applied with a finger-pinch grip or a hand grip.
- The force required to manipulate controls, levers, or hand tools is unacceptably high.
- Repetitive pushing of buttons or manipulation of toggle switches on equipment.

Examples of risk factors in an electrician's job:

- Direct contact of tool handles to the palm and fingers while applying high amounts of force to the handle (e.g., utility blade handle, ridges on screw driver handle, short handles of pliers).
- Force from hitting, pushing, and pulling with the butt of the hand (e.g., hitting end of the wrench to turn it).

- Static grip and squeezing motion required to operate hand tool and support the weight of the hand tool with its power cord (e.g., cutting cable, power tools).
- Repetitive twisting motion of the hand and forearm to prep the work area while manually removing protective covers or threaded fasteners as well as replacing them after the electrical work is completed.

3.3.2 Survey and Observation Tools

More detailed analyses can be conducted by using one of the observation methods for tasks that involve use of the hands and arms. These methods are approaches for ranking tasks according to relative risk and are not measures of risk. In other words, a task scoring 4 according to this method is judged to contain more risk than a task scoring 2, but the task scoring a 4 need not have twice the risk of the task scoring 2.

Many methods are available for scoring tasks, but the best option is to choose a tool shown to be both valid and reliable for detecting risks related to CTDs. A valid tool reasonably identifies the risks related to the five CTD risk factors, the risk factors listed in the previous section. A reliable tool produces relatively similar results when tasks are scored by different raters.

One of these methods is the rapid upper limb assessment (RULA) method, developed by McAtamney and Corlett (1993) for analyzing upper-limb-intensive tasks (see Figure 3-8). RULA is easy for an untrained individual to apply to rate priority for tasks involving the hand, arm, and shoulder.

3.3.3 Hand Tools

Use of hand tools potentially involves the five CTD risk factors as well as potential risk for acute injury, such as a finger pinch. Work areas and hand tools used by personnel with jobs where hand tools are used extensively, such as pipefitting, electrical, mechanical, and shop areas, should be evaluated for features that may contribute either to cumulative or acute injuries. The features of the hand tool that should be evaluated are

Wrist Postures And Scores:

-0° 1. No wrist flexion or extension

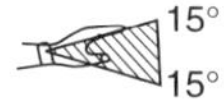

2. Wrist flexed or extended up to 15°

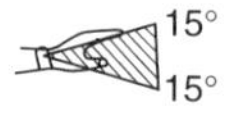

3. Wrist flexed or extended beyond 15°

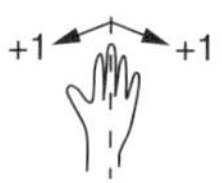

Note:
Add 1 if the wrist is bent away from the midline

Wrist Twist And Scores:

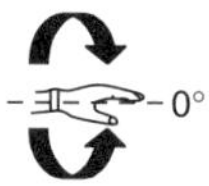

1. Twisting at wrist is absent or only in the mid-range

2. Twisting is extreme, at or near the end of the range

Figure 3-8 Excerpt of RULA observation worksheet for wrist postures (McAtamney and Corlett, 1993).

- Weight and balance.
- Grip characteristics.
- Trigger or other controls.
- Vibration (described in Chapter 4).
- Posture while using the tool.

In general, a tool should be easy to grasp and operate. The grip area in contact with the hand should minimize hard edges. Vibration should be minimized where possible. Heavy tools should be equipped for two-handed use. The following is a summary of tool guidelines based on detailed specifications found in Eastman Kodak Company (1983).

The recommended upper weight of a hand tool is about 4.5 lb (2 kg) with a maximum of 6 lb (2.7 kg) supported with one hand. If the tool is heavier, then provide an external device to counter or support the tool weight. A tool designed to be operated with two hands distributes forces, provides a more stable grip, overcomes torque, and provides the user better control over the tool. The hand or arm muscles should not have to be used to counteract a tool that does not have the weight balanced correctly.

A grip shape that is circular or oval with a diameter of 1.25 to 1.75 in. (3.1 to 4.4 cm) is recommended. If the grip requires a span across a lever (such as the handles of pliers), then the span should be in a range 2–3 in. (5–7.5 cm). All the hand and fingers should comfortably fit onto the grip. Operation of the tool should not require a restricted portion of the finger (such as the tip of the finger) or hand to provide prolonged static pressure on the tool. The thumb and index finger should slightly overlap around a closed grip for maximum strength. The grip is where the hands and fingers apply force to the tool, so it should be covered with a soft material that reduces hard edges and pressure points. The covering should protect the hands from the temperature of the tool (e.g., hot or cold surface) and reduce the slipperiness of the tool in the hand. Some coverings also claim to attenuate vibration exposure to the hand.

The manner in which the tool is used also influences the potential for injury. Hand and wrist postures are important indicators. Tools should be able to be used without extreme bending of the wrist or awkward arm posture. The work should be arranged at about elbow height to avoid awkward arm and shoulder posture. The work should also be designed to minimize the need for static posture, such as prolonged gripping with the hands or holding out an outstretched arm. A correct glove fit is also important. Tight-fitting gloves can put pressure on the hands, while loose-fitting gloves reduce grip strength and dexterity.

3.4 BEHAVIOR

Individual work tasks can, of course, be assessed as needed with one or a few of the tools described in the previous sections of the chapter. The actual challenge of assessing work tasks, however, is to assess multiple work tasks and relative risk so that intervention efforts and resources can be focused at high-priority conditions—the aspects of work tasks identified as the highest potential contributors to a MSD injury.

The procedures described in this chapter are used to assess the risk associated with job and task characteristics. Risks associated with the work environment, human behavior, and management systems can be identified and addressed with other tools.

Conditions of the work environment that could contribute to a MSD injury, such as a back injury resulting from a slip, trip, or fall, include good housekeeping and adequate illumination of work areas. Examples of environmental issues that could contribute to an injury are

- Poor housekeeping (e.g., obstacles, such as hoses, left on the floor or in areas traversed by personnel carrying loads).
- Slippery walking surfaces (e.g., ice, rain, or slick substances).
- Dark areas, where the walking surface is not adequately illuminated.

These types of issues should be identified during safety inspections, job safety analyses, and task risk assessments.

Individuals can contribute to controlling MSD risk by proper behavior while doing physical tasks. Human behaviors that can help prevent MSD injuries are, but not limited to,

- Proper lifting technique.
- Proper pushing and pulling technique.
- Use of lift assist devices when indicated.

Proper lifting technique involves

- Planning the lift.
- Positioning the feet.
- Using good posture.
- Getting a firm grip.
- Keeping close to the load.
- Lifting the load smoothly.
- Moving the feet.
- Putting down, then adjusting the load.

Proper pushing and pulling technique involves

- Where possible, pushing instead of pulling.
- Where possible, locating the hands at about elbow height when standing erect.

- Staggering the feet, one foot forward.
- Not using sudden movements.
- Leaning slightly forward.

The results of studies on the effectiveness of back belts in preventing back injuries are equivocal, including the most recently published ones (Krause et al., 1996; NIOSH, 1994; Magnusson and Pope, 1996). The use of back belts as injury prevention devices should not be considered a substitute for a comprehensive ergonomics program. To prevent back injuries, it is necessary to train individuals to lift properly, provide adequate engineering controls, and ensure that injuries are reported early, so they can be treated and the situation ameliorated before any further, more serious, injury might occur.

To address behavioral risk factors, the site behavior-based safety program can incorporate critical behaviors that help prevent MSDs. In this type of program, critical MSD injury prevention behaviors can be identified and evaluated as safe or at risk during routine observations conducted as part of the program and are addressed as needed. As an example, for one program, the observers watch for the following critical lifting, bending, and twisting behaviors during a safety observation of warehouse employees:

- While lifting, is the employee using primarily the legs, keeping a curve in the back, the breastbone facing forward, and the load close to the body?
- Are the shoulders in line with the hips?
- Is the load held isometrically with the arms, as much as possible?
- Is the lift balanced and controlled?
- Does the employee avoid lifting or moving heavy or bulky objects without assistance?
- Is the load held close to the body?
- Does the employee keep frequently lifted material at a level between the waist and shoulders?

- While bending, is the employee keeping a curve in the back, and shoulders in line with hips?
- Does the employee bend from the hips or knees rather than from the waist, using a wide base of support as much as possible?
- Does the employee avoid bending where possible?
- Does the employee avoid twisting the spine at all times, keeping the shoulders in line with the hips?
- Does the employee turn by moving the feet and taking steps?

3.5 ERGONOMICS PROGRAM

An ergonomics program provides a process to assess, evaluate, and control the risks associated with work-related MSD injuries. The following basic principles are recommended as a basis for any ergonomics program:

- Ergonomic processes are supported at every level of the organization.
- Ergonomic processes are proactive, designed with input from all levels of the organization and aligned with the culture, complexity, and level of risk of the site.
- Ergonomics applications are integrated into corporate safety practices, standards and management systems.
- Management is actively involved through participation in goal setting, training, and stewardship.
- Ergonomic processes support risk identification, assessment and interventions to reduce risks.
- The program complies with local ergonomics regulation.

The ergonomics program recommended in this chapter is a specific application of the general program model presented in Chapter 1. There are other sources of recommendations for ergonomics program structures, such as those by NIOSH (1997), which grew from the proposed ergonomics regulations in the United States.

3.5.1 Planning an Ergonomics Program

The stages of ergonomics program implementation are planning, training, assessment, interventions, control, and continuous improvement.

Planning the program involves the following steps:

1. Develop the site plan and documentation (e.g., reference materials, manual).
2. Develop the training package.
3. Identify key site personnel and contacts.
4. Define the goals of the ergonomics program.

Training, awareness, and expertise in ergonomics should be developed and maintained. This includes the initial orientation and awareness training and periodic ongoing and refresher training for managers, supervisors, professionals, and other employees as appropriate. Training can be integrated into existing or ongoing safety and health training courses.

The level of training depends on the employee's role in the program. (Table 3-5 is an example of assigned roles and responsibilities for an ergonomics program.) General awareness training for all employees should include

- Common MSDs, signs and symptoms.
- The importance of reporting MSDs early.
- How to report MSDs, signs, and symptoms.
- Risk factors, jobs, and work activities associated with MSD hazards.
- Proper lifting and material handling techniques.
- A description of relevant local regulations and standards.

Safety meetings may be used to provide supplemental job specific ergonomic training, such as use of checklists, worksheets, and calculations. Management, supervisors, medical, and safety directors and the site skilled contacts should receive training on risk identification, setting risk priorities, and assessment. The site skilled contacts or subject matter experts should receive training on how to use detailed risk assessment tools for ergonomics agents (e.g., how to assess lifting, pushing, pulling,

TABLE 3-5

Example Roles and Responsibilities for an Ergonomics Site Program

Management sponsor	Ensure implementation, execution, and maintenance of program Participate in an annual review of the program, facilitating improvements to the program and manual as necessary Assure that the management and local program manual administrators are properly trained and qualified Identify the responsible and accountable resources (administrator and skilled contacts) and ensure they are trained
Program administrator	Oversee the program Conduct periodic evaluations of the effectiveness of the ergonomics control program for the management sponsor Periodically review the program manual for necessary changes Assure that applicable personnel are properly trained Assure appropriate records are maintained, including key performance indicators Notify the employee and his or her supervisors of the results of the workplace evaluations
Line supervisor	Assure the program is implemented for their work areas Promote participation in the program Ensure proposed changes that affect the program are addressed through the management of change process Consider ergonomic principles when designing or reviewing job tasks Ensure employees receive ergonomics training Present job-specific ergonomics topics at safety meetings Promptly refer employees to medical personnel if they report symptoms and initiate incident investigation procedures as necessary Request workplace ergonomic evaluations as needed Support implementation of interventions and controls
Site skilled contact	Implement the program at each field location Ensure site-specific procedures are written Ensure worker training is conducted Maintain records and training Arrange general awareness training for site personnel Evaluate work environment or job tasks as indicated by risk assessment or referral (e.g., after an incident) Identify, prioritize, and assess risks Work with other personnel to develop and implement controls Report key performance indicators

continued

TABLE 3-5

Example Roles and Responsibilities for an Ergonomics Site Program *Continued*

	Serve as the subject matter expert (SME) for field-related questions Review injury, illness, and incident reports for ergonomic trends
Health clinic or medical personnel	Evaluate persons reporting discomforts or concerns with MSD signs and symptoms Report ergonomics injuries and illnesses to safety and management personnel for workplace evaluation follow-up Evaluate employee's fitness for work as needed
Trainer	Maintain documentation of ergonomics training
Ergonomist	Assist with development of training packages Train site contacts and other key personnel to assess risks Assure quality of data and analyses Provide technical expertise as needed Assist with intervention and controls as needed
Site engineer	Assist development of controls Include ergonomic principles in new design and modification projects Ensure project employees receive ergonomics training Include ergonomics considerations in design execution, including design reviews
Employees	Attend and participate in scheduled ergonomics training Use tools and equipment as instructed Use proper work practices as instructed in ergonomics training Report symptoms of injury or illness at their first appearance Report workplace hazards or conditions

and upper extremity repetitive tasks). The project engineers and design specialists should receive training on how to consider minimizing design features that affect ergonomic hazards or risks.

3.5.2 Risk Assessment Process

The risk assessment process should cover the following:

- Reactive (e.g., response to incidents) and proactive (preventative) risk assessment.
- Screening using the following data to set priorities on tasks that need assessment: safety incident data, health illness data, absenteeism data,

behavior-based safety observations on ergonomics (e.g., following lifting procedures, using proper lifting technique).

- Detailed assessments (e.g., use of NIOSH lifting equation, biomechanical model) on specific high-risk tasks.
- Review of projects for new facilities or modified plant areas.

There are two levels of risk assessment for proactively assessing existing operations: screening and detailed. The first level is a screening process that may be used to proactively assess a specific job or task for potential risk factors to prevent injuries. The screening process is used to qualitatively assess a job or task, identify obvious concerns or conditions that can be quickly fixed, and assess priorities on interventions to reduce risk or initiate further detailed assessment.

When using these tools, a higher score indicates that a task contains a relatively high number of key MSD risk factors. Management should consider these jobs or tasks at the highest priority for risk reduction. High- or moderate-risk tasks that have no simple or quick solution are quantitatively assessed with the best exposure assessment methods available. Tools such as simple checklists, postural observation, and task analysis worksheets are appropriate for screening and have already been described in this chapter. Data such as injury reports, medical, management, and employee concerns should also be considered in the screening process to assign priority to tasks for follow-up.

The second level is a detailed assessment process, which is used when quick fixes to high or moderate risks are not immediately evident or the identified solutions are considered costly or resource intensive. Detailed assessments take more effort and skill to conduct, but they provide quantitative insights that can help target the intervention for effective use of resources. The detailed assessment process is quantitative where possible and based on the best-available exposure assessment methods. Quantitative assessment provides insight on which portions of a job or tasks contribute the most risk, therefore guiding the cost-effective solution that most effectively reduces risk. In addition, quantitative assessments can be used to conduct sensitivity analyses, in which the tasks or jobs can be modeled with the proposed changes and the risk reduction estimated before any design changes actually take place. This process is illustrated in Figure 3-9.

More comprehensive analysis approaches encompass risk identification of factors beyond the physical risk factors. One approach, described by

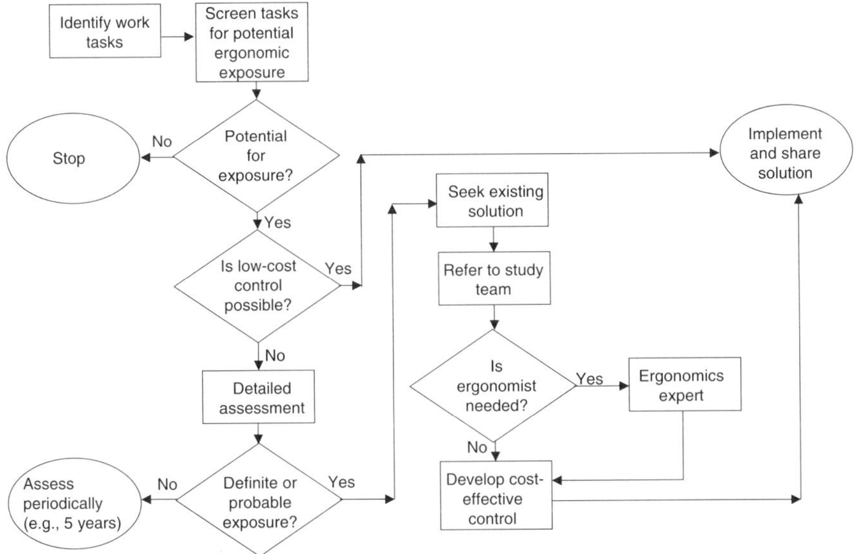

Figure 3-9 Flowchart on risk assessment process.

the Canadian Association of Petroleum Producers (2000), involves the use of a series of forms that guide the risk identification and assessment process. Several potential risk factors are reviewed on the forms, including a signs and symptoms questionnaire and posture, force, repetition, organizational, and environmental factors. The scores on each form are summarized and the tasks are categorized into low-, medium-, and high-risk categories based on the aggregate task score.

Another comprehensive approach is by NIOSH (1997) as part of proposed ergonomics regulations. This approach is based on criteria, called *triggers*, that initiated aspects of an ergonomics program based on certain conditions.

3.5.3 Solutions

Timely, successful solutions, regardless of size, build credibility, which is the foundation on which to support a longer-term process. In the solution development phase of the project, the emphasis is to "design in" ergonomic improvements in selected workplaces and implement appropriate engineering, work-practice, and administrative controls. These solutions take the form of modifying equipment, tools, or materials; changing work procedures; or modifying work schedules, where appropriate, by distributing difficult tasks among employees and shifts or job rotation.

In some cases, simple solutions are available through modifications in purchasing specifications, procedural changes, and the like. In other cases, solutions are not as obvious and require further development. Where solutions are costly or complex, it may be useful to request input from other similar operations, consultants, or equipment designers.

Engineering controls are physical changes to a job that reduce MSD hazards. Examples of engineering controls include changing or redesigning workstations, tools, facilities, materials, or work processes (refer to Chapter 6 for more detail). Examples of tools that reduce hazards include

- Box tables to raise loads.
- Tools to move loads (pallets, drums, boxes, or furniture).
- Hoists to assist lifting.
- Scissor lifts to keep loads at an appropriate height.
- Tilt tables to improve access.
- Portable stairs to prevent stretching.

Administrative controls are changes in the way work in a job is assigned or changes to schedules that reduce the magnitude, frequency, or duration of exposure to ergonomic risk factors. Examples of administrative controls include

- Rotating employees in positions to vary physical activity.
- Recombining jobs or tasks to adjust the physical workload.
- Providing alternative tasks.
- Changing the work pace (if the work pace is not under the control of the employee, such as a conveyor belt with product moving past).

3.5.4 Evaluating the Ergonomics Program

The ergonomics program should be assessed periodically to evaluate its effectiveness and identify improvement areas. Some indicators of effectiveness may include

- Percentage of incidents related to MSD injury.
- Evaluation of local work area.
- Percent training completed.
- Relative percentage safe and at-risk behavior observed.

The program evaluation checklist in Appendix 3 can used to conduct a gap assessment on existing programs.

3.6 CASE STUDY

This is a study on how to apply the musculoskeletal risk identification tools available from the Canadian Association of Petroleum Producers (CAPP, 2000). The risk identification process that it recommends consists of two levels: (1) screening and (2) priority setting and detailed assessment.

The CAPP provides a series of worksheets that support the screening and detailed assessment processes. The CAPP worksheets are used to screen and assess a series of work tasks to change out the catalyst in a reactor. The work tasks are

- Remove the 50-lb nuts from each bolt that secures the cover at the top of the reactor, then replace them and close the cover when finished changing the catalyst. This task requires hoisting an impact gun to about waist height and reaching upward to support it while it rests vertically over the nut to loosen and tighten it. The handles of the impact gun are above shoulder height.
- While sitting at the bottom of the reactor, push a long pole vertically up through the individual tubes within the reactor to loosen the catalyst so that it falls out of the reactor. Each push with the rod requires about 30 lb of upward force with one hand. The rod handle is a contact stress to the hand.
- While sitting near the top of the reactor, use a pneumatic drill to remove the caked catalyst from each tube within the reactor, so that the drilled pieces fall through the tubes to the bottom floor. Pushing the drill through the caked material requires about 20–35 lb of force depending on the psi used for the drill.

- Move the drums of catalyst to the top of the reactor—400-lb drums of catalyst are moved from an elevator to the reactor area. Drums are staged at a height of about 41 in. from the floor.
- Pour the new catalyst in the drums into the top of the reactor by pushing with 95 lb of force on the drum to tilt it so the contents pour into the reactor.

The level 1 screening process worksheet is shown in Figure 3-10.

The CAPP screening worksheet indicates a lower potential of issues associated drilling out the used catalyst and pouring in the new catalyst. The screening worksheet indicates that detailed assessments may be useful on the other tasks to help further assess risk. Level 2 detailed assessments consist of biomechanical analyses of the removal of the reactor cover, prodding the catalyst, and moving the drums.

	Task name	Perceived risk						Frequency and Duration Task Is Performed
	In distinct work activity comprised of several steps or actions	Moderate or Severe Body Part Discomfort: (Y/N)	Awkward Work Postures? (Y/N)	High Effort or Force? (Y/N)	High Repetition or Work Rate? (Y/N)	Contact Stress or Skin? (Y/N)	High Mental Stress? (Y/N)	1/month: 1/week: <1/week 1/day, and 1/hour
1.	Remove /replace reactor head	N	Y	Y	N	N	N	1/month
2.	Loosen catalyst with rod	N	Y	Y	N	Y	N	1/month
3.	Drill caked catalyst	N	N	N	N	N	N	1/month
4.	Move drums	N	N	Y	N	N	N	1/month
5.	Tilt drums to pour contents	N	N	N	N	N	N	1/month
6.								

Figure 3-10 Level 1 summary form from CAPP risk identification process.

Biomechanical assessment indicates that the recommended weight limit for the nuts on the reactor is about 44 lb for the current height of about chest height. A platform is installed to bring the operator up in height so that the work is at about 31 in. from the floor, resulting in a recommended limit of 50 lb. Loosening the catalyst by prodding the reactor with a vertical rod requires moderate shoulder and elbow strength and has no excessive compression force on the spine. For handling the drums, biomechanical assessment indicates that most people lack the shoulder strength to safely perform this task, although compression forces on the spine are not excessive. Drum lifting and handling equipment is acquired to minimize the amount of manual drum movement required to maneuver the drums over to the reactor.

APPENDIX

APPENDIX 3-1

Manual Handling Screening Checklist: Risk Identification and Priorities (Deeb, 1994)

Work Location ______________________ Date ________________

Task Description

____

Task Code ______________________________

Assessed by ____________

Please answer the questions in all three sections. You can have multiple answers to some questions. An asterisk next to a question means that multiple answers are possible. Please answer *all* questions to the best of your ability, leave no question unanswered. If you are unsure about an answer to a question, go to highest answer category. For example, in Question 2.3, if the object weighs 30 lb (14 kg), choose answer number 4.

Name: ____________________ Age: ________

Height: __________

Weight: ____________________ Male/Female

Left/Right Handed

For Authorized Use Only SCORE

Section 1

1. Task Characteristics
   1.1 The operator performs the task:
   1. Sitting 2. Standing 3. Squatting or kneeling

   Other comments:

   1.2 When performing the task, the upper body is: (multiple answers are possible)*
   1. Straight/erect 2. Bent forward 3. Bent backward 4. Twisted

   Other comments:

   1.3 Performing the task involves: (multiple answers are possible)*
   1. Normal arm reaches
   2. Awkward postures, such as deviation from stable body position
   3. Reaching above shoulders
   4. Uneven distribution of body weight on the feet, such as one foot is lower than the other

   Other comments:

   1.4 The task is performed at the pace of the operator
   1. Yes
   2. No

   Other comments:

For Authorized Use Only SCORE

1.5 Performing the task involves: (multiple answers are possible)*

L 1. Lifting/lowering (answer 1.6, 1.7, 1.8, and go to Section 2)

P 2. Pushing/pulling (answer 1.9, 1.10, and go to Section 2)

C 3. Carrying (answer 1.11 and go to Section 2)

Other comments:

L 1.6 This question has two parts, please answer Part A and Part B individually:

A. The initial lifting/lowering height is at
   1. Waist or elbow height
   2. Near the floor
   3. Above the shoulders

B. The location of object in relation to the worker's body is:
   1. Close to the front of the body
   2. Far from the body

Other comments:

L 1.7 The object is moved (lifted/lowered): (multiple answers are possible)*

1. At the same level
2. From waist to shoulder level/shoulder to waist level
3. From floor to waist level/waist to floor level
4. From floor to shoulder level/shoulder to floor level

Other comments:

L 1.8 This question has two parts, please answer Part A and Part B individually:

A. The task of lifting/lowering objects is performed
   1. For less than half an hour at a time
   2. Between half an hour and 1 hour at a time
   3. Between 1 hour and 2 hours at a time
   4. More than 2 hours

B. The frequency of lifting/lowering of objects is performed
   1. Less than once per minute
   2. Between one and six times per minute
   3. More than six times per minute

Other comments:

For Authorized Use Only SCORE

P 1.9 In what direction is the pushing/pulling applied?

1. In front of the body

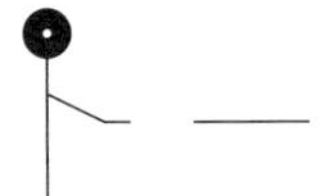

2. Across the front of the body

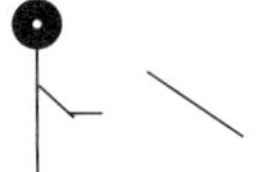

Other comments:

P 1.10 The force applied to push/pull the object is sustained for

1. Less than 10 seconds
2. Between 10 and 20 seconds
3. Between 20 and 30 seconds
4. More than 30 seconds

Other comments:

C 1.11 Carrying the object requires a holding grip or position for

1. Less than 5 seconds
2. Between 5 and 10 seconds
3. More than 10 seconds

Other comments:

1.12 The task (please answer parts A, B, C, and D individually)

A. Requires unusual capability to be performed (i.e., beyond expectations of the normal worker)
   1. No
   2. Yes

B. Aggravates those with a health problem (i.e., back and muscular problems)
   1. No
   2. Yes

C. Aggravates those who are pregnant
   1. Not applicable (for men)
   2. No
   3. Yes

D. Requires special training/individual received training
   1. No/not applicable
   2. Yes/yes
   3. Yes/no

Other comments:

Total of Section 1

Section 2

For Authorized Use Only SCORE

2. Material/object characteristics:
   2.1 The type of object can best be described as a
      1. Bucket
      2. Box
      3. Bag
      4. Other (e.g. hose, drum, pump) ____________________

      Other comments:

   2.2 The geometric shape of the object can best be described as
      1. Small
      2. Medium
      3. Large, bulky, or unwieldy

      Other comments:

   2.3 The weight of the object handled is
      1. Below 10 lb (4.5 kg)
      2. Between 10 and 20 lb (4.5 and 9 kg)
      3. Between 20 and 30 lb (9 and 14 kg)
      4. Between 30 and 40 lb (14 and 18 kg)
      5. Between 40 and 60 lb (18 and 27 kg)
      6. Above 60 lb (27 kg)

      Other comments:

   2.4 The object's load
      1. Is handled using one hand (e.g., bucket, hose)
      2. Is shared equally between the hands
      3. Is not shared equally between the hands (i.e., one side is heavier than the other)
      4. Can shift suddenly (i.e., unstable contents such as liquid, powder)

      Other comments:

   2.5 The object has
      1. Handles
      2. No handles

      Other comments:

For Authorized Use Only SCORE

2.6 The area of contact between hands/arms and object
1. Presents no hazard/difficulty
2. Is cold/hot
3. Has sharp edges

Other comments:

Total of Section 2

Section 3

3. Workplace characteristics
3.1 The task is performed in
1. An unrestricted space
2. A restricted space (places constraints on posture)
3. A very restricted space (places severe constraints on posture)

Other comments:

3.2 The task is performed in
1. A closed environment with no strong air movement
2. An open environment with strong air movement

Other comments:

3.3 The ambient temperature and humidity of the workplace is
1. Within the comfortable zone (temperature between 19° and 26°C [66° and 79°F] and humidity between 35 and 65%)
2. Outside the comfortable zone (hot/cold/humid conditions)

Other comments:

3.4 The workplace is considered to have (consider day and night work)
1. Adequate lighting for the job
2. Inadequate lighting for the job

Other comments:

3.5 The floor surface (multiple answers are possible)*
1. Is even
2. Has variation in levels (i.e., uneven, steps)
3. Is slippery

Other comments:

	For Authorized Use Only SCORE
3.6 The workplace has 1. Good housekeeping 2. Poor housekeeping, such as crowded or waste on the floor Other comments:	
3.7 The conditions where manual materials handling tasks are performed require 1. Standard protective clothing 2. Additional protective clothing (e.g., chemical suits, respirators) 3. Special protective clothing such as thermal coverall and gloves Other comments:	
Total of Section 3	

Summary

Relative Scores

	Total (max)	*% Severity*
Section 1	(116)	
Section 2	(22)	
Section 3	(20)	
Total	(158)	

Comments/recommendations:

APPENDIX 3-2

Musculoskeletal Ergonomics Program Gap Analysis Checklist

Checklist Item	*Confirmed (✓)*
Program Feature	
Implementation	
The program is implemented throughout the site	
Employees/local unions support the program	
A system is in place for employees to report concerns	
Feedback is given to employees regarding reported concerns	

continued

APPENDIX 3-2

Musculoskeletal Ergonomics Program Gap Analysis Checklist *Continued*

Checklist Item	*Confirmed* (✓)
Concerns are addressed in a timely manner	
The program includes the following components:	
Planning	
Selection/priority assignment of worker groups for screening	
Screening for detailed assessment	
Detailed risk assessment	
Training for the general population and assessors	
Control development	
Measurement/continuous improvement	
Changes to tasks triggers re-evaluation	
Members of worker group are interviewed	
Jobs are divided into tasks for assessment	
Training and qualifications of ergonomics resources are commensurate with role and level of involvement	
A program is in place to address office ergonomics	
The program is part of every day practice/awareness	
Management Support	
The program has visible support by site management and line management	
Roles in the program are clearly defined	
Enough personnel to effectively implement and sustain the program	
Enough time to effectively implement and sustain the program	
There is enough budget to effectively implement and sustain the program	
The program is linked to other site systems such as incident investigation, near miss reports, procedures development and training, risk assessments, management of change and medical system	

continued

APPENDIX 3-2

Musculoskeletal Ergonomics Program Gap Analysis Checklist *Continued*

Checklist Item	*Confirmed (✓)*
Training	
All employees are trained with an introduction to ergonomics	
Supervisors/management/medical directors trained in the screening process (overview)	
Training includes consideration for equipment design that affects musculoskeletal ergonomics	
Refresher training scheduled as needed	
Training given new employees	
Verification method in place to determine if training is adequate and understandable	
General awareness training includes identification of symptoms and signs	
Training addresses procedures for reporting work-related musculoskeletal disorders and concerns	
Workers who change program roles have sufficient training to fill new role	
Training enables managers, supervisors, and employees to identify aspects of a job that increase risk of developing work-related musculoskeletal disorders	
Assigning Priorities	
Jobs assigned priority prior to the screening process	
Symptom surveys used across the site to assess trends	
Health/illness data used to assess trends and set priorities	
Absenteeism data used to assess trends and set priorities	
Safety incident data used to assess trends and set priorities	
Communication with similar sites used to identify potentially high-risk tasks	
Outcome of employee concerns are used in assigning priorities	

continued

APPENDIX 3-2

Musculoskeletal Ergonomics Program Gap Analysis Checklist *Continued*

Checklist Item		*Confirmed (✓)*
Screening Tools		
	Screening tools are easy to use and understand	
	Screening tools identify tasks that require detailed assessment	
	Person conducting screening process is qualified	
	Screening based on medical data and concern of management, employees, and evaluator	
	All field jobs are screened based on prioritization process	
Detailed Assessment Tools		
	Whole body posture assessed using a standardized tool	
	Tool/method: ______________________	
	Upper extremity exposure assessed using a standardized tool	
	Tool/method: ______________________	
	Vehicle operation assessed using a standardized tool	
	Tool/method: ______________________	
	Lifting and lowering assessed using a standardized tool	
	Tool/method: ______________________	
	Whole body pushing and pulling forces assessed using a standardized tool	
	Tool/method: ______________________	
	Energy expenditure assessed using a standardized tool	
	Tool/method: ______________________	
	Results for each tool mapped to an action level rating to identify relative risk	
	Trained evaluator observes the worker performing the task	
	Tasks are videotaped for later detailed analysis	
	Workstation measurements taken (surface heights, clearance space, etc.)	

continued

APPENDIX 3-2

Musculoskeletal Ergonomics Program Gap Analysis Checklist *Continued*

Checklist Item		*Confirmed (✓)*
	Forces and hand positions measured	
	Exposure to heat, cold, and vibration considered	
	Repetition and duration of task considered	
	Hand tool usage assessed	
	Coefficient of friction on walking surface (slippery floor) considered	
	Person conducting the detailed assessment is qualified	
Intervention and Controls		
	Solutions evaluated with considerations of relative risk and relative cost/benefit analysis	
	Effectiveness of solutions implemented measured for musculoskeletal risk factors and work efficiency	
	Solution development documented	
	Solutions developed and implemented in a timely manner, depending on severity of issue	
	Solutions and learning effectively communicated throughout the organizational levels and shifts/work groups, across similar sites, and a central resource	
	Employees empowered to make changes to their work environment	
	Employees empowered to make suggestions for changes to their work environment	
	Solution evaluated before implementation by a qualified skilled resource	
	Control strategies reported through management of change procedures	
	Changes in work procedures or required work effort (force, shift length, container size, etc.) trigger re-evaluation	
	Assessed tasks given priority for improvement according to relative risk	

continued

APPENDIX 3-2

Musculoskeletal Ergonomics Program Gap Analysis Checklist *Continued*

Checklist Item	*Confirmed (✓)*
Continuous Improvement/Measurement	
The program stewarded on a regular basis	
The program has been continuously improved	
Benchmarking data (i.e., health illness data, absenteeism data) collected	
Measures of program performance in place and include safety incident data, health illness data, absenteeism data, and restricted work.	
Internal and external quality assurance processes in place to evaluate data and the program	

3.7 REVIEW QUESTIONS

Test your understanding of the material in this chapter.

1. Which of the following is not a type of cumulative trauma disorder?
   (a) low back pain.
   (b) broken wrist.
   (c) carpal tunnel syndrome.
   (d) tendinitis.
2. What five risk factors are associated with cumulative trauma disorders?
3. How would you determine whether the weight of a box of spare parts is too heavy to be safely lifted by the warehouse attendant?
4. Which dimension, distance of the load away from the body or vertical location of the load, is the more significant contributor to the effect of the load?
5. What general instructions would you give someone on how to lift safely?

6. How would you identify which tasks need a detailed ergonomic assessment?
7. What topics should be covered in a general employee awareness course on preventing musculoskeletal injuries?

REFERENCES

Ayoub, M. M., Mital, A., and Nicholson, A. S. (1992) *A Guide to Manual Materials Handling*. London: Taylor and Francis.

Canadian Association of Petroleum Producers (CAPP) and Canadian Petroleum Products Institute (CPPI). (2000) *Ergonomic Risk Identification and Assessment Tool*. Calgary, Alberta: CPPI.

Deeb, J. (1994) Personal Communication.

Dempsey, P. G., Burdorf, A., Fathallah, F., Sorock, G., and Hashemi, L. (1991) "Influence of Measurement Accuracy on the Application of the 1991 NIOSH Equation." *Applied Ergonomics* 32, pp. 91–99.

Eastman Kodak Company. (1983) *Ergonomic Design for People at Work*. New York: Van Nostrand Reinhold Company.

Hignett, S., and McAtamney, L. (2000) "Rapid Entire Body Assessment." *Applied Ergonomics* 31, pp. 201–205.

Karhu, O., Harkonen, R., Sorvali, P., and Vepsalainen, P. (1981) "Observing Working Postures in Industry: Examples of OWAS Application." *Applied Ergonomics* 12, no. 1, pp. 13–17.

Karhu, O., Kansi, P., and Kuorinka, I. (1977) "Correcting Working Postures in Industry: A Practical Method for Analysis." *Applied Ergonomics* 8, no. 4, pp. 199–201.

Kivi, P., and Mattila, M. (1991) "Analysis and Improvement of Work Postures in the Building Industry: Application of the Computerized OWAS Method." *Applied Ergonomics* 22, no. 1, pp. 43–48.

Kraus, J. F., Brown, K. A., McArthur, D. L., Peek-Asa, C., Samaniego, L., Kraus, C., and Zhou, L. (1996) "Reduction of Acute Low Back Injuries by Use of Back Supports." *International Journal of Occupational and Environmental Health* 2, pp. 264–273.

Magnusson, M., and Pope, M. (1996) "Does a Back Support have a Positive Biomechanical Effect?" *Applied Ergonomics* 27, no. 3, pp. 201–205.

Mattila, M., Karwowski, W., and Vilkki, M. (1993) "Analysis of Working Postures in Hammering Tasks on Building Construction Sites Using the Computerized OWAS Method." *Applied Ergonomics* 24, no. 6, pp. 405–412.

McAtamney, L., and Corlett, N. (1993) "RULA: A Survey Method for the Investigation of Work-Related Upper Limb Disorders." *Applied Ergonomics* 24, no. 2, pp. 91–99.

NIOSH. (1981) *Work Practices Guide for Manual Lifting.* Cincinnati, OH: U.S. Department of Health and Human Services.

NIOSH. (1994) *Workplace Use of Back Belts.* DHHS (NIOSH) Publication No. 94-122. Cincinnati, OH: U.S. Department of Health and Human Services.

NIOSH. (1997) *Elements of Ergonomics Programs.* DHHS Publication No. 97-117. Cincinnati, OH: U.S. Department of Health and Human Services.

Occupational Safety and Health Administration. (1995) *OSHA Proposed Ergonomic Protection Standard.* Washington, DC: Bureau of National Affairs.

Scott, G. B., and Lambe, N. R. (1996) "Working Practices in a Perchary System, Using the OVAKO Working Posture Analysing System (OWAS)." *Applied Ergonomics* 27, no. 4, pp. 281–284.

Silverstein, B. A., Fine, L. J. and Armstrong, T. J. (1987) "Occupational Factors and Carpal Tunnel Syndrome." *American Journal of Industrial Medicine* 11, pp. 343–358.

Snook, S. H., and Ciriello, V. M. (1991) "The Design of Manual Handling Tasks: Revised Tables of Maximum Weights and Forces." *Ergonomics* 34, no. 9, pp. 1197–1213.

University of Michigan, Center for Ergonomics. (1995) *3D Static Strength Prediction Program Manual.* Ann Arbor: University of Michigan.

Waters, T. R., Putz-Anderson, V., and Garg, A. (1994) *Applications Manual for the Revised NIOSH Lifting Equation.* DHHS (NIOSH) Publication No. 94-110. Cincinnati, OH: U.S. Department of Health and Human Services.

Waters, T. R., Putz-Anderson, V., Garg, A., and Fine, L. J. (1993) "Revised NIOSH Equation for the Design and Evaluation of Manual Lifting Tasks." *Ergonomics* 36, no. 7, pp. 749–776.

Wiktorin, C., Selin, K., Ekenvall, L., Kilbom, A., and Alfredsson, L. (1996) "Evaluation of Perceived and Self-Reported Manual Forces Exerted in Occupational Materials Handling." *Applied Ergonomics* 27, no. 4, pp. 231–239.

Woldstad, J. D., McMulkin, M. L., and Bussi, C. A. (1995) "Forces Applied to Large Hand Wheels." *Applied Ergonomics* 26, no. 1, pp. 55–60.

4

Environmental Factors

4.1 INTRODUCTION

This chapter discusses four environmental factors or features that arise from the surroundings (i.e., illumination and temperature) and mechanical features of the workplace (i.e., noise and vibration) that could be encountered at work and affect behavior. Other examples of environmental features that could be encountered at work and may affect behavior are dangerous chemicals, radioactive materials, atmospheric pressure, and the like. These other environmental features are not discussed in this chapter.

This chapter discusses the four environmental features that can affect people in three ways: (1) health, (2) performance, and (3) comfort. The affects of these three aspects are usually combined. For example, poor health can lead to both poor performance and reduced comfort and, thus, reduced work satisfaction. Stressors, arising from illumination, temperature, noise, vibration, or any other aspect of the environment can adversely affect people when they reach a certain level, although the effect may not be apparent either to the person being affected or an observer. The ideal range for performance and comfort is narrow. Therefore, trying to adapt to conditions outside the ideal range can make people use more effort, which can lead to reduced performance and comfort; for example, a person trying to see fine details when illumination levels are too low or too high.

This chapter ends with a case study examining the lighting quantity and quality in a control room. In addition, a set of review questions are included to help you check your understanding of the material covered in this chapter.

4.2 ILLUMINATION

Light stimulus is characterized by its intensity and wavelength. The intensity of light is expressed in terms of the amount of luminous flux (energy) it generates. Flux is measured in candlepower (cd) or lumen, where 1 candle (or candela) is equal to 1 lumen. The wavelength of light is measured in terms of nanometer (nm). One nanometer is one-billionth of a meter (10^{-9}). For example, the eye discriminates between different wavelengths in the range of 380 and 780 nm by the sensation of color.

In our daily environment, the amount of illumination (also called *illuminance*) that falls on a surface (e.g., desk) depends on the energy of the light source, its distance from the surface, and the angle of the surface to the light source. Most light falls on a surface perpendicular to the light. Illumination is described in terms of the rate of flux (lumens) produced by the source and the surface area over which it spreads. One lumen per square meter is defined as 1 lux and one lumen per square foot is defined as a foot-candle (FC). Illumination meters are usually calibrated in both measures. One foot-candle is about 10.76 lux (there are 10.76 square feet in a square meter). Light may fall on a surface from multiple sources or, for example, as reflections from walls, ceilings, and other objects in the immediate environment. These reflections can be considered individual light sources, and the efficiency of each depends on the angle at which it reflects light on the surface. In summary, the amount of light falling on a surface depends on the:

1. Light source and its luminous intensity.
2. Distance between the light source and the surface.
3. Angle of the light source to the surface.
4. Number of light sources and reflecting sources in the immediate environment.

We do not see the light falling on a surface but we see the light reflected from a surface. Luminance is a measure of the light reflected from the surface and is associated with the subjective sensation of brightness. Keep in mind that, depending on their surface characteristics, different bodies absorb and reflect different amounts and qualities of light. Luminance is expressed in terms of candela per meter squared (cd/m^2).

The subjective sensation of brightness (that is, the luminance qualities of an object) is perceived after the light has stimulated the retinal cells of the eye and the information passed on to the optic cortex in the brain. Then, the person reports the intensity of the object's brightness. Therefore, the intensity is determined by the luminance of the object, the amount of light passing through the eyes and to the brain, and the subjective experiences of the observer. In turn, the subjective experiences depend on the observer's past experiences and the brightness of other bodies in the visual field. This process in illustrated in Figure 4-1. This section discusses how

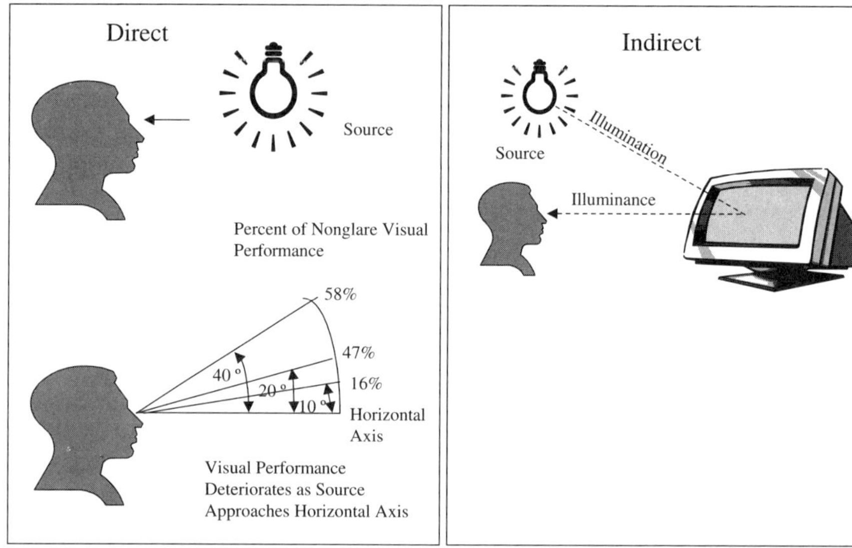

Figure 4-1 Quality of illumination—glare.

illumination affects performance in terms of lighting quantity and quality and provides guidelines on improving lighting in the workplace.

Visual performance in the workplace is primarily a function of the overall quantity and quality of illumination, task factors, age of the individual, glare, and contrast.

4.2.1 Lighting Quantity

Studies in the literature demonstrate that increasing illumination can certainly lead to an increase in visual performance in terms of productivity and efficiency. However, the increase depends highly on the type of task performed (Barnaby, 1980; Jaschinski, 1982; Knave, 1984) and age of person (Bennett, Chitlangia, and Pangrekar, 1977; Ross, 1978). One must be careful in prescribing higher and higher levels of illumination: a continuous increase in the level of illumination results in smaller and smaller improvements in visual performance, until that performance levels off.

4.2.2 Task Factors

The level of illumination needed generally depends on the task performed. The more visually intense the task (i.e., smaller details or lower contrast), the higher is the illumination level required, up to a maximum, where performance levels off no matter how much more the illumination is increased. The point at which this leveling off occurs is different for different tasks. Another approach to enhance visual performance is through changing the features of the task; for example, increasing the size of the object seen or the contrast between the object and its background (discussed in Chapter 2). The visual tasks that people perform can vary in difficulty from simple ones, such as finding a car in a parking lot at night, to the highly demanding, such as calibrating an instrument. Within this wide range of task difficulties are wide ranges of illumination levels that can help increase performance. For example, Table 4-1 presents

TABLE 4-1

Illumination Levels for Different Types of Activities

Activity	*Illumination Range (lux)*
Public area navigation	20–50
Occasional visual tasks	100–200
Visual tasks with large objects and high contrast, reading printed material	200–500
Reading poorly printed material with medium contrast or small objects	500–1000
Reading very poorly printed material with low contrast and very small objects for a short period	1000–2000
Reading very poorly printed material with low contrast and very small objects for a long period	2000–5000
Difficult visual tasks such as inspection or fine assembly	5000–10,000
Difficult visual tasks with very low contrast and very small objects	10,000–20,000

Note: To obtain the values in foot-candles, simply divide the values in lux by 10.76; for example 20 lux = 1.86 or about 2 foot-candles.

ranges of illumination levels (in lux) recommended for different types of visual work activities and areas. The more demanding the visual task and the longer it is sustained, the higher is the illumination level required. It is recommended that local task lights be used to provide illumination levels above 1000 lux. Table 4-2 lists illumination levels for selected tasks in process areas of a petrochemical environment.

4.2.3 Age Factors

The aging process has its affects on the visual system. As we grow older, in general, over 40 years old, we experience the following changes (Hughes and Neer, 1981; Wright and Rea, 1984):

TABLE 4-2

Illumination Levels for Selected Tasks in Process Areas of a Petrochemical Environment

Process Area	*Illumination Level (lux)*
Control Room	
Desk area and console	325
General	215
Manual Sampling Points	215
Compressor Houses	160
Loading racks	
Loading point	110
General area	32
Platforms	
Main operating area	55
Ordinary area	22
Stairways and ladders (frequently used)	55
Exchanger area	32
Streets	4

Note: To obtain the values in foot-candles, simply divide the values in lux by 10.76; for example, 215 lux = 19.98 or about 20 foot-candles.

- The lens of the eye atrophies and hardens, decreasing its ability to conform to the shape necessary for quick and efficient focus on objects at different distances from the eyes. These changes are also the contributors to the improvement of the far vision (farsighted) and the deterioration of the near vision (nearsighted).
- The muscles controlling the diameter of the pupil also atrophy, reducing the size of the pupil and decreasing the range and speed of adjusting and adapting quickly and efficiently to different levels of illumination.
- A lower level of illumination reaching the retina leads to a reduction in the level and rate of dark adaptation.
- As the lens hardens, it yellows, leading older people to make errors in sorting or matching colors in the blue-green and red regions.
- The changes in both the lens and pupil flexibility and efficiency lead to a decrease in visual acuity; the ability to see fine details.

These changes should encourage us to account for the elderly population in the design of lighting to ensure visual comfort and efficient visual performance. For example, we should provide:

- *More light for elderly people.* This can be achieved through providing adjustable personal light.
- *Higher contrast.* For example, for older people to see as well as people 20–35 years old (baseline), the following multiplication factors are needed to ensure higher contrast and maximize visual performance (Blackwell and Blackwell, 1968; Grandjean, 1988):
  40 years old, 1.17.
  50 years old, 1.58.
  65 years old, 2.66.

4.2.4 Lighting Quality

A number of factors have a direct effect on task visual performance and subjective impressions of people. These are the lighting color, glare, and illumination ratio. Each factor is discussed briefly next.

4.2.4.1 Lighting Color

The objects in the visual field and the various types of lamps we interact with daily differ in color. In terms of lighting, color is also referred to as *characteristic spectral distribution*. The color of an object is determined by the:

- Wavelengths of visible light that it absorbs and reflects. For example, a red object absorbs most colors in the incident light except red. Incident light is the light falling or striking on something.
- Color of the incident light. Two properties of the incident light determine the resulting color of the object: the color rendering index (CRI) and the color "temperature" of the light.

The color rendering index was developed to quantify the accuracy with which the light source renders (reproduces) the natural colors in an object. The index varies from 0 to 100. As the CRI increases, color judgment errors tend to decrease. Sources with CRIs below 60 should be avoided when accurate color judgment is important (Collins and Worthey, 1985; Wotton, 1986). However, CRIs greater than 80 indicate good color rendering.

Color temperature refers to the numerical assessment of a lamp's color. The numerical assessment is based on a comparative color appearance of glowing of light sources known as *blackbodies*. This is referred to as *color temperature*, expressed in Kelvins (K). For example, color temperatures up to 3000 K are "warm" and make reds appear more vibrant. Color temperatures between 3000 and 4000 K enhance most colors equally. Color temperatures greater than 4000 K enhance greens and blues. Table 4-3 provides CRI, color temperature, and life values for various types of common lamps.

4.2.4.2 Glare

Glare occurs whenever one part of the field of vision is brighter than the level to which the eye has become adapted, causing decrease in visual performance, annoyance, and discomfort. Glare can be characterized as direct or indirect (refer to Figure 4-1). Direct glare is caused when the light source appears directly in the field of view (discussed in Chapter 2), such as a lightbulb beside a hard to read instrument, the sun during the

TABLE 4-3

Most Common Lamp Types and Their Color Rendering Index (CRI), Color Temperature (K), and Life in Hours

Lamp Type	*Color Rendering Index (CRI)*	*Color Temperature (K)*	*Life (hr)*
Incandescent filament lamp	100	2600–3000	1000–2000
Cool white deluxe fluorescent	89	4200	10,000–20,000
High-pressure sodium	22–75	1600	10,000–15,000
Warm white deluxe fluorescent	73	2900	10,000–20,000
Metal halide	65–80	3300–5700	6000–20,000
Cool white fluorescent	66	4200	10,000–20,000
Warm white fluorescent	52	3000	10,000–20,000
Low-pressure sodium	40	1600	16,000
White fluorescent	60	3500	10,000–20,000
Tungsten halogen lamp	100	3000–3400	100–2000
Mercury lamp	15–55	55–60	12,000–24,000
Sunlight at sunrise	—	1800	—
Sunlight at noon	—	5000	—

day, or car headlights, especially high beam, at night. Indirect glare, also referred to as *reflected* or *specular glare*, is caused by bright reflections from polished or glossy surfaces, such as screens of visual display terminals (VDTs), glossy paper, or white walls. In addition, glare is commonly classified into disability and discomfort glare.

Disability glare occurs when there is direct interference with visibility and visual performance and often is accompanied by discomfort. Discomfort glare produces discomfort, annoyance, irritation, or distraction but does not necessarily affect visibility or visual performance. The distinction between the terms *disability* and *discomfort* is often unclear. The same lighting conditions can produce disability and discomfort glare simultaneously.

Reducing glare is an important ergonomic consideration when designing workplaccs. The guidelines presented in Table 4-4 should be considered.

4.2.4.3 Luminance Ratio

The ratio of the brightness (luminance) of the object of interest to the luminance of the surrounding visual field is termed the *luminance ratio.* Our eyes adapt to the luminance of the immediate visual field. The surroundings can have different distribution of luminance and, therefore, create high and low luminance ratios, which can cause transient adaptation problems. Transient adaptation occurs when the visual system moves from a bright to a dark area or vice versa (e.g., entering a dark movie theater on a bright sunny day or coming back outside after the movie). The greater the difference in luminance level (ratio of change), the greater the problem of transient adaptation, which can affect visibility, visual comfort, and performance. An example of large differences in luminance levels is found in the control room. The VDT screen has low luminance

TABLE 4-4

Reducing and Controlling Glare

Reduce or Control Direct Glare	*Reduce or Control Indirect Glare*
Light source should not be directly in the field of vision or line of sight	Position the light source or work area so reflected light is not in the direction in which the person normally needs to look to achieve maximum visibility
Use shades, hoods, or glare shields to reduce the brightness (luminance) of the light	Avoid using reflective colors and materials, such as bright metal, glass, glossy paper
Use several low-intensity lamps in place of few high-intensity lamps	Use surfaces that diffuse light, such as textured finishes, nonglossy surfaces, flat paint
When using fluorescent tubes, they should be at right angles to the line of sight	Provide several low-intensity lamps in place of few high-intensity lamps and use indirect lighting
Use indirect lighting	The brightness of the lamps (luminaires) should be kept as low as feasible

level as compared with background windows but about the same luminance as a background wall. So, it is best to mount the VDT against the wall rather than against the window.

Contrast (or relative luminance) is defined by specifying the ratio of the two luminances and is calculated using the following formula:

$$C = (L_t - L_b)/L_b$$

where C = contrast, L_t = target luminance, L_b = background luminance.

The following guidelines are widely accepted:

- The distribution of luminance on the major surfaces in the field of view should be uniform with the working visual field brightness—brightest in the middle and darker toward the edges.
- The luminance ratio (brightness : contrast) between the middle of the visual field and the immediate surroundings should not exceed 1 : 3, where 1 is the task or object viewed (middle of the visual field) and 3 is the immediate surroundings. The ratio between the middle of the visual field and remote areas should not exceed 10 : 1.
- The maximum luminance ratio within an office-like setting (includes control room) should not exceed 40 : 1.

Reducing high luminance ratios can be accomplished by eliminating bright walls, reflective and glossy surfaces, and dark or black furniture, machines, or filing cabinets. For example, Grandjean (1988) recommends the following reflectance for an office environment:

Ceiling, 80–90%.

Walls, 40–60%.

Furniture, 25–45%.

Machines and equipment, 30–50%.

Floor, 20–40%.

4.3 TEMPERATURE

The human body's thermal regulation system tries to maintain a relatively stable internal (core) temperature of between 36.1 and 37.2°C (97 and 99°F). The core temperature must stay within a narrow range to prevent serious damage to health and performance. When physical work is performed, additional body heat is generated. If we add relatively high humidity to the ambient temperature, the resulting condition can lead to fatigue and potential health risks.

The human body maintains heat balance by increasing blood circulation to the skin; therefore, we sweat on hot days. When it is cold, the body reduces the blood circulation to the skin and we shiver to keep the extremities warm. The body generates heat through metabolism and physical work. To maintain the internal heat balance, the body exchanges heat (gain or loss) with the environment in four ways:

1. *Convection.* This process depends on the difference between the air and skin temperatures. If the air temperature is hotter than the skin, then the skin absorbs heat from the air, which becomes heat gain to the body. If the air temperature is cooler than the skin, then the body loses heat.

2. *Conduction.* This process relates to the difference between the temperature of the skin and the surface contacted directly. For example, touching a hot stove, the skin gains heat and may burn.

3. *Evaporation.* This process depends on the difference between the water vapor pressure of the skin and water vapor in the environment (or relative humidity).

4. *Radiation.* This process refers to the difference between the temperature of the skin and surfaces in the environment. For example, standing in sunlight, we receive radiation from the sun.

This section discusses how temperature affects performance and health. It also looks into the comfort and discomfort zones and, finally, discusses the tools used to measure energy demands and how to reduce heat stress. Excellent information on this subject can be found in the literature (ISO, 1994, 1995; Parson, 1995).

4.3.1 Effects of Heat on Performance

Many studies over the last five decades investigated the effects of thermal conditions (heat) on performance of cognitive and physical tasks.

No firm conclusions can be drawn. For example, while many studies have shown some performance decrement, others demonstrated no differences or even improvements in performance (Kobrick and Fine, 1983). The reasons may be related to the following interacting factors:

- Most studies occur in a laboratory environment.
- Subjects are mainly young healthy men.
- Different techniques and exposure conditions are used.
- Wide differences among the tasks performed.
- Subjects have different skill levels.
- Individual differences (biologically and physiologically) not controlled.

These reasons create difficulty generalizing the results to the general working population. However, despite the differences, some reasonable generalized conclusions can be used in the work environment. These are summarized next. A good review of task performance in heat is provided in Ramsey (1995).

4.3.1.1 Cognitive Tasks

The effect of heat on performing cognitive (higher mental function) activities depends on the type of tasks at hand. For example, simple cognitive tasks, such as visual or auditory reaction time, mathematical problem solving, and short-term memory tasks, respond differently to heat exposure than complex cognitive tasks that process control operators typically do, such as monitoring DCS screens for out-of-limit conditions (Ramsey and Kwon, 1988). In simple cognitive tasks, heat does not affect performance negatively. In fact, an improvement in performance was obtained when these simple cognitive tasks were performed during a brief exposure to heat. On the other hand, in complex cognitive tasks, heat negatively affects performance, starting between 30 and 33°C or 86 and 91.4°F (Ramsey and Kwon, 1988).

4.3.1.2 Physical Activities

Physical activities, especially heavy activities, performed in hot, humid conditions may cause fatigue and exhaustion much sooner than in more moderate conditions. For example, sitting at rest, men can tolerate 38°C (100°F) for 3 or more hours. However, the tolerance is reduced to 30 minutes (Lind, 1963) if working moderately at 4.67 kilocalories per minute (kcal/min). Kilocalories is the heat required to raise the temperature of 1 kg water from 15 to 16°C. In another study by Wyndham (1974) on men shoveling rocks, the productivity level was maintained until the temperature exceeded 28°C (82.4°F). At about 30°C (84°F), productivity dropped to 90% and to 50% when the temperature reached 34°C (93°F). It is worth adding here that an increase in temperature can affect people's safe behavior. A large study observing the behavior of over 17,000 people at work found that the incidence of unsafe behavior increased as the climatic conditions fell outside the range 17–23°C or 63–73°F (Ramsey et al., 1983).

4.3.2 Effects of Cold on Performance

As an increase in core body temperature can lead to heat stress (hyperthermia), so a small decrease in core body temperature can produce cold stress (hypothermia). However, the human body is better able to tolerate increased heat than compensate for heat loss. Examples of jobs where people work in a cold environment are oil and gas extraction, electric line repair, gas line repair, sanitation, trucking, warehousing (especially cold storage), and military (Sinks et al., 1987). Similar to the effects of heat on performance, the studies of the effects of cold on performing different functions are not well understood and therefore lead to no firm conclusions. The reasons may be related to the following interacting factors:

- Differences in air flow, temperature, and humidity.
- Differences in the duration of exposure.
- The difference in exposure level to the different parts of the body.
- Individual differences (biologically and physiologically).

In spite of the interacting factors and differences in the results of the studies, some generalized conclusions can be used from the literature. These conclusions are summarized next.

4.3.2.1 *Cognitive Tasks*

The effect of cold on cognitive tasks has been detected in studies on reaction time and complex mental activities. However, the effect depends on the severity of the temperature, task performed, and skill and experience of subject with the task and with performing in the cold. The following is a summary of the results:

- Mental performance decreases when performing complex, demanding tasks requiring high levels of concentration and significant use of the short-term memory (Bowen, 1968; Enander, 1987, 1989).
- Cold does not affect the cognitive efficiency of well-motivated subjects (Baddeley et al., 1975).
- Cold acts as a distractor that interferes with some types of mental performance tasks (Enander, 1989).

4.3.2.2 *Physical Activities*

Cold significantly affects physical work. The reduction in limb or whole body temperature reduces physical capacity. For example, reduced limb temperature (i.e., arm and hand) affects motor ability, causes a loss of cutaneous (touch) sensitivity, affects the limb's muscular control, and reduces dexterity, muscular strength, and endurance abilities (Lockhart, 1968; Morton and Provins, 1960; Enander, 1989). The reduction in whole body temperature reduces the rate of metabolism within the muscles of the extremities and, in turn, the performance due to shivering. Indeed, Lockhart (1968) suggested that part of the reason performance decreases in tasks involving dexterity activities may be due to shivering.

In practice, gloves and heavy thermal boots and clothing can protect operators in cold weather. In severely cold temperatures, the clothes and gloves can affect the ability of operators to perform tasks that require strength or dexterity by limiting body motion.

4.3.3 Effects of Heat on Health

4.3.3.1 *Hot Environment*

Exposure to high levels of heat can affect a person's health in two ways. First, increased temperature on the skin can lead to tissue damage from burning (e.g., over 45°C or 113°F). This situation is usually manageable because it is observable and the people react to it by distancing themselves from the heat. The second way heat can affect health is more serious and dangerous. It deals with increased core body temperature to about 42°C (108°F), where heat stroke (hyperthermia) occurs. This rise in core body temperature results in an increase in metabolism, which in turn produce heat that needs to be dissipated. If the cycle is allowed to continue, death can ensue. This cycle is referred to as Q10 effect (for every 1°C increase in deep body temperature, metabolism rises 10%). There are two possible reasons why the heat generated does not leave the body: The body is exposed to heat and humidity that reduces the sweat evaporation; and some protective clothing acts as insulation.

The continuous exposure to hot environmental conditions and the heat stress it generates can lead to several forms of heat illnesses, with heat stroke being the most serious:

- Heat rash, also known as *prickly heat*—where the skin erupts with red pimples with intense itching and tingling caused by inflammation around the sweat gland ducts.
- Heat cramps—spasms of the muscles result from loss of salt.
- Heat exhaustion—results from dehydration, with a general feeling of headache, loss of performance through muscular weakness, dizziness, possible vomiting.
- Heat stroke—a significant and alarming increase in core body temperature and the failure of the body to loose the heat. The general symptoms are headache, dizziness, vomiting, shortness of breath, strange behavior, and eventually loss of consciousness and death.

Heat stroke is caused mainly by strenuous physical work or exercise in hot conditions, duration of exposure to extreme heat sources, and highly motivated and competitive individuals (i.e., sports).

In addition, the exposure and tolerance to heat stress is a function of individual differences (Burse, 1979; Strydom, 1971; Sanders and McCormick, 1993):

- *Gender.* Generally, women are less tolerant to hot environmental conditions and, therefore, more vulnerable to heat illnesses than men.

- *Age.* Older people are more vulnerable to heat illnesses than younger people.
- *Physical fitness.* The more the person is physically fit, the longer is the tolerance to hot environmental conditions, up to a point.
- *Body fat.* Body fat acts as an insulating layer blocking the release of heat. It also acts as an extra weight to move around, increasing the level of physical activity leading to higher body heat and energy expenditure.
- *Alcohol.* Alcohol affects peripheral nervous functioning and produces dehydration.

4.3.3.2 Cold Environment

Exposure to cold can lead to a drop in core or deep body temperature, likely to produce a risk to health. When the core body temperature falls below 35°C (95°F), the state is referred to as *hypothermia* or *cold stress.* With core body temperature below 35°C (95°F), the risk of disorientation, hallucination, and unconsciousness increases, leading to cardiac arrhythmia as the core temperature drops even further and subsequent death from cardiac arrest.

As the body is exposed to cold and the core body temperature starts to drop, the body automatically and rapidly uses its regulation system and produces two physiological reactions to cold stress:

- Shivering—Shivering is characterized by increased muscular activities and contractions to produce heat rapidly; the colder it is and the faster the core body temperature drops, the faster and more intense shivering becomes, leading to exhaustion, until core body temperature reaches 30–33°C (86–91°F), where shivering gradually is replaced by muscular rigidity.
- Constriction of blood vessels, also known as *vasoconstriction*—Vasoconstriction is characterized by an increase in the constriction of the peripheral (skin and extremities) blood vessels and an eventual shut off of blood flow to these regions as the core body temperature drops further, leading to an increase in blood pressure. This process

serves two main purposes: First, the warm blood is diverted from the cold skin, such as hands, fingers, and face, and distributed to the internal organs, so less body heat is lost. Second, the insulating capacity of the skin is increased up to six times by cutting off the flow of blood (Sanders and McCormick, 1993). As the constriction of blood vessels increases and blood flow is shut off, the temperature of all exposed areas (hands, fingers, toes, face, nose) rapidly approaches the air temperature, causing freezing of body tissues (cold injury) or frostbite.

In addition, the exposure and tolerance to cold exposure is a function of individual differences (Timbal, Loncle, and Boutelier, 1976):

- Body composition—amount of subcutaneous (under the skin) fat around the body, which acts as good insulating material. This is especially important when blood vessels constrict in response to the cold and blood is diverted away from the body surface toward the internal organs for survival.
- Individual size and weight—since the degree of heat loss is proportional to the body's surface area, the bigger an individual is in terms of height and weight, the greater is the amount of heat that can be generated by shivering. Note that the heat generated through shivering depends on the mass of active muscular tissues in the body.
- Physical fitness—the better the physical fitness, the more efficient the production of heat through the shivering process and the longer it is maintained with delayed physical exhaustion.

4.3.4 Comfort and Discomfort Zones

Comfort in the comfort zone is perceived when the body's physiological responses to the environmental temperature and humidity are within the normal regulatory responses. However, when those physiological responses exceed the normal regulatory responses, individuals perceive discomfort. People's perception of comfort, in the comfort zone, is influenced by the following environmental factors:

Air temperature—comfort zone, 20–25°C (68–78°F).

Relative humidity—comfort zone, 30–70%.

Air velocity—comfort zone, 0.1–0.3 m/sec (20–60 ft/sec).
Other factors that contribute to the perception of comfort are

- Workload. The heavier the work load, the lower the upper limits of the temperature and humidity must be.
- Radiant heat. Examples are furnaces, sun, ovens, radiant-heat lamps.
- Clothing. The type and amount of clothing worn and its insulation value determine how much the upper limits of temperature and humidity must be lowered.
- Age. Elderly people prefer higher temperatures than younger people, since the metabolism rate decreases slightly with age.
- Gender. Women have a lower metabolic rate than men and, in general, might prefer a warmer environment.
- Color. Color of the surroundings can influence an individual's thermal comfort. This influence is psychological in nature; for example, when red predominates, individuals feel warmer. The color blue elicits the opposite response.

4.3.5 Work Tolerance in a Hot Environment

A key factor to examine and quantify a warm working environment is the energy that the human body consumes to perform the job. Through the metabolism process, the body uses oxygen to convert food and body fat into mechanical energy and heat. In turn, the body consumes energy to stay alive. The total energy consumption (measured in kilocalories, kcal, per unit time) is composed of three components:

1. Basic metabolism—the number of kcal used during sleeping, breathing, and the like to keep the organs alive.
2. Leisure calories—the number of kcal used when performing activities such as reading or playing a musical instrument.
3. Work calories—the number of kcal used when performing different physical movements to accomplish a task, such as typing a

TABLE 4-5

Examples of Energy Expenditure for Some Occupations

Occupation	*Energy Expenditure (kcal/min)*
Light assembly work (seated)	2.2
Work in laboratory	2.5
Walking on level (4.5 km/h)	4.0
Walking on level (6 km/h)	5.2
Chopping wood	8.0
Mowing the lawn	8.3
Shoveling (16 lb weight)	10.2

manuscript on the computer, lifting objects, walking up or down stairs (Table 4-5).

A tool to predict energy expenditure was developed by the University of Michigan (1988). The tasks this tool analyzes are lifting and lowering; holding, walking, and carrying; horizontal arm work forward; lateral arm work; pushing and pulling; general hand and arm work; throwing loads; and climbing steps.

Energy consumption is certainly affected by (1) the ambient temperature and the workload, (2) the age of the individual, and (3) rest duration.

The interaction of temperature and workload is illustrated in Table 4-6.

The effect of age on permissible energy consumption is given in Grandjean (1988). If we set the maximum energy consumption of a 20–30 year old as the 100% level, then the relative maximum energy that can be consumed by older workers is given by the following:

Age (years)	*Maximum energy consumption*
40	96%
50	80%
60	80%
65	75%

TABLE 4-6

Maximum Working Temperature by Workload

Energy Consumption, kcal/min	*Type of Work*	*Maximum Effective Temperature, °C (°F)*
6.7	Heavy manual work (walking with a 66-lb load (30 kg)	26–28°C (78.8–82.4°F)
4.2	Moderate heavy work at 2.5 miles/hr (4 km/hr)	29–31°C (84.2–87.8°F)
1.7	Light sedentary work	33–35°C (91.4–95°F)

The rest duration is determined in terms of the energy consumed to perform a particular task. For example, if we take an individual who works in a warehouse, lifts and stacks bags of chemical powder over the day, and consumes energy at an estimated rate of 7.5 kcal/min, then using the following equation we can determine the rest time needed to be able to continue the job without fatigue:

$$\text{Rest time (as \% of work time)} = (\text{kcal/min} - 1) \times 100\% = 87.5\%$$

The results suggest that, for 10 minutes (600 sec) of steady work, he or she may need about 525 sec (about 9 min) rest.

4.3.6 Recommendations to Improve Working Conditions

The following recommendations for improving both working conditions in hot and cold environments are summarized from several sources in the literature (Grandjean, 1988; Sanders and McCormick, 1993; Eastman Kodak, 1983; Konz, 1990).

4.3.6.1 Guidelines for Heat Conditions

The recommended guidelines to achieve a level of comfort working in hot conditions follow a systematic approach by making changes to the ambient conditions, the task, the personal protective equipment, and finally the individual:

- A worker, new to a hot environment, should be acclimatized to heat by initially spending only 50% of the working time in the heat and increasing this by 10% each day.
- Train workers to recognize signs of heat illness, and provide and locate drinking water close to workers and the task performed.
- The greater is the heat load and the greater the physical effort performed under heat stress, the longer and more frequent should be the rest or cooling periods.
- If the limit of heat tolerance is being exceeded, the working day must be shortened.
- Where the radiant heat is excessive (i.e., blast furnaces), the worker must be protected by special goggles, screens, and special protective clothing against the risk of burning the eyes and hands. Special protective clothing includes such equipment as ice vests and vortex cooling suits.
- Where radiant heat is excessive (sun heat in the middle of July in traditionally warm states and countries), provide cooling facilities (cool rooms) for individuals to use for their work breaks or when needed.
- Match the workload to ambient temperature and relative humidity, where possible.
- Increase the flow of air in the workplace, where possible, using, for example, fixed or portable fans.

4.3.6.2 Guidelines for Cold Conditions

The recommended guidelines to achieve a level of comfort working in cold conditions are

- Design to ensure that the work is performed in a protected environment (e.g., provide windproofing around a task normally performed outside).

- Block the source of air flow with a shield.
- Change the work area location, if possible, away from the air flow source.
- Increase the ambient temperature where possible. If not, provide adjustable auxiliary heaters to accommodate individual differences. For continuously moving and changing locations, this may not be possible.
- Recommend and provide proper warm clothing to extend the tolerance to cold conditions.
- In extremely cold environments, if the limit of cold tolerance is being exceeded, the working periods must be shortened.
- Provide warming facilities (warm rooms) for individuals to use for their work breaks or when needed.
- Use proper gloves to protect hands and fingers. When finger dexterity is required, use partial gloves that cover the hands but expose the fingers.
- Train people to recognize signs of cold stress.

4.4 NOISE

Noise is an aspect of our work and daily living environment produced by equipment, machines, and tools. Noise is commonly referred to as *unwanted sound.* Burrows (1960) clearly defines *noise* as "an auditory stimuli bearing no informational relationship to the presence or completion of the immediate task." However, it depends on the individual's subjective reaction whether to consider a sound source as noise or not. In addition, sound labeled *noise* on one occasion may not be noise in a different occasion or environment.

Sound is defined in terms of two parameters, frequency and intensity, because a sound is a vibration stimulus experienced through the air. Frequency is an acoustic energy reflecting the number of vibrations per second, expressed in Hertz (Hz). One Hertz is equal to 1 cycle per second. The range of frequency between 2 and 20,000 Hz defines the frequency limits of the ear. People subjectively perceive frequency as pitch. Intensity is related to the sound pressure variation caused by the sound source. People subjectively perceive intensity as loudness. Intensity is specified in terms of sound pressure level (SPL) and expressed in the logarithmic units of decibels (dBA). Being logarithmic, a 10-fold increase in power (intensity) occurs with each 10 dBA increase. For example, 60 dBA on the decibel scale is not twice as intense as 30 dBA but 1000 times as intense as 30 dBA. Table 4-7 presents some examples of sources of noise. This section discusses how noise affects performance and health. It also provides guidelines on how to reduce and control noise in living and working environments.

4.4.1 Effects of Noise on Performance

The effects of noise on performance have been studied for many years with mixed results. While some studies showed performance decrements, others showed no effects or even an improvement in performance (Kryter, 1970; Gawron, 1982). Similar to the studies on the effects of temperature

TABLE 4-7

Examples of Sources of Noise

Sound	*dBA Level*	*Sound Power Increase*
Rock concert 30 m from speakers	120	1,000,000,000,000 (10^{12})
Pneumatic hammer	100	10,000,000,000 (10^{10})
Average traffic	80	100,000,000 (10^{8})
Vacuum cleaner (10 ft)	70	10,000,000 (10^{7})
Conversational speech	60	1,000,000 (10^{6})
Quiet office	40	10,000 (10^{4})
Bedroom at night	30	1000 (10^{3})
Normal breathing	10	10 (10^{1})

on performance, the mixed results of noise effects are related to the following interacting factors:

- Most studies are in a laboratory environment.
- Different techniques used to produce noise (e.g., office noise to rocket noise) and different exposure conditions (i.e., definition of quiet ranged from 0 to 90 dBA).
- Wide differences among tasks, from reaction time and number checking all the way through to accuracy of shooting; because of the different nature of difficulty, these tasks require relatively different demands on memory, cognitive, and motor capabilities of those tested.
- Individual differences (biologically and physiologically) in the way humans respond to noise or a noise and task combination.
- Different techniques of experimental design used to test the effects of noise.

Despite the mixed results, reasonable conclusions can be made about the effects of noise on performance. These are summarized next. It is safe to say at this point that noise has little or no effect on manual work, and this is not discussed further.

4.4.1.1 Speech and Communication

Speech and communication depend highly on the listener's ability to receive and decode the sounds. This means that the listener must hear the signal (speech or auditory display) and understand the message. Therefore, the speech and communication comprehension depends on the loudness of the voice as well as the level of background noise.

Hearing and understanding the signal may be masked (blocked) by a noisy environment. If the signal intensity remains constant and the masking intensity increases, then more masking occurs. The reverse is also true. Similarly, as the frequency of the masker approaches that of the signal, more masking occurs. From a practical side, it is important to know the intensity and frequency of both the signal and masker. For example, if the intensity and frequency of an auditory display are much higher than those of the noise, then masking does not occur. Otherwise, performance reductions (not hearing and understanding) due to masking occurs and corrective measures must be considered (this is discussed in detail in the section on how to control noise).

4.4.1.2 Cognitive Performance

The relationship between noise and cognitive performance is similar to that between heat or cold and cognitive performance. The results are again mixed. Even so, some reasonable conclusions can be made about the effects of noise on cognitive performance for specific types of tasks:

- Generally speaking, noise clearly has an effect on overall performance, especially when it reaches a level over 90 dBA.
- Noise levels over 100 dBA can affect monitoring performance over relatively long duration (Jerison, 1959).
- Tasks with large short-term memory component (i.e., mathematical calculations) are affected by high noise levels (Poulton, 1976, 1977). The reasoning here is that we can not hear ourselves think in noise.
- Noise produces more errors in complex rather than simple tasks (Boggs and Simon, 1968).
- Tasks performed continuously without rest pauses between responses are affected by noise (Davies and Jones, 1982).
- Tasks requiring high perceptual or information processing capacity (i.e., high mental demands, complexity, and considerable detail) are affected by noise (Eschenbrenner, 1971).
- Noise does not reduce performance on simple routine tasks or motor performance (i.e., manual work). Positive noise, in the eyes of the beholder, such as music, may facilitate an improvement in performance and satisfaction. The reasoning here is that music can aid in reducing boredom and fatigue associated with these tasks (Fox, 1971).

4.4.1.3 Nuisance and Distraction

Noise is considered a nuisance (annoying or unpleasant) when it interferes with our ability to carry out an activity. Thus, the response is subjective. This subjective nature indicates that what is annoying to one individual may not be to another. Furthermore, what is a nuisance on one occasion may not be at another time. For example, people are not disturbed by the noise generated by their own activities. A few observations can be made regarding this point:

- Loud, high-frequency noises are considered a nuisance or annoying. They contribute to performance decrements by creating a distraction and affecting concentration and effective communication.
- More familiar noises are perceived as less annoying (Grandjean, 1988).
- Sound features that contribute to the different degrees of "noise perceived annoyance" are intensity, frequency, duration, and the meaning to the listener. For example, the annoyance level increases as the intensity and frequency of sound increases, the longer it remains, the less meaningful it is to the listener (Kryter and Pearsons, 1966).
- Overhearing conversations is also considered, by most people, annoyance (Waller, 1969).
- Intermittent noise can reduce performance on mental tasks more than continuous noise.

4.4.2 Effects of Noise on Health

The most important and obvious effect of noise on health is hearing loss and deafness. Hearing loss and deafness can have two causes: conduction deafness and nerve deafness. Conduction deafness is caused by the inability of the sound waves to reach the inner ear and vibrate the eardrum adequately. Conditions that may lead to conduction deafness are wax in the ear canal and infection or scars in the middle ear or eardrum. Conduction deafness is not the product of a noisy environment and does not result in complete hearing loss. Nerve deafness is caused by reduced sensitivity in the inner ear due to damage or degeneration of nerve cells. The hearing loss here is in the higher frequencies, whereas conduction deafness is more even across frequencies.

Establishing a strong relationship between hearing loss and exposure to a specific noise is not easy. This is simply because during the normal aging process we are exposed to nonoccupational noises (vehicles, TV, vacuum cleaners, etc.) as well as occupational noises. The combined effects and the cumulative nature of the process make it difficult to point to a specific cause. The degree to which hearing is affected depends on the type, intensity, frequency of noise, exposure duration and, individual sensitivity and susceptibility. Two types of hearing loss are discussed here: aging, which is a normal process, and noise-induced temporary and permanent hearing loss.

4.4.2.1 Aging Hearing Loss

As a result of the normal aging process, hearing loss can occur due to changes in the structure of the auditory system. The change is more noticeable in men than in women. This hearing loss is also more noticeable at the higher frequencies. The decline in hearing loss is not consistent in all individuals simply because of individual biological and physiological differences and susceptibility. Taking a frequency of 3000 Hz as standard, Grandjean (1988) reports the loss of hearing to be expected at various ages as

Age (years)	*Expected Hearing Loss (dBA)*
50	10
60	25
70	35

4.4.2.2 Noise-Induced Hearing Loss

Noise-induced hearing loss or deafness can be related to continuous or noncontinuous exposure to noise. This noise-induced deafness can be temporary (temporary threshold shift, TTS) or permanent (permanent threshold shift, PTS). Even though the initial TTS is not damaging, continuous experiences of TTS lead to PTS (Kryter, 1970). This is because the TTS and PTS affect the same areas of the ear.

Temporary hearing loss is characterized by the ability to recover normal hearing in a few hours or days after an exposure to continuous high-intensity noise (hence, the term TTS). For example, being exposed to 90 dBA noise can cause about a 10-dBA shift in the hearing threshold. However, exposure to 100 dBA can cause about a 60-dBA shift. Permanent hearing loss or deafness is characterized by an irreversible deafness after repeated exposure to high-intensity noises (hence, the term PTS). This PTS gradually appears as the number of years of noise exposure increases, and it occurs first at 4000 Hz, which is within the region of frequencies to which the ear is most sensitive.

The United States National Institute of Occupational Safety and Health (NIOSH, 1972) indicates that a continuous exposure, about 8 hr, to noise levels above 85 dBA can cause hearing loss. Therefore, personal protective equipment (i.e., hearing protection) and engineering controls must be implemented in the workplace. This is discussed later. The U.S. Occupational Safety and Health Administration (OSHA, 1983) established permissible levels of noise exposure depending on the level of noise and the duration of the exposure. This is shown in Table 4-8.

4.4.3 Guidelines to Control Noise

Despite the recommendations by NIOSH and OSHA, in some work areas, workers must perform their jobs in high-noise situations for long hours. This section presents an overview of noise control methods. Noise control methods can be applied at the noise source, in the path of noise transmission and the surrounding area, and to the individual receiving the noise. In many cases, a combination of methods must be applied. Haslegrave (1995) provides a good review of noise assessment.

TABLE 4-8

Permissible Noise Exposure According to OSHA

Permissible Time (hr)	*Sound Level (dBA)*
16	85
8	90
6	92
4	95
2	100
1	105
0.5	110

Before embarking on solving a noise issue, we need to determine the level, location, and extent of the noise issue confronting the human and the task to be accomplished. The following are some principles to follow:

- Use proper, well-calibrated equipment.
- When doing measurements, keep away from reflecting surfaces (i.e., machines, walls) and hold the sound pressure level meter away from your body, at about arm's length, in the hearing zone.
- Sketch the area where noise level measurements are taken. This may not be necessary for simple noise sources.
- Map the noise level measurements on sketched area.
- Depending on the human activities performed at and around the noise source, determine what noise level is acceptable according to local regulations.

4.4.3.1 Noise Control at the Source

Controlling noise at its source is the ultimate goal. However, this method can prove expensive. The following are some examples on reducing noise at the source:

- Enclose the source area with sound-absorbing material. This can reduce radiated noise by 20–30 dBA. High-frequency noise can be better contained and absorbed by barriers than low-frequency noise; for example, barriers around the flare of an offshore platform, baffles for air intakes such as in furnaces, and mufflers such as on steam vents with high-volume air.
- Routinely maintain, lubricate, and align equipment to reduce vibration and therefore noise levels.
- Choose and use tools and equipment that generate low-frequency noise. This may not always be possible because such tools may not exist commercially and a special order may be costly.

4.4.3.2 Noise Control in the Path of Noise Transmission

This method can be used either in addition to or after all the possibilities of controlling noise at the source have been exhausted. Some of the applied principles here are

- Place a barrier, such as a shield or wall, between the human and noise source to deflect and absorb noise. Please note that the reduction of low-frequency noise (below 1000 Hz) is not possible, since it travels over and around barriers.
- Cover walls, ceilings, and floors with sound-absorbing panels to reduce noise reflection back into the work area. It is worth noting that the overall noise reduction, using this technique, is rather limited; however, it does further reduce noise.
- Take advantage of distance from the source, since the level of noise reduces with the square of the distance traveled.

4.4.3.3 Noise Control at the Receiver

If the two preceding methods fail to reduce noise to a safe level, then this final method involves the use of personal hearing protection. The applied principles here are

- Make available and properly train individuals on the use of hearing protection devices. These can include ear plugs, which can reduce noise level by up to 30 dBA and are suggested use for noise levels of 85 dBA and above; ear muffs, which can reduce noise level by about 40–50 dBA and are suggested use for noise levels over 100 dBA; a combination of the use of ear plugs and muffs; and soundproof helmets, which cover the head and ears.

- Provide instructions and train people on how to use hearing protection.
- Rotate jobs to reduce exposure time.

4.5 VIBRATION

Vibration is any regular or irregular movement of a body about its fixed position. Regular vibration (such as experienced in machinery like pumps, compressors, or power saws) is known to have a predictable waveform that repeats itself at regular intervals. Random vibration is irregular, unpredictable, and the most common type encountered in the real world; for example, vibration experienced in trucks, tractors, trains, and airplanes.

Vibration is characterized by its direction and amount (frequency and acceleration). The body can vibrate and move in one or more directions. The movement direction is described in terms of three orthogonal components: X, Y, and Z, where

X direction is back to chest.

Y direction is right side to left side.

Z direction is foot or buttocks to head.

For example, looking at Figure 4-2, if the person is standing or sitting on an up and down vibrating platform, the vibration is in the Z direction. If the person is lying on his or her back, the vibration is in the X direction.

Frequency of movement expresses the speed of vibration. Frequency is measured in Hertz (Hz), where 1 Hz = 1 cycle per second. For example, 1 cycle is the body movement from its fixed reference point to its highest point, to its lowest point, back to its reference point. Therefore, the higher the speed and the faster the body vibrates, the more cycles take place per second.

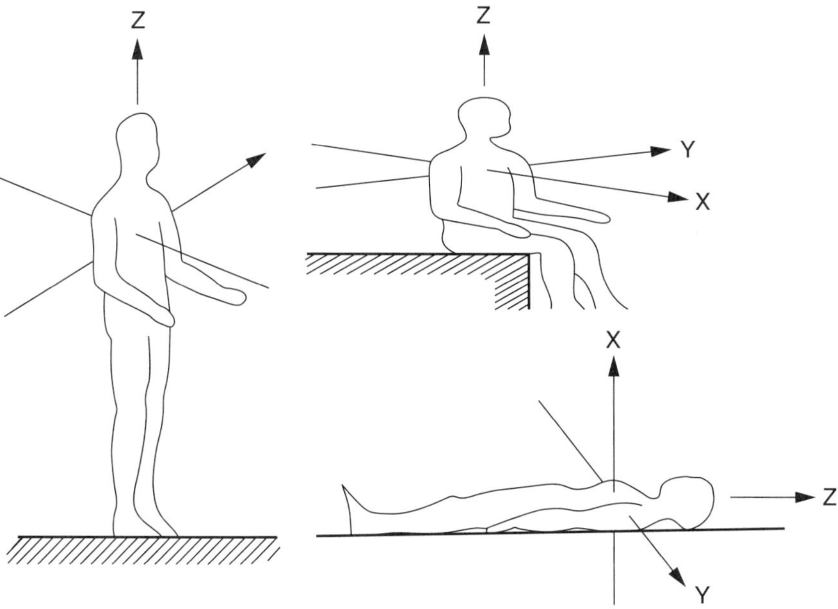

Figure 4-2 Direction of vibration relative to torso.

Intensity of vibration is expressed in terms of the maximum amount (amplitude or displacement) the body moves from its fixed reference point, measured in inches or meters. Intensity can also be specified in terms of acceleration. Acceleration can be expressed in terms of inch per second squared (in./s^2) or meter per second squared (m/s^2). Acceleration can also be expressed in terms of gravity (g) where 1 g is the amount of acceleration needed to lift the body off the earth's surface. The metric system defines 1 g as equal to 9.81 m/s^2. This section discusses how vibration affects performance and health. It also provides guidelines on how to reduce and control vibration. Griffin (1990) provides a full review of the subject.

4.5.1 Effects of Vibration on Performance

The principal effects of vibration on performance are reduced motor (i.e., hand steadiness) and visual (i.e., eye fixation) control. No strong evidence exists to suggest vibration affects information processing, reaction time, pattern recognition, or any intellectual (cognitive) tasks.

4.5.1.1 Motor Control

Considerable research has consistently demonstrated the effects of vibration on motor performance (Hornick, 1973; Levison, 1976; Lewis and Griffin, 1978, 1979a). A summary of the findings follows:

- Vibration has residual effects that may last up to half an hour after exposure.
- At low vibration frequencies (less than about 15 Hz), performance reduction is related to the amount of vibration experienced by the controlling limb.
- As the vibration intensity transmitted to the related limb increases, tracking performance decreases.
- Manual performance degradation depends primarily on the intensity of the vibration.
- Manual performance degradation occurs where the task demands movement in the axis in which the individual is being vibrated. For example, performing vertical tracking is affected more by the vibration in the Z axis than the Y or X axis.

4.5.1.2 Visual Performance

Examples of the effects of vibration on visual performance are observed in work situations such as driving trucks or tractors or riding trains. The most commonly reported problems associated with these situations are blurry images in the visual field and reduced visual acuity (ability to see fine details).This is probably due to the instability of the image on the retina. A summary review of the findings (Griffin, 1976; Meddick and Griffin, 1976; Barnes, Benson, and Prior, 1978; Griffin and Lewis, 1978; Lewis and Griffin, 1979b; Moseley and Griffin, 1986) follows:

- When both object and person are vibrating, the two May be vibrating at different frequencies and intensities.

May or may not be moving in phase.
May or may not be moving in the same axis.

- The effects of vibration on visual performance depend on the size of object observed and its distance from the observer.
- Visual performance degradation is determined mainly by the vibration frequency. For example, visual performance impairment occurs in the frequency range 10–25 Hz.
- Low frequency vibration affects the ability of the eye to track targets and maintain a stable image on the retina (e.g., reading a book on a train ride).
- High-frequency vibration creates blurry images.

4.5.2 Effects of Vibration on Health

Exposure to vibration, especially long term, leads basically to two health issues: damage to body organs due to whole-body vibration, which is a function of high-intensity vibration; and damage to blood vessels, body tissues, and nervous system to the fingers, hands, and arms due to segmental vibration, which is a function of high-frequency vibration. The majority of studies in the literature focus on tractor drivers, truck drivers, and workers using power tools (Rosseger and Rosseger, 1960; Ramsey, 1975; Abrams and Suggs, 1977; Whitham and Griffin, 1978; Seidel and Heide, 1986; Grandjean, 1988; Sanders and McCormick, 1993).

Whole-body vibration is mainly the product of vertical vibration (Z axis), when standing or sitting (Figure 4-2), absorbed through the feet when standing and the buttocks and feet when sitting. The most common situations where individuals are exposed to long-term, high-intensity, whole-body vibration are driving trucks, buses, tractors, and operating construction and heavy equipment. The most common physiological effects and complaints of whole body vibration are

- Stomach or abdominal pain and spinal disorders at the low back and neck regions. In addition, traces of blood in the urine have been found, probably due to kidney damage.

- Chest and abdominal pain from operating within the frequency range 4 10 Hz.
- Back and neck aches within the range 8–12 Hz.
- Headaches, muscular tension, and irritation in the intestines and bladder within the range 10–20 Hz.

Segmental vibration is mainly the product of long exposure to high-frequency vibrating objects absorbed through the fingers, hands, and arms. The most common situations where individuals are exposed to long-term, high-frequency segmental vibration come from examples of using hand-held power tools such as grinders, chain saws, pneumatic road drills and hammers, or polishing machines. Further, the design, size, and weight of the power tool also affects its vibration level (Abrams and Suggs, 1977). The most common physiological affects and complaints of segmental vibration are

- Structural damage to blood vessels and nerves, mainly in the fingers and hands, with vibration frequencies between 40 and 500 Hz. The arms could also be affected over time.
- Damage to bones, joints, tendons, and muscles of hands and arms with vibration frequencies less than 40 Hz.
- Symptoms associated with these situations are referred to as *white fingers condition, Raynaud's disease,* or *vibration-induced white fingers* and consist of
  - Pain, stiffness, and numbness of the fingers and hands.
  - Blanching (white or bluish color) of either all or part of the hands and fingers.
  - Loss of muscular control and strength.

4.5.3 Guidelines to Reduce or Control Vibration

There are basically two ways to reduce or control the effects of vibration on the individual: engineering and administrative controls (Eastman Kodak, 1983; Grandjean, 1988, Sanders and McCormick, 1993). Engineering controls can basically be achieved through two approaches:

- Reducing vibration at the source by
  Proper equipment maintenance.
  Mounting equipment on pads or springs.
  Modifying equipment to reduce speed and acceleration.

- Preventing vibration from reaching the individual by
  Placing special spring suspensions under seats and platforms to absorb vibration.
  Using rubber mats for standing tasks.
  Designing hand-operated power tools with damping-isolating materials, especially between the hand and hand-grip of the tool.
  Using special vibration-absorbing gloves and, reducing the time spent working in cold conditions.
  Designing the size and weight of the hand-operated power tools (i.e., using counterweight tools) to minimize the force needed to grip, balance, and operate them.

Administrative controls focuses on three approaches to reduce exposure to vibration:

1. Train individuals on the issues associated with vibration in terms of performance and health effects.

2. Provide the proper personal protection equipment, such as special gloves, and invest in properly designed equipment.

3. Rotate individuals among work tasks to control and reduce exposure. For example, Bovenzi, Petroniv, and DiMarino (1980) recommend the use of a chipping hammer (used to remove caked material from metal surfaces) should not exceed 4 hours per person per shift.

4.6 CASE STUDY

The proper quality and quantity of lighting in a control room environment is critical to the efficiency and safety of the petrochemical plant. This case study deals with the evaluation of an existing lighting system in a process control center, and recommendations for follow-up.

4.6.1 Method

The evaluation methods include

- Recording the illumination levels at each writing surface of the control room (with the brightness adjusted to a "personnel preference" level by each control room operator). The levels are typically selected to minimize glare reflected from the screens and the stainless steel panels.
- Discussion with individual operators about the lighting in their areas.
- Photos of qualitative HF concerns.

4.6.2 Results

We determine the following:

- Illumination levels are too low for reading tasks.
- Stainless steel panels are highly reflective. The ratio of luminance from stainless steel panels to the luminance of the screens in most cases is greater than the maximum recommended luminance ratio, 10:1.
- Small cell parabolics (Figure 4-3) are not located or used correctly:
  Control equipment is located beneath the luminaires that provide illumination for paper-and-pencil tasks. The operators decrease the brightness of these lights to reduce glare.
  The luminaires located around the room periphery cause less screen glare, but they are not lit, since they would cast shadows on the writing surfaces.
- The large circular light (Figure 4-4) structure in the center of the room is a glare source for each console. It is composed of "egg crate" filters over recessed fluorescent lights. The cut-off angle for the egg crates is poor, resulting in spillover of illumination.

4.6.3 Recommendations

The recommendations for the new lighting system are

Figure 4-3 Parabolic light.

Figure 4-4 Large circular light.

- *Task lights.* Ceiling mounted spotlights should be installed to light the writing surfaces in each of the console areas.
- *Ambient illumination.* Drop-down indirect lighting units are recommended to eliminate glare from overhead fixtures.

Recommendations for existing lighting systems are

- Replace the large circular illuminated area in the center of the control room by small-cell parabolics (at the current location of the egg crates) to eliminate screen glare, ceiling-mounted track lights, or inexpensive portable task lights located on the work surfaces that surround the center pole in the room.
- Diffuse the glare on the control system stainless steel panels by sandblasting the existing panels; painting them with a flat, matte-finish paint (an electrostatic paint process may work here); or install a matte finish covering over the panels.

4.6.4 Installation of a Pilot Lighting System

The goal here is to design and install a retrofit lighting system for evaluation. The system consists of two components.

The first component is an indirect lighting luminaire that hangs about 24 in. from the ceiling. The luminaire is inverted and throws its light up, onto the ceiling, from which it is reflected onto the surfaces below. The intensity of the light is continuously adjustable from off to full intensity.

Owing to the even dispersion of the light on the ceiling, no "hot spots" reflect on the panels below, and the light from the luminaire is evenly dispersed across the control space.

The second component of the lighting system is a "spot" task (track) light recessed into the ceiling. The light is equipped with a lens that shines a focused, 3-ft diameter spot of light on the worktable located in the center of the control space. The brightness of the light is continuously adjustable from off to full intensity.

The illumination levels provided by the new lighting system are obtained at several locations around the control space. Each reading is recorded with the indirect light source at full intensity. Readings are obtained with the spotlight off or the light adjusted to full intensity.

4.6.5 Final Results

The illumination levels provided by the combination of an indirect, ceiling-hung luminaire and track lights are acceptable for the operators' tasks. Levels at each console location are very similar, 2–3 FC (20–30 lux), whether the spotlight is on or off. With the spotlight adjusted to full intensity, the levels for reading at the operator's worktable are higher than required.

The overhead indirect light causes little glare on the console and the task (track) light is so directional that it provides virtually no spillover from a spot of light about 3 ft in diameter on the reading table.

Operators report being very pleased with the result and management has decided to install the indirect lighting system throughout the space. The number and size of the indirect luminaires will be determined from lighting surveys by the contractor. The number and location of the track lights will be determined by where the operators want reading levels of illumination.

REVIEW QUESTIONS

1. Light stimulus is characterized by its intensity and wavelength. Define light intensity and wavelength.
2. What lighting factors affect performance in the workplace?
3. How is age affected by the light quantity and quality?
4. What types of glare do people experience?
5. What is the luminance ratio and why is it important?
6. How does heat affect physical performance?
7. How does cold affect physical performance?
8. What are the forms of heat illness?
9. What does shivering do to the human body?
10. Define the human comfort zone.
11. How do you calculate rest duration when working in the heat?

12. Define noise and its parameters.
13. How does noise affect people's speech and communication?
14. How can noise lead to a hearing disorder?
15. How can noise be controlled?
16. What is vibration?
17. What are the signs of Raynaud's disease (also known as *white fingers disease*)?
18. What are the effects of vibration on visual performance and motor control?
19. How can vibration be controlled?

REFERENCES

Abrams, C. F., and Suggs, C. W. (1977) "Development of a Simulator for Use in the Measurement of Chain Saw Vibration." *Applied Ergonomics* 8, no. 2, pp. 130–134.

Baddeley, A. D., Cuccaro, W. J., Egstrom, G. H., Weltman, G., and Willis, M. A. (1975) "Cognitive Efficiency of Divers Working in Cold Water." *Human Factors* 17, pp. 446–454.

Barnaby, J. F. (1980) "Lighting for Productivity Gains." *Lighting Design and Application* 10, no. 2, pp. 20–28.

Barnes, G. R., Benson, A. J., and Prior, A. R. J. (1978) "Visual-Vestibular Interaction in the Control of Eye-Movement." *Aviation, Space and Environmental Medicine* 49, pp. 557–564.

Bennett, C., Chitlangia, A., and Pangrekar, A. (1977) "Illumination Levels and Performance of Practical Visual Tasks." Proceedings of the Human Factors Society 21st annual meeting, Santa Monica, CA, pp. 322–325.

Blackwell, H. R., and Blackwell, O. M. (1968) "The Effects of Illumination Quantity upon the Performance of Different Visual Tasks. *Illumination Engineering* 63, pp. 143–152.

Boggs, D. H., and Simon, J. R. (1968) "Differential Effect of Noise on Tasks of Varying Complexity." *Journal of Applied Psychology* 52, pp. 148–153.

Bovenzi, M. L., Petroniv, L., and DiMarino, F. (1980) "Epidemiological Survey of Shipyard Workers Exposed to Hand-Arm Vibration." *International Archives of Occupational and Environmental Health* 46, pp. 251–266.

Bowen, H. M. (1968) "Driver Performance and the Effects of Cold." *Human Factors* 10, pp. 445–464.

Burrows, A. A. (1960) "Acoustic Noise, and Informational Definition." *Human Factors* 2, no. 3, pp. 163–168.

Burse, R. (1979) "Sex Differences in Human Thermoregulatory Response to Heat and Cold Stress." *Human Factors* 21, no. 6, pp. 687–699.

Collins, B., and Worthey, J. (1985) "Lighting for Meat and Poultry Inspection." *Journal of the Illuminating Engineering Society* 15, pp. 21–28.

Davies, D., and Jones, D. (1982) "Hearing and Noise." In W. Singleton (ed.), *The Body at Work.* New York: Cambridge University Press.

Eastman Kodak Co. (1983) *Ergonomic Design for People at Work*, vol. 1. New York: Van Nostrand Reinhold Company.

Enander, A. (1987) "Effects of Moderate Cold on Performance of Psychomotor and Cognitive Tasks." *Ergonomics* 30, no. 10, pp. 1431–1445.

Enander, A. (1989) "Effects of Thermal Stress on Human Performance." *Scandinavian Journal of Work and Environmental Health* 15, supplement 1, pp. 27–33.

Eschenbrenner, A. J. (1971) "Effects of Intermittent Noise on the Performance of a Complex Psychomotor Task." *Human Factors* 13, no. 1, pp. 59–63.

Fox, J. G. (1971) "Background Music and Industrial Productivity—A Review." *Applied Ergonomics* 2, pp. 70–73.

Gawron, V. (1982) "Performance Effects of Noise Intensity, Psychological Set, and Task Type and Complexity." *Human Factors* 24, pp. 225–243.

Grandjean, E. (1988) *Fitting the Task to the Man.* London: Taylor and Francis.

Griffin, M. J. (1976) "Eye Motion during Whole-Body Vibration." *Human Factors* 18, pp. 601–606.

Griffin, M. J. (1990) *Handbook of Human Vibration.* London: Academic Press.

Griffin, M. J., and Lewis, C. H. (1978) "A Review of the Effects of Vibration on Visual Acuity and Continuous Manual Control: Part I. Visual Acuity." *Journal of Sound and Vibration* 56, pp. 383–413.

Haslegrave, C. M. (1995) "Auditory Environment and Noise Assessment." In J. R. Wilson and E. N. Corlett, *Evaluation of Human Work.* London: Taylor and Francis.

Hornick, R. (1973) *Vibration in Bioastronautics Data Book.* Second Ed. NASA Sp-3006. Washington, DC: National Aeronautics and Space Administration.

Hughes, P., and Neer, R. (1981) "Lighting for the Elderly: A Psychobiological Approach to Lighting." *Human Factors* 23, pp. 65–85.

ISO. (1994) *Moderate Thermal Environment—Determination of the PMV and PPD Indices and Specification of the Conditions for Thermal Comfort.* ISO 7730. International Standards Organization.

ISO. (1995) *Hot Environment—Estimate of the Heat Stress on Working Man, Based on the WBGT-Index (Wet Bulb Globe Temperature).* ISO 7243. International Standards Organization.

Jaschinski, W. (1982) "Conditions of Emergency Lighting." *Ergonomics* 25, pp. 363–372.

Jerison, H.J. (1959) "Effects of Noise on Human Performance." *Journal of Applied Psychology* 43, pp. 96–101.

Knave, B. (1984) "Ergonomics and Lighting." *Applied Ergonomics* 15, no. 1, pp. 15–20.

Kobrick, J. L., and Fine, B. J. (1983) "Climate and Human Performance." In D. J. Oborne and M. M. Gruneberg (ed.), *The Physical Environment at Work.* Chichester, UK: John Wiley.

Konz, S. (1990) *Work Design: Industrial Ergonomics.* Third Ed. Worthington, OH: Publishing Horizon.

Kryter, K. D. (1970) *The Effects of Noise on Man.* New York: Academic Press, pp. 545–585.

Kryter, K. D., and Pearson, K. S. (1966) "Some effects of spectrum content duration on perceived noise level." *Journal of the Acoustical Society of America* 39, pp. 451–464.

Levison, W. H. (1976) *Biomechanical Response and Manual Tracking Performance in Sinusoidal, Sum of Sines, and Random Variation Environments.* Army Medical Research Laboratory Report AMRL-TR-75–94.

Lewis, C. H., and Griffin, M. J. (1978) "A Review of the Effects of Vibration on Visual Acuity and Normal Control. Part II. Continuous Manual Control." *Journal of Sound and Vibration* 56, pp. 415–457.

Lewis, C. H., and Griffin, M. J. (1979a) "Mechanics of the Effects of Vibration Frequency, Level, and Duration on Continuous Manual Control Performance." *Ergonomics* 22, pp. 855–889.

Lewis, C. H., and Griffin, M. J. (1979b) "The Effect of Character Size on the Legibility of a Numeric Display during Vertical Whole-Body Vibration." *Journal of Sound and Vibration* 67, pp. 562–565.

Lind, A. R. (1963) "Tolerable Limits for Prolonged and Intermittent Exposure to Heat." In J. D. Hardy (ed.), *Temperature—Its Measurement and Control in Science and Industry*, vol. 3. New York: Reinhold.

Lockhart, J. M. (1968) "Extreme Body Cooling and Psychomotor Performance." *Ergonomics* 11, pp. 249–260.

Meddick, R. D. L., and Griffin, M. J. (1976). "The Effect of Two-Axis Vibration on the Legibility of Reading Material." *Ergonomics* 19, pp. 21–33.

Morton, R., and Provins, K. A. (1960) "Finger Numbness after Acute Local Exposure to Cold." *Journal of Applied Psychology* 15, pp. 149–154.

Moseley, M., and Griffin, M. J. (1986) "Effects of Display Vibration and Whole-Body Vibration on Visual Performance." *Ergonomics* 29, pp. 977–983.

NIOSH. (1972) *Criteria for a Recommended Standard—Occupational Exposure to Noise.* Cincinnati OH: U.S. Department of Health, Education, and Welfare. National Institute for Occupational Safety and Health.

OSHA. (1983) "Occupational Noise Exposure: Hearing Conservation Amendment." *Federal Register* 48, pp. 9738–9783.

Parson, K. C. (1995) "Ergonomics Assessment of Thermal Environments." In J. R. Wilson and E. N. Corlett, *Evaluation of Human Work*. London: Taylor and Francis.

Poulton, E. C. (1976) "Continuous Noise Interferes with Work by Masking Auditory Feedback and Inner Speech." *Applied Ergonomics* 7, pp. 79–84.

Poulton, E. C. (1977) "Continuous Intense Noise Masks Auditory Feedback and Inner Speech." *Psychological Bulletin* 84, pp. 977–1001.

Ramsey, J., Burford, C., Beshir, M., and Jensen, R. (1983) "Effects of Workplace Thermal Conditions on Safe Work Behavior." *Journal of Safety Research* 14, pp. 105–114.

Ramsey, J., and Kwon, Y. (1988) "Simplified Decision Rules for Predicting Performance Loss in the Heat." In *Proceedings on Heat Stress Indices*. Luxembourg: Commission of the European Communities. Cited in M. S. Sanders and E. J. McCormick, *Human Factors in Engineering and Design*. Seventh Ed. New York: McGraw-Hill, 1993, pp. 570–571.

Ramsey, J. D. (1975) "Occupational Vibration." In C. Zenz (ed.), *Occupational Medicine: Principles and Practical Applications*, pp. 553–562.

Ramsey, J. D. (1995) "Task Performance in Heat: A Review." *Ergonomics* 38, no. 1, pp. 154–165.

Ross, D. (1978) "Task Lighting—Yet Another View." *Lighting Design and Application* pp. 37–42.

Rosseger, R., and Rosseger, S. (1960) "Health Effects of Tractor Driving." *Journal of Agricultural Engineering Research* 5, pp. 241–275.

Sanders, M. S., and McCormick, E. J. (1993) *Human Factors in Engineering and Design*. Seventh Ed. New York: McGraw-Hill, pp. 551–588.

Seidel, H., and Heide, R. (1986) "Long-Term Effects of Whole-Body Vibration: A Critical Survey of the Literature." *International Archives of Occupational and Environmental Health* 58, pp. 1–26.

Sinks, T., Mathias, C., Halpern, W., Timbrook, C., and Newman, S. (1987) "Surveillance of Work-Related Cold Injuries Using Workers' Compensation Claims." *Journal of Occupational Health and Safety* 29, pp. 505–509.

Strydom, N. (1971) "Age as a Causal Factor in Heat Stroke." *Journal of the South African Institute of Mining and Metallurgy* 72, pp. 112–114.

Timbal, J., Loncle, M., and Boutelier, C. (1976) "Mathematical Model of Man's Tolerance to Cold Using Morphological Factors." *Aviation, Space and Environmental Medicine* 47, pp. 958–964.

University of Michigan. (1988) *Energy Expenditure Prediction Model*. Ann Arbor: University of Michigan.

Waller, R. A. (1969) "Office Acoustics—Effect of Background Noise." *Applied Acoustics* 2, pp. 121–130.

Whitham, E. M., and Griffin, M. J. (1978) "The Effects of Vibration Frequency and Direction on the Location of Areas of Discomfort Caused by Whole-Body Vibration." *Applied Ergonomics* 9, no. 4, pp. 231–239.

Wotton, E. (1986) "Lighting the Electronic Office." In R. Lueder (ed.), *The Ergonomics Payoff, Designing the Electronic Office.* Toronto: Holt, Rinehart and Winston, pp. 196–214.

Wright, G., and Rea, M. (1984) "Age, a Human Factor in Lighting." In D. Attwood and C. McCann (ed.), *Proceedings of the 1984 International Conference on Occupational Ergonomics.* Rexdale, Ont., Canada: Human Factors Association of Canada, pp. 508–512.

Wyndham, C. (1974) "Research in the Human Sciences in the Gold Mining Industry." *American Industrial Hygiene Association Journal* pp. 113–136.

5

Equipment Design

5.1 HUMAN/SYSTEM INTERFACE

One main way that operators interact with plant equipment is by monitoring instrument displays and manipulating the process via controls or switches on control panels or a keyboard. Therefore, good design of the

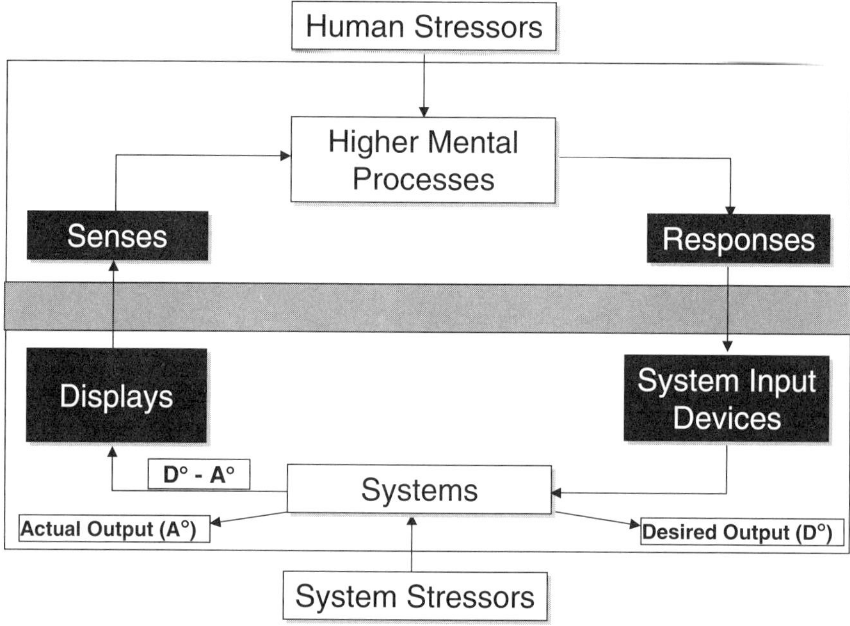

Figure 5-1 Human/system interface.

displays that provide information to the operator about the status of the process and the control devices that are available to the operator for manipulating the process can minimize the potential for error. The operator's interaction with the controls and displays is represented by the shaded portions of the human system interface shown in Figure 5-1.

This chapter discusses how displays and controls should be designed to minimize the potential for human error. Appendix 5-1 contains a checklist for displays, controls, control panels, and process control displays for reviewing aspects of this equipment that may contribute to human error at a site.

5.2 CONTROLS

Controls are physical devices used to send signals to plant equipment. Some controls require little manual effort to operate, such as panel-mounted pushbuttons, rocker switches, toggle switches, knobs, and slide

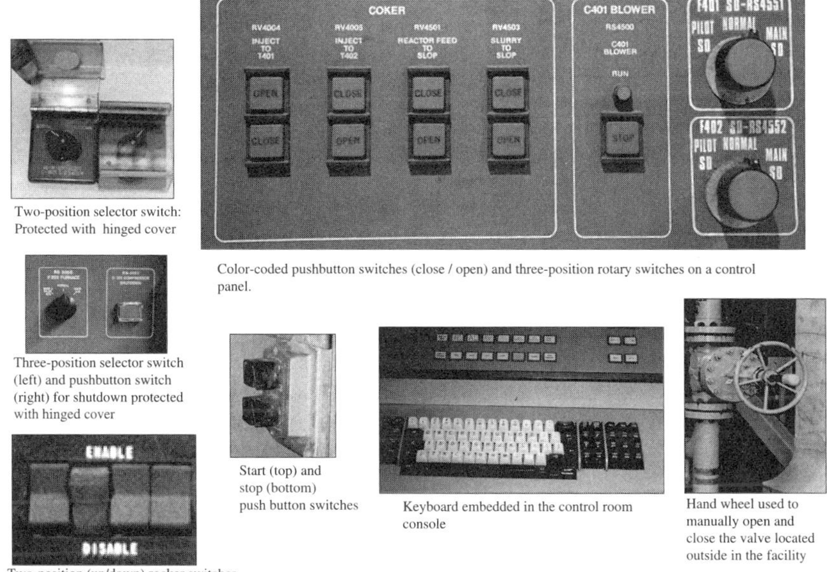

Two-position selector switch: Protected with hinged cover
COKER
C401 BLOWER
Color-coded pushbutton switches (close / open) and three-position rotary switches on a control panel.
Three-position selector switch (left) and pushbutton switch (right) for shutdown protected with hinged cover
ENABLE
DISABLE
Two-position (up/down) rocker switches
Start (top) and stop (bottom) push button switches
Keyboard embedded in the control room console
Hand wheel used to manually open and close the valve located outside in the facility

Figure 5-2 Types of controls.

switches. Other controls require muscular effort to operate, such as hand-operated valve wheel actuators, levers, and handles. Input devices such as keyboards, keypads, mice, and trackballs are also controls (see Figure 5-2).

Many displays are accessed or manipulated by a related manual control. Near misses and incidents may be related to an operator's misunderstanding or confusion about the operation of a control. Unclear operation of controls may result from poorly worded labels; hard-to-see labels; missing labels on the control; poorly designed controls; inappropriate location of a control on a panel; or controls that do not operate as expected. In these cases, change the design of the control or its settings so that it operates as most people would expect.

Controls can be improved by the following simple quick fixes:

- Replace missing labels and improve them, if necessary.
- Replace or repair broken controls or parts of controls.

- Group controls (e.g., on a panel) by function or sequence with labeling schemes, such as color-coded escutcheon plates or line demarcation.
- Remove controls that are permanently out of service.
- Replace controls that operate in a manner inconsistent with similar controls.

Several issues should be considered when selecting controls for use in a plant or evaluating a work situation that involves the use of controls:

- Physical design of the control itself—types of controls appropriate for various tasks (e.g., dimensions of the control).
- Physical requirements related to musculoskeletal injury risk factors (force, repetition, posture and contact stress).
- Appropriate labeling to clearly identify the controls.
- The way the controls operate matches operators' expectations.
- Physical access to operate the control.
- Guarding against inadvertent operation.
- Relationship of the control to the equipment it operates and the related displays of equipment status.

5.2.1 Physical Requirements of Operating Controls

The physical requirements to operate the control should be considered. Controls should not require excessive force to operate; for example, strenuous exertion to turn a valve hand wheel or excessive force by leaning down with the whole body on a lever to activate it. If excessive force is needed, then the contact force of the control on the hands and fingers (e.g., pressure points and sharp edges) may be a concern. Details on the recommended maximum activation forces for different types of controls are available in Eastman Kodak (1983).

A force gauge is used to measure the actual force required to operate the control; then the actual force can be compared to the recommended maximum force. However, a simpler qualitative approach is for operators to identify the controls that need to be examined further as those that require strenuous exertion or that dig into the fingers or hand during repetitive control operation. Other aspects of the physical requirements to watch include repetition and awkward body posture. Repetitive control manipulation and awkward body postures, with an extended and outreached arm, may be involved in repetitive or prolonged operations, such as operating a crane or operating a batch process via a local control panel.

5.2.2 Types of Controls

The type of control to install depends on what the control must do; that is, does the operator use the control to select a discrete setting or is the control used to select a setting from a continuous scale? A problem in this area is often related to the type of control or switch being inappropriate for the function it is supposed to perform. The following features should be considered when selecting the controls to be installed on a control panel (Eastman Kodak, 1983):

- The speed and accuracy of signal the operator needs to send to the plant or equipment via the control.
- The surface area available on the equipment or control panel for installation of the control.
- Ease of use—how intuitively clear is the operation of the control to the operator?
- The consistency of the operation of the control with other controls in the plant with a similar function.

Push buttons are suitable for two-position controls. They can be operated quickly and, depending on size, with the fingers, thumb, or palm of the hand. A push button needs some means to indicate its position, either a physical displacement or indicator lights. Examples of the use of push buttons are on/off and start/stop switches.

Rocker switches and toggle switches are most often used for two-position controls, although thcy can be used for three-position controls as well. They are fast and easy to operate. Their position can be used to indicate their setting. They are not suitable for continuous control. Examples of the use of rocker switches and toggle switches are on/off, open/close, and up/down controls.

Slide switches and thumb wheels are used for continuous control. They offer a wide range of settings and usually have no indents or indicators to indicate position of the switch. Since they require fine motion, the hand should be supported when they are used. Examples of the use of slide switches and thumb wheels are to increase flow, temperature, or pressure controls.

Rotary switches and lever-type handles are used for both discrete and continuous settings. When used for discrete settings, detents should be provided. Ensure that the number of knob discrete choices match the number of selections that can be made. Do not use a knob with three positions if only two choices available, such as on/off. The rotary knob is suitable for a large range of continuous motions, whereas the lever is suitable for only a smaller range. Rotary controls operated by the fingers should have knurled or textured surfaces. For rotary controls, avoid selection points that are 180° apart. Examples of the use of rotary switches and lever-type handles are controls to change flow lines and multiposition controls.

More details and specifications on various types of controls can be found in Kodak and McCormick Controls are used to communicate with equipment. If the design does not clearly indicate the position of the control, then the operator may make an error in either interpreting the current setting or making changes by setting the control in a different position. Figure 5-3 shows the difference between good and bad designs for a selector switch.

Dimensions of controls are important, especially when multiple controls, such as thumb wheels, buttons, or switches, are located on a single panel. Clearance between the controls reduces the potential for inadvertent operation. The size of the controls must be large enough to fit the fingers and hands of the operators. The use of gloves should be considered when designing the space required for effective operation of controls. The dimensions of the control and the clearance required between

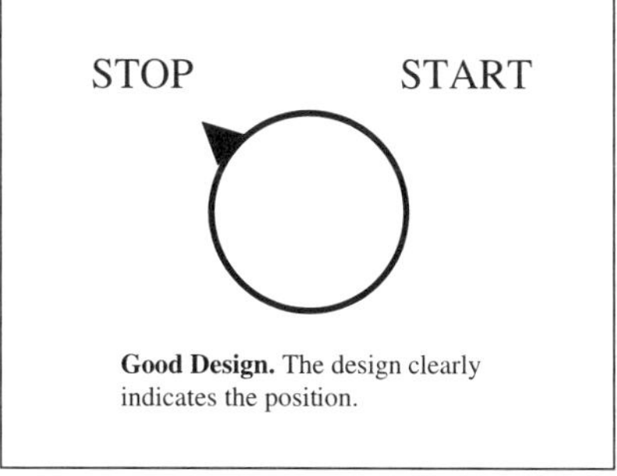

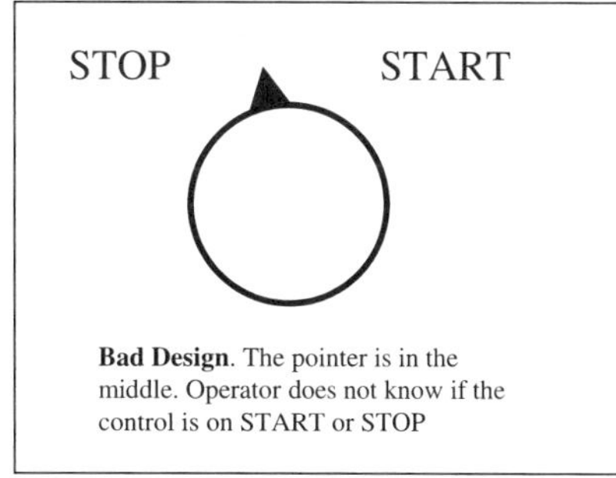

Do this

Not this

Figure 5-3 Good and bad designs for a selector switch.

controls are determined by finger anatomy dimensions, and are available from Eastman Kodak (1983).

The following points should also be considered for selecting controls to use in the plant:

- Controls should operate in the way that the operator expects.
- Controls that perform the same function should be of the same design and operate in the same manner.
- Discrete controls should have a positive indicator that the control has activated. This can be an indent, a click (only in quiet environment), an indicator light, or a large displacement of the switch.

5.2.3 Control Labels and Identification

All controls and associated displays should be labeled. The content of the labels should be clear and use complete words (e.g., Surge Pump Pressure) instead of acronyms or abbreviations (i.e., SPP). Labels of controls that are important, such as those that identify emergency equipment, should also be color-coded to facilitate quick identification. The label should be located above the control, or between a control and its associated display as shown in figure 5-4.

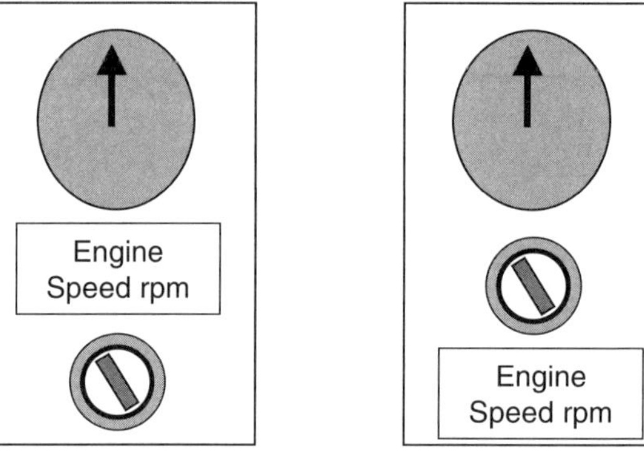

Figure 5-4 Placement of control labels.

5.2.4 Stereotypes

It is important that operators are able to manipulate specific controls in a manner consistent with their expectations. For example, the expected behavior for operating valve hand wheels are counterclockwise turns to "open" the valve and clockwise turns to "close" the valve. The operator's expectation is called a *stereotype*. A strong stereotype is one in which 85% of a user population has the same expectation of how a control operates. Equipment operated by plant personnel should operate consistently with their stereotypes. If the design of the equipment conflicts with a stereotype, there is more potential for error when operating it. Although it is possible to educate people to operate systems that do not follow the stereotypes, their performance may deteriorate, especially when in an emergency situation. Such controls should be simple to use without the operator's hesitation and require minimal decision making. Individual controls should be designed so that the direction in which the control is moved to achieve a desired setting or result is clear to the operator and does not violate any expectations.

Stereotypes for expectations on how controls should operate may differ among countries. Therefore, the design of a control panel arrangement in one country may not be effective for operators of a plant in a different country, especially for emergency operations. Stereotypes also apply to the meaning of color coding for signs and labels. For example, the color red is associated with danger in both the United States and in Hong Kong, but in the United States this stereotype is very strong, while in Hong Kong it is weaker. Color stereotypes in the United States for colors associated with Caution and Warning also are not as strong as the color red for Danger.

If controls are difficult to use because they are the wrong type for the situation or if controls operate in an unexpected manner, then these pose serious problems. These controls should be replaced or redesigned so that they operate according to expectations.

5.2.5 Access to Operate

A properly designed panel locates controls within the functional reach of smaller operating personnel, specifically within the functional reach of the smallest 5% of the personnel expected to operate the plant. This means that 95% of the population can reach a control comfortably and 5% can reach the control but not so comfortably. For a smaller operator to reach the control, it may require an undesirable working posture by extending the upper half of the body and while perhaps standing on tiptoes. Emergency controls, such as emergency shutdown valves and emergency isolation valves, should be within comfortable reach of all operators.

5.2.6 Preventing Accidental Operation

Once the location of controls has been set, the design should consider whether those locations prevent the controls from being accidentally activated. For example, if too little space is provided between adjacent controls or if the controls are arranged so that the operator's limb movements are such that they pass over (with the danger of touching) another control, the "wrong" control may be activated. Even if controls are in ideal positions, they can still be activated accidentally. A list of techniques to guard against accidental activation follows:

- Recess the control.
- Cover the control with a hinged cover (plastic see-through covers).

- Lock the control.
- Increase control resistance, the force required to activate the control.
- Operationally sequence a set of controls; for example, ensure that control 2 cannot be operated until control 1 has been activated.

5.2.7 Valves

Hand wheel actuators that operators grasp and turn to manually open and close valves are a major type of control device used in a petrochemical plant (see Figure 5-5). For valves that must be manually operated, the force that the operator must exert on a hand wheel to open, or unseat, a

Figure 5-5 Example of good valve stem orientation.

closed valve (also known as *cracking* a valve) is an important consideration. The force that an operator can apply to a hand wheel to crack a valve depends on the location of the hand wheel, the orientation of the valve stem, the diameter of the hand wheel, and the quality of grip on the hand wheel (Shih, Wang, and Chang, 1997). If the force required to crack open a valve is greater than the force the operator can safely exert, then there is a potential risk factor for musculoskeletal injury due to overexertion. Another concern is that, in a critical situation, an operator may not be able to exert the strength to move the valve quickly enough.

A study on 57 male operators reported the range of force that could be applied by the 5th percentile male operator (meaning that most, 95%, of the men could provide more force; Attwood, et al., 1999). Five "zones" of force capabilities are identified, based on the orientation and location of the valve for 5th percentile male operators. More detail on the study is described in chapter 2, section 2.4, a case study. These five zones are shown in Figure 2-7 of the case study, and are tabulated in Table 5-1.

The operators could generate the most force on the hand wheels positioned at waist height on a vertical stem (refer to Table 5-1, zone 1). The next most favorable positions (zone 2) were: shoulder height on a horizontal stem and knee height on a vertical stem. A hand wheel positioned at waist height was slightly less favorable (zone 3). The operators could

TABLE 5-1

Valve Force Zones for the 5th Percentile Male Operator

Zone	*Force Capability (lb)*	*Hand Wheel Height*	*Stem Orientation*
1	97 (highest)	43.5 in (waist height)	Vertical
2	87	28 in (knee height)	Vertical
	83	58 in (shoulder height)	Horizontal
3	78	43.5 in (waist height)	Horizontal
4	74	58 in (shoulder height)	45° angle
	70	25 in (knee height)	45° angle
	70	25 in (knee height)	Horizontal
5	62 (lowest)	58 in (shoulder height)	Vertical
	55	72 in (overhead)	Vertical

generate the least force on hand wheels that were positioned overhead or at shoulder heights on a vertical stem (zone 5). Hand wheels positioned at knee and shoulder heights on a 45° angle stem, and at knee height on a horizontal stem were only slightly better (zone 4) than the worst case.

Shih et al. (1997) reported that operators could exert more force on knurled and textured than smooth hand wheels for small valves. In addition to the type of hand wheel and its location and orientation, the availability of clearance for the operator to assume an advantageous posture and the working environment also affects the force that can be applied to the hand wheel (Amell and Kumar, 2001). An advantageous posture is determined by the distance required to reach the valve and the clearance in front of the valve for the operator to assume a preferred position to exert force on the hand wheel.

Operators should be able to reach the hand wheel to turn it from a maximum horizontal distance of 24 in. (610 mm), with the preferable range between 16 and 20 in. (410–510 mm). The recommended minimum clearance behind a hand wheel for a valve that must be manually operated is 30 in. (760 mm), with the preferable range between 40 and 48 in. (1010–1220 mm), depending on the height of the hand wheel location. Use the higher end of the preferable range of clearance if access requires the operator to squat or kneel on one knee.

For existing plants, a hand wheel actuator survey can be used to identify valves that may require too much force for safe manual operation, valves that may cause overexertion due to high force requirements, and valves that require a high number of repetitions to manually open or close. These valves may contribute to a musculoskeletal injury, be difficult to quickly close during an emergency, or require the use of a wrench to provide the torque necessary to move the valve. Plant surveys indicate that few modern continuous facility petrochemical operations require highly repetitive operation of valves by a single operator. At the high end of the range was a batch unit operation at a single plant that required up to 70 valves to be operated over a shift by an individual operator. For that operation, a unit survey was conducted to identify and replace valves with excessive force requirements and high numbers of hand wheel revolutions required to open and close them.

For screening purposes, operators can point out the valves that obviously require very high exertion to open or close. These valves can be replaced during equipment upgrades and turnarounds. After these valves are identified, the next in priority are valves with hand wheel actuators located below knee height or above shoulder height and that exceed recommended force. The force required to operate these valves can be measured with a torque wrench and compared to the force for the zones listed in Table 5-1. Depending on the frequency at which the valve is opened or closed and the critical nature of the valve, remaining valves can be assigned priorities for replacement as equipment is upgraded or modified.

Other considerations for valve operation include labels and hand clearance around the hand wheel to avoid hand or finger pinches. Hand clearance should be a minimum of 4 in. (1010 mm) extended from the rim of the hand wheel and a minimum of 2 in. (50 mm) behind the hand wheel. Severe hand clearance situations can also be reviewed in the survey and fixed by adjusting the heights of the hand wheels to alternate higher and lower to provide clearance.

There are many documented incidents in which operators out in the plant thought they were turning a hand wheel that moved the valve that they intended to operate but in error moved a different valve (Kletz, 1998). Each valve should be clearly labeled to avoid this type of error. The valve labels should match other references to the valve, such as those in process diagrams, process displays, control panels, or procedures documents.

One method for identifying and correcting problem valves is to incorporate the identification and follow-up actions into an existing management system, such as the maintenance system. Criteria for identifying problem valves can include one or combinations of the following: high rim pull force, high repetitions, awkward or difficult access, confusing layout, and lack of or poor labels.

5.3 VISUAL DISPLAYS

To display the information required by the operator, the goal is to choose the correct type of display to minimize error when reading and interpreting information acquired from visual displays. To accomplish this goal, one needs a full analysis of the task that the operator has to undertake, which can be matched against the display's ability to carry out these requirements.

5.3.1 Types of Displays

There are many types of displays in a plant. These include digital displays of pressure, temperature, and flow from indicators out in the plant; status on/off lights; and annunciator tiles. Visual displays generally take one of two forms: digital and analog. A digital display presents the information directly as numbers; for example, a pocket calculator or digital watch. With analog displays, the operator has to interpret the information from a pointer's position on a scale; or from some other indicator analogous to the real state of the process. For example, a clockface with a dial for the face and hour/minute hands is an analog display and a warning light is an analog display since the state of the light (on/off) is analogous to the state of the process (danger/safe) in the real world. Figure 5-6 shows examples of visual displays.

Visual displays may be used

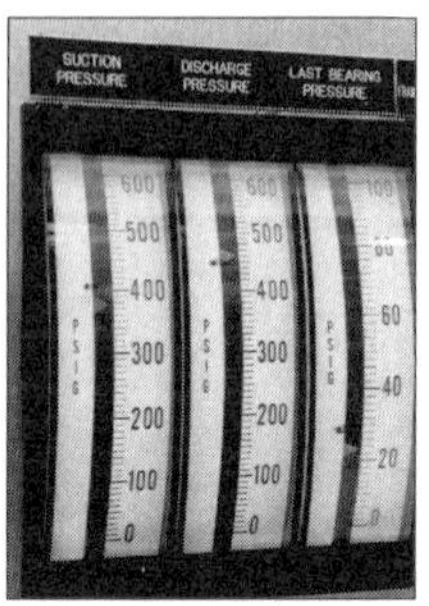

Moving pointer pressure scales on a control panel

Video Display Terminal (VDT), part of the console located in the control room

Moving needle on gauge outside in the facility

Annunciator tile light split in two: Pump running (left) and pump stopped (right)

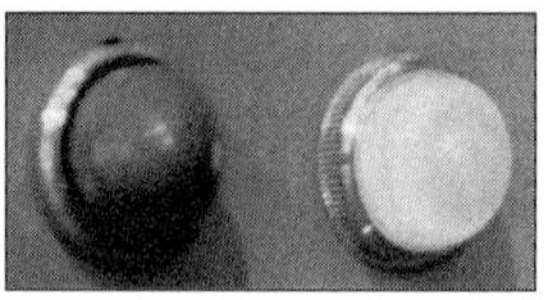

Off (left) and On (right) status lights on a control panel

Digital readout display on a meter

Figure 5-6 Types of visual displays: analog moving pointer, annunciator, status lights, digital display, and VDT.

- To make quantitative readings, to read the state of a machine or process in numerical terms. For example, a digital display of production volume provides the fastest and most accurate source of information to the operator.
- To make qualitative readings, to infer the "quality" of the machine state, such as temperature and pressure ranges. A moving indicator can provide qualitative temperature ranges cold, warm, or hot rather than the precise temperature. Changes are easy to detect.
- To check readings, to determine if parameters are within some "normal" bounds or that several parameters are equal, such as Hi/Lo pressure or excessive revolutions per minute (rpms). For example, check reading would draw a response such as "All pressures are normal." For check readings, zone markings can be used to help quickly determine if the check reading is within normal range.
- To make adjustments, such as setting an indicator to a desired value, to modify set points on control information.
- To provide status indicators. For example, a High Pressure status light indicates that a specific discrete condition, high pressure, is occurring.
- To monitor a status indicator in a mimic display, to determine condition and location; for example, an Emergency Shutdown (ESD) indicator in a mimic panel.

Specific types of displays are best for certain types of information; for example (Eastman Kodak Company, 1983).

- A moving pointer is best for qualitative or check reading and some adjustments.
- A digital readout is best for numerical readings.
- A graph (pen recording or trend plot) is best for detecting trends and qualitative readings.
- An annunciator light, best for giving operating instructions on a control panel where many functions are monitored.

Depending on the circumstance and the nature of the information needed by the operator, displays are designed to present information both in quantitative and qualitative forms. Quantitative information provides precise numeric values, such as "the temperature is 85°F." Qualitative information is used to read an approximate value, a trend, rate of change, or change in direction, such as a moving pointer that indicates that "the pressure is rising." Another qualitative type of information is the use of color codes for status, such as the color red for "hot" and blue for "cool." Once these guidelines have been adopted, they should serve the goal of being consistent in designing displays that present formation of the same type. For example, a digital counter for quantitative information, moving pointers and color coding for qualitative or status information.

5.3.2 Mounting Displays

Displays that are read during operator surveillance rounds in the plant should be readable from the surveillance path. This means that they should be oriented so that the operator can accurately read them.

The letters and numbers on displays, such as gauge dials and digital instrument readouts, need to be readable from normal working positions or surveillance paths. The operator should be able to read the display from head on, perpendicular to the operator's line sight. This avoids parallax error, a misreading due to looking at the gauge or level indicator from a side angle. Safety-related displays should be in the operator's primary field of vision (see Figure 5-7).

The required height of the displayed characters to be readable depends on the viewing distance and the criticality of the information. If a display is to be viewed at a greater distance from normal working positions, the features have to be enlarged to maintain the same visual angle. The following formula may be used:

$$\text{Height of letters or figures in inches} = \frac{\text{viewing distance in inches}}{200}$$

For example, if the viewing distance is 28 in. (0.7 m) the height of a character should be between 0.1 and 0.2 in. (2 and 5 mm). For a viewing distance of 6 feet, the character height should be between 0.3 and 0.5 in. (7 and 13 mm). If the viewing conditions are suboptimal, divide by 120 instead of 200 to increase the character height.

Figure 5-7 Displays visible from surveillance path.

These indicators for the discharge pressure, and for pressure, temperature and liquid level in the seal fluid drum face the operator's normal surveillance path.

In addition, the symbols or numerical displays on the device should be visible (e.g., not covered by shadow) and legible (with characters that can be read accurately from a distance). Displays should be mounted so that they can be read when the operator needs the information, such as when starting pumps. If the operator makes a local adjustment in the process and it appears on a related display such as a pressure indicator, the operator should have it in the field of view. Where possible, displays should be in consistent units so operators need not do mental arithmetic to interpret the data. Measurement units (e.g., psi, bar) should be clearly visible and large enough to be read. Adjacent displays with similar functions should have the same layout of graduation marks and characters. Also, where practical, the display, such as a gauge or level indicator, can be marked to indicate the range of normal operation parameters for check readings.

Displays should be mounted to maximize readability. Mount the displays to avoid glare from nearby lights or sunlight. Displays with numerical readouts, such as LCD displays or LED status indicator lights, may be difficult to see if in direct sunlight. It may help to shield these types of displays so the operator can read them in bright light. Some displays may also be difficult to read under low light conditions, such as nighttime darkness or in shadow. If possible, avoid unwanted shadows that shroud displays. Also avoid optical distortion of the display caused by domed covers by using flat glass covers instead.

5.4 RELATIONSHIP BETWEEN CONTROLS AND VISUAL DISPLAYS

If a control and a display are related both in the machine and the operator's mind, it is important to ensure that the directions of their moving parts are compatible with each other and conform to the operator's movement expectations. Further, the direction in which controls are moved to achieve a given result should be consistent throughout those parts of the control system used by the same operator.

It should also be clear to the operator which equipment or status is affected by the control being operated (see Figures 5-8 and 5-9). For example, if a control panel is for pumps used to fill three separate tanks, then it should be clear which control operates each pump that corresponds to the appropriate tank, as shown in figure 5-9.

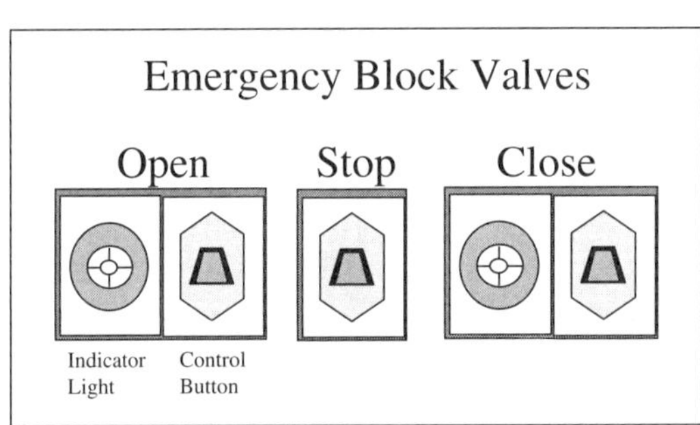

(a) A variation of a control panel layout for operating a set of emergency block valves

Figures (a) and (b) are variations of a control panel layout for operating a set of emergency block valves. Displays and control that perform similar functions should be laid out in the same way. It is preferable to use consistent layouts within a facility—that is, consistently use either layout (a) or layout (b), but not both.

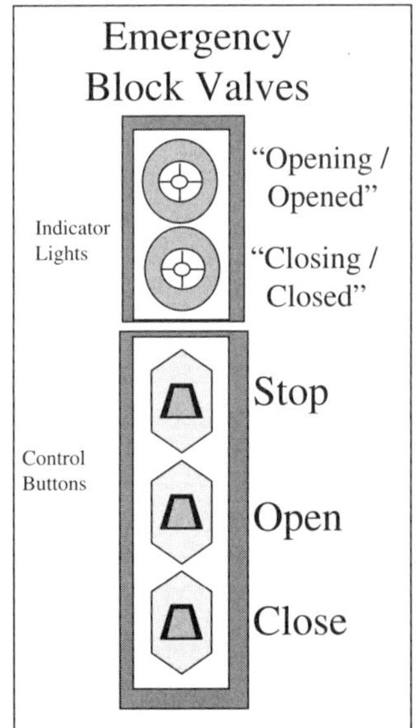

(b) Another variation of a control panel layout for operating a set of emergency block valves

Figure 5-8 Consistency of motor operated valve controls and indicators.

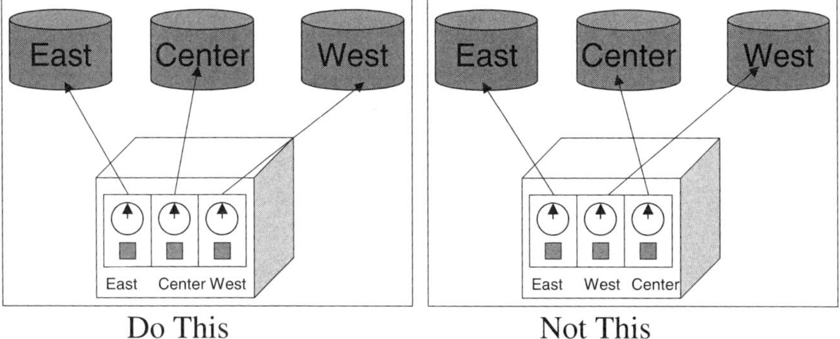

Figure 5-9 Consistency of control location in relation to plant equipment.

A classic example of the relationship between controls and displays is the burners on a kitchen stove and the knobs that turn each of the burners on or off. Many studies have been conducted to understand which individual knobs people associate with each of the burners. Some important learning from these studies is that the arrangement people prefer may not be the one that results in the fastest performance or fewest errors (Sanders and McCormick, 1987). In addition, there are population differences in preference for the relationships of the knobs to the controls. And, the results appear to be affected by whether the subjects are shown a drawing of a stove on paper or a physical model.

Therefore, just asking personnel for their opinions of how controls and displays should be arranged on an emergency control panel is not enough. The design effort should be augmented to include evaluation of a physical prototype of the panel that measures the number of correct responses of operators being able to operate the correct sequence of controls on the panel. The prototype panel can be mimicked by using a paper and pencil survey with a rendering of the panel as a drawing or sketch, but the best approach appears to be a physical mock-up.

5.5 AUDITORY ALARMS

The advantage of using an auditory signal is that the operator need not be looking at the display to notice it. Auditory alarms should be used to attract attention to a condition that requires action and not to announce information that requires no action. Auditory alarms are used to attract attention to local plant conditions (e.g., field panel alarms), attract and send messages to plantwide personnel (e.g., fire alarms, evacuation alarms, or mustering alarms), and attract the attention of operators within the control room.

Our perception of an alarm is determined by its frequency, amplitude, and modulating pattern. An audible signal that modulates is a signal with fluctuating frequency or amplitude over time. Sirens and warble tones are common examples of modulating alarm signals. For plant alarms, modulating signals are best because they tend to attract human attention better than signals made from a single tone (Eastman Kodak Company, 1983).

Each audible plant alarm that requires a unique response should have a clearly distinct signal. Studies at a nuclear power facility indicate that personnel are able to identify 12 distinct signals when they can be compared to each other. But, if a person must choose to respond based on hearing a single alarm, limit the number of alarms to at most five distinct audible signals (Sanders and McCormick, 1987). The best alarm signals are those that

- Elicit fast response times from people who hear them.
- Are readily recognizable and the response action needed is clear.
- Can be heard over ambient noise (about 10 dB above ambient noise is recommended for emergency alarms) (Eastman Kodak Company, 1983).

The following are characteristics for different types of plant alarms based on frequency (Sanders and McCormick, 1987):

- Best for quick response times (Sanders and McCormick, 1987) are "Yeow" (descending from 800 to 100 Hz every 1.4 sec) and "Beep" (425 Hz, on for 0.7 sec and off for 0.6 sec).
- Best frequencies for signal to travel through the plant are 500–1000 Hz.
- Best for noise penetration and attracting attention are sirens and horns.

For control rooms, it is important to avoid alarms that cause "startle" responses, as some research indicates that startle responses degrade the performance of the desired response and may cause mental stress (Sanders and McCormick, 1987). The best frequencies for control room alarms are 200–1000 Hz, and if there is ambient noise, a warbling signal is recommended (Eastman Kodak Company, 1987).

5.6 FIELD CONTROL PANELS

Field control panels contain displays and controls used to monitor and control the operation of process equipment in a local area of the plant. The purpose of a field control panel is to provide important local information to the operator and local access to the plant equipment for safe process operation.

Important aspects of field panels include

- Completeness and accuracy of labels, signs, and instructions.
- All lights and indicators working, no "live" readings from disconnected equipment.
- Good arrangement of controls and displays on the panel.
- Emergency panels and components are clearly visible and readily accessible.

One indicator of opportunity for the improvement of existing control panels is the presence of handwritten notes or labels that give instructions or reminders, placed on the panels by operators or maintenance personnel. These notes may be evidence of a history of confusion that someone attempted to clarify by labeling the panel for others. This type of hand labeling has problems in that it may be inaccurate, difficult to read, unclear, and nonpermanent.

5.6.1 Field Panel Layout

When an operator walks up to a control panel, the function and identity of each control on the panel should be obvious. This involves knowing what each control does to the status of the plant, and how the manipulation of the control results in that action. Some of this is achieved by good labels on the panel. An important contributor to the operator's understanding of how the panel should be operated is the arrangement of controls in relation to the displays or indicators and plant equipment affected. Another important consideration is how the arrangement of the controls and displays aligns with what the operator might expect.

When deciding how controls and displays should be arranged in front of the operator, the goal is always that they can be used quickly, accurately, and without physical fatigue. For this reason an attempt is normally made to ensure that complementary sets of components are arranged so that they suggest to the operator how they should be used.

Decisions on how to group the controls and displays for a field control panel are based on five criteria (Sanders and McCormick, 1987):

1. *Emergency use.* Locate emergency controls in easily identified and easily reached positions.
2. *Functional grouping.* Related functions are located together in an easily identifiable fashion.
3. *Sequence of use.* Determine the more common operations performed at the control panel, then position the controls so those operations can be performed by sequentially moving from one control to the next (for North American operations, the standard sequence is from left to right and top to bottom).
4. *Frequency of use.* Displays with more frequent use should be located in the primary visual area and controls placed closer to the hands.
5. *Relation of the panel itself to the equipment it operates.* The panel layout of controls and displays be should be arranged on the control panel in a layout similar to the geographical locations of the equipment in the plant (see Figure 5-9).

The sequence of use can be divided into time sequence and functional sequence. Time sequence suggests that, if controls or displays are normally operated in some sequence, for example, switching a machine on, increasing rpm, moving spindles together, then they should be arranged in that sequence. All the operator needs to do is follow the arrangement provided, rather than work in an apparently random fashion.

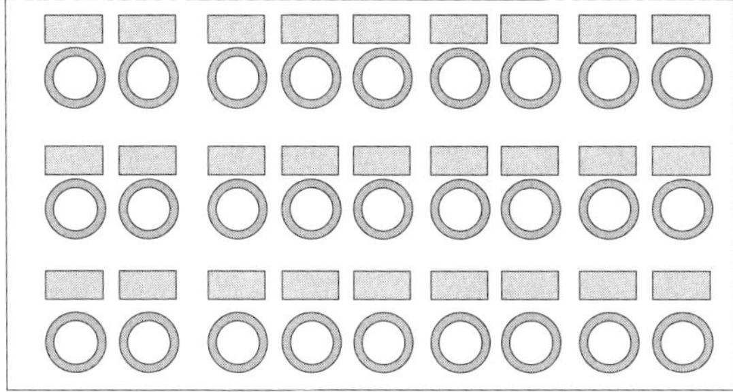

(a) Emergency Shutdown (ESD) Panel without controls grouped by function

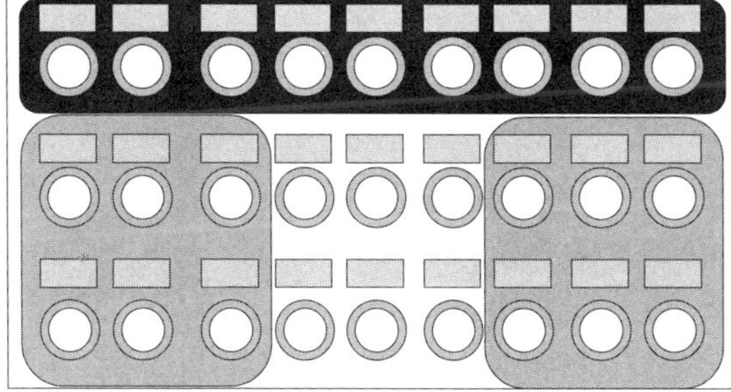

(b) Emergency Shutdown (ESD) Panel with controls grouped by function by placing shaded regions behind the buttons

Figure 5-10 Functional grouping on a control panel.

Function sequence suggests that controls and displays can be arranged in terms of function, either within the temporal sequence or in terms of a temporal sequence of different functions. For example, the workspace around an operator may include components concerning the start/stop, temperature, pressure, status, shutdown, and the like. It is suggested that all the controls and displays related to any one of these functions be grouped together. Also, the different functions may be used in a temporal sequence. For example, the start/stop, then the temperature, then pressure, then status, and so forth. In such a case, group components not only according to their function but also according to the order in which they are used.

The frequency of use principle suggests that controls and displays should be arranged in terms of how frequently they are used by the operator. More frequently used components should be placed within easy sight and reach of the operator. On the other hand, some situations may arise in which rarely used but extremely important components, such as emergency controls, might be positioned well outside of the operator's effective area if we rely only on the frequency of use principle. By their nature, they are operated infrequently, but when they are used, they need to be operated quickly and accurately. The frequency of use principle, therefore, must be tempered by importance.

The design, layout, arrangement, and installation of visual displays and controls affect the performance of the operator. Frequency of use, sequence of use, and importance are principles used to guide the designer to the best design. No single principle applies to all control panel designs. However, the principles should be used to create control and display layouts that are consistent from one location to another. This consistency enhances human performance and reduces the potential for error (see Figure 5-8).

Other guidelines for field control panel design include

- For large groups of controls use mimics instead of traditional panels. Mimics are control panels in which the displays and controls are imbedded within a schematic of the process flow depicted graphically on the panel.

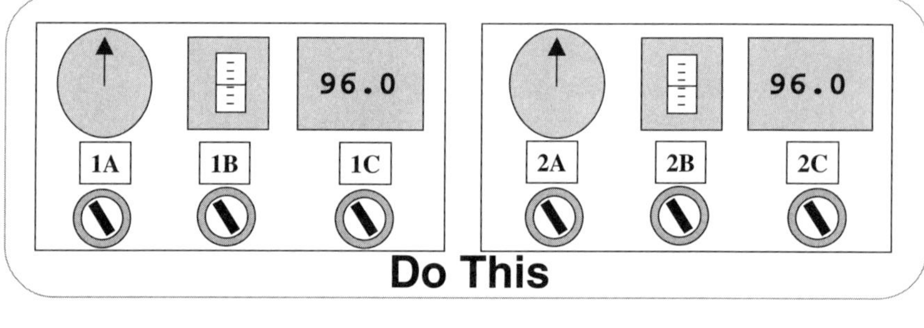

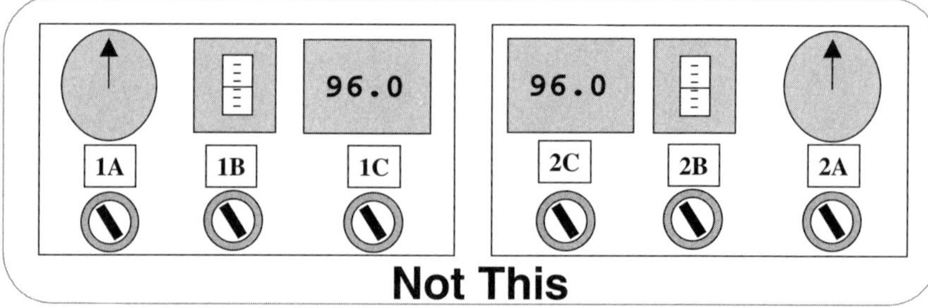

Figure 5-11 Do not use "Mirror Image" layouts such as the arrangement of controls and displays shown in the bottom pair of panels.

- Avoid mirror image layouts (See figure 5-11).
- Locate displays above their associated controls. Displays may be located beside their controls if the association is clear controls and the displays they affect should be as close together as possible.
- Ensure that the panel can be accessed by the user population (see Figure 5-12).
- Ensure adequate control instructions and feedback that the control has been activated.
- Select and mount the various panel components so that cleaning instrument displays, recorders, and control devices should be possible without affecting the process or requiring much dismantling.

5.6.2 Field Panel Labels

Each control panel should be identified with a human readable label, such as "title," located at the top-center of the panel. The various components of the control panel should also be labeled: subsections or dedicated areas of the panel, displays, and controls. For the subsection labels, center the labels with the title for the panel subsection at the top of the section and within the demarcation lines. The label placement, terminology, and text should be consistent within the control panel and across the plant (see Figure 5-13).

Avoid using one large label at the top or bottom of a series of controls or displays. Label each set of controls and displays separately. Labels for the controls and displays should include a functional description on the top line and the process tag number of the equipment below the functional description.

Control labels should be centered above the control so the worker's hands do not obstruct the label while manipulating the control. For controls, the labels that indicate the position of the control should indicate the functional result of the movement; for example, ON, OFF, PUSH TO TEST, and INCREASE (see Figure 5-14).

Labels for displays should be centered on or under the display. For displays of measurements, such as gauges, dials, or digital readouts, the label text should indicate what is measured, such as level, voltage, or pressure.

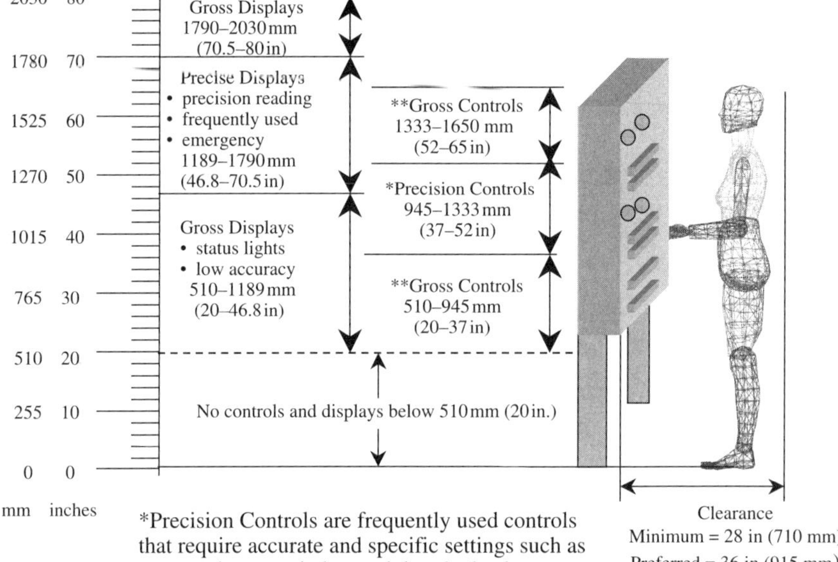

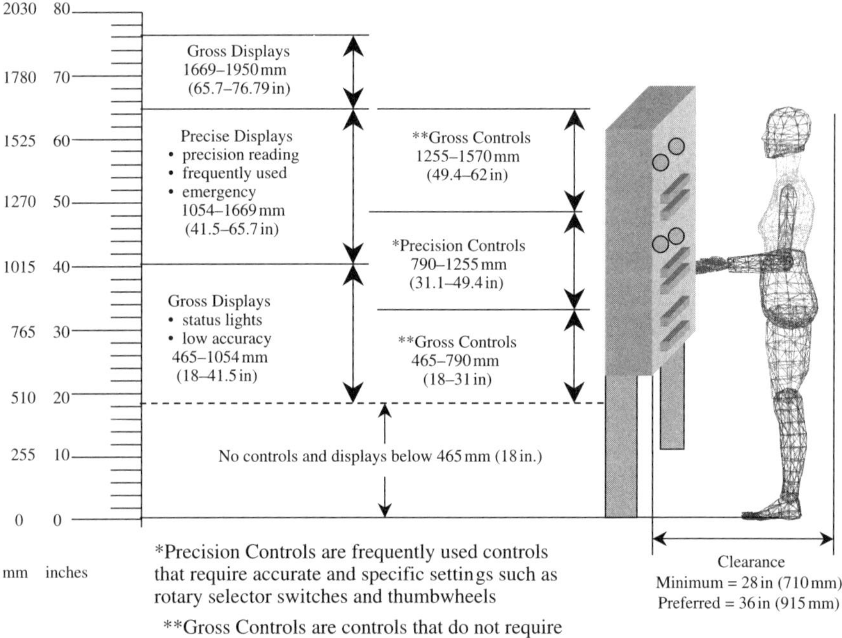

Figure 5-12 Panel design for populations with different physical sizes: (A) North American, (B) Singaporean.

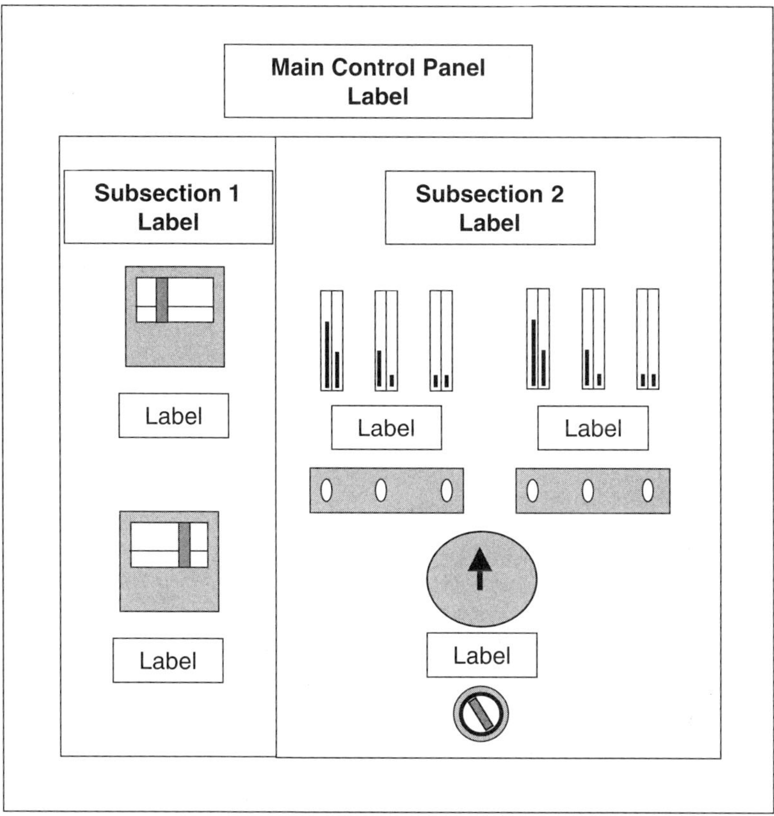

Figure 5-13 Placement of labels on a field panel.

The units of measure (meters, volts, psi) should appear on the face of the display not on the label. If the label is for a sensor, then it should indicate the equipment it is sensing; for example,

Ballast Pump Pressure Safety Low PSL-1200

The text on the labels should be readable from the operator's viewing distance; for example, while using the panel or scanning the panel to select a display or control. The label should be mounted horizontally where possible, so that the text can be read from left to right. Black text on white

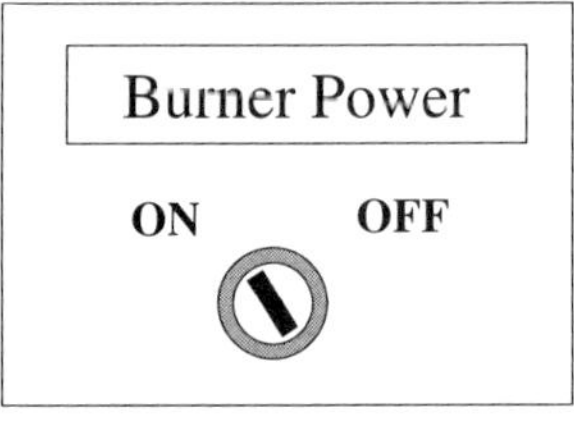

Figure 5-14 Design of label content for control operation.

labels is recommended for most situations. Use all uppercase letters if the label consists of three or fewer words, and use mixed upper- and lowercase for labels consisting of more than three words.

Labels should be made out of a durable material, such as engraved plastic or UV-protected vinyl. If engravings become filled with dirt, they can be covered with clear plastic. Attach labels using a permanent method, such as adhesive or screws. Do not place labels on controls that can turn.

Escutcheon plates are large labels used to show groupings of displays and controls. The escutcheon plates may be simple shapes or designed as graphic illustrations of the component being labeled.

5.6.3 Improving Field Control Panels

Many improvements can be made to field panels without redesigning them from scratch. In some cases, just clarifying the labels and adding markings to clarify groupings is a good enough fix. In cases where there is confusion about which controls operate which equipment or which indicators link to which equipment, the controls should be fixed so that confusion is eliminated.

Finally, redesign of the control panel may be indicated if several of the above improvements are needed and some controls must be moved to locate them within reach, out of the range for accidental activation, within a functional group, or close to the corresponding display. Any related displays should also be considered in the redesign effort.

The general steps to design or modify a control panel are

1. Identify type of information that should be displayed and adjustments to be made.
2. Select instrument types for controls and displays that best match the types of information displayed and the adjustments to be made.
3. Specify the design of each control and display based on human factors design guidelines (refer to Control Panel Layout in Appendix 5-1). Determine strategy to prevent accidental activation of controls that could result in serious misoperation.
4. Select a scheme for organizing the panel elements. The preferred way to organize panel elements is by function, layout consistency, and critical instruments. Secondary criteria are sequence and frequency of use. If a mimic is used, ensure that it accurately follows the process.
5. Work out the overall layout of the panel and determine spacing requirements between elements and groups, placement of labels, supporting demarcation or details to define the groupings, and labeling or coding techniques.
6. Specify access (physical and visual) requirements for maintenance and operation and lighting requirements.

5.7 PROCESS CONTROL DISPLAYS

Process control displays are the screens that a control room operator uses to monitor the process via a computerized distributed control system (DCS). These displays are typically on video display unit (VDU) monitors; for example, desktop computer monitors. Some control rooms may have additional displays that are "hardwired" emergency systems. These systems may be composed of annunicator tiles and status lights with controls, similar to a "field" panel.

With advances in technology, petrochemical processes are controlled with computerized digital distributed systems that have more information and complexity than ever before. Although an increase in available information appears to be an advantage for the operator in managing the process, there is the potential to overload the operator with information. Because of the large amount of information and process complexity, the digital control room potentially provides less process overview to the operator and less feel for process dynamics, possibly resulting in the operator's loss of ability to handle upsets (Bollen and van der Schaaf, 1993). Several factors contribute to operator workload: design of the displays and alarms, characteristics of the process, fitness for duty, environmental factors, control room design, operator interactions with personnel, and operator training and experience (Connelly, 1995). The focus for this section of the book is to optimize the design of the process display screens to ensure that the control room operator's primary interface to the process does not exacerbate his or her overall workload (see Figure 5-15).

Figure 5-15 Process control human/system interface.

The optimal interface takes best advantage of the technology, helps the operator effectively manage normal and abnormal operations, yet does not overload the operator during abnormal operations. The operator interface consists of information and alarms displays in the screens as well as the console unit (e.g., keyboard, touch screen, control panels). It is the major tool used by the operator to sense what is happening with the process and make adjustments. Therefore, the design of the operator interface is critical because it influences how the operator perceives, understands, and acts on the process.

The guidelines summarized in this chapter describe the features of effective process control display screens. The content of the displays should be designed with the approach described in Section 5.7.2.

5.7.1 Process Control Display Interface

Examples of potential problems with process control interfaces include

- Nuisance alarms and alarm flooding.
- Lack of coding for alarm priority.
- Alarm sounds that have been disengaged because they startle the operators.
- Inconsistent or overuse of color coding for data and especially on graphics.
- Information displayed in groups that are not functionally related.
- Graphics that are inconsistent with the actual process.
- Individual displays that exceed the recommended density of information packed onto a single screen.
- Critical values or equipment that are unlabeled.
- Instrumentation that is out of service but displays a "live" inaccurate value.

The operator is usually able to compensate for any one of these situations in normal and smooth operating conditions, but the compensation strategy may break down during periods of high stress, such as unusual or upset situations, or if the operator must compensate for multiple and simultaneous problems.

The International Standards Organization has standards for control display interfaces (ISO, 1999a, 1999b), as do several other organizations (International Instrument Users' Associations, 1998a, 1998b, 1998c; Instrument Society of America, 1985, 1986, 1992, 1993, 1995a, 1995b; Mulley, 1994; Gilmore, Gertman, and Blackman, 1989). Other sources of control system guidelines include industry and government regulations for fossil and nuclear power plants (O'Hara et al., 1994; O'Hara, 1993; Rankin et al., 1985; U.S. Nuclear Regulatory Commission, 1981; Electric Power Research Institute, 1981). These sources include lessons learned and recommendations that have been published on disasters such as Three Mile Island, Piper Alpha, and Bohpal. Because the operator interacts with a digitally controlled plant through a computer, some human/computer interface guidelines are also appropriate (U.S. Nuclear Regulatory Commission, 1982).

5.7.1.1 Display Hierarchy

Levels of hierarchy help the operator navigate the process control displays. The top level is a system overview that contains key process and emergency alarms, fire and gas alarms, and key process information. In addition, this level could also display locations of open hot work permits. The next level of displays provides process overviews for each area of the plant. These displays include schematics of process flow, equipment, and integrated alarms within the process. Within this level of hierarchy also should be a start-up display that includes the parameters the operator needs to monitor the start-up process. A third level of hierarchy contains more detail about subsections of the plant or equipment (e.g., detail of trays in a tower or multiple temperatures in an exchanger). Ready access to the shutdown controls should be provided without forcing the operator to dig down into the hierarchy to find them.

5.7.1.2 Contents of Displays

Display contents should be based on the operator's tasks. Contents of display screens consist of both static and dynamic components. Static information remains the same on the screen and does not change, such as the process flow lines, titles, and equipment symbols. Dynamic information is the portion of the display that changes, such as displays of real-time instrument readings and status of equipment. Examples of dynamic information include pressure, temperature, and flow data and the status of valves and pumps. A single display screen may include a schematic of the process, with a table of critical values, a trend chart for one of the critical values, and a selectable pop-up of temperatures within a reactor. Given the increasing display capabilities of distributed control systems, contents of displays need to be thought out to avoid producing cluttered displays that are poorly structured and increase the operator's mental workload.

It is important to systematically select and organize the information the operator needs on the display screens, as there are thousands of data points from which to choose. Connelly (1995) recommends that display systems avoid overloading short-term memory by minimizing the number of displays and placing more functionally related information on each display. It is better to have operators comparing information on a single screen than to try to make comparisons of data across screens by holding the data in short-term memory. Displays that are dense but well structured can be suitable and clear to expert operators (Connelly, 1995). To support high-density displays, ensure that alarm signals are salient and key process variables and alarms are available in an overview. To increase display density while minimizing a "cluttered" effect, use effective color coding and "pop-up" supplementary data in windows. Consider providing an overview that displays status of key process variables and other indications.

5.7.1.3 Display Layout

The components of each display, such as the display title, alarm list, dedicated navigation targets, and main area for graphics, should be located consistently on the screen. For example, the title always is at the top left, the list of areas of the plant with active alarms is always at the top right of the screen, and the navigation targets (back, next, overview) are located below and to the left of the main screen area.

5.7.1.4 Abbreviations and Labels

Abbreviations and labels should be standardized and used consistently within the system. Equipment should be labeled within the graphic area that defines the equipment. Exceptions are pumps, filters, heaters, and valves, which have shapes and corresponding tag numbers that identify them to the operators.

5.7.1.5 Alarms

An alarm requires operator action; therefore, only the conditions that require operator action should be assigned an alarm priority. An alarm priority scheme ranks each alarm level and ensures that the alarms are consistently assigned appropriate priority. One example of a scheme for a nuclear power plant is based on the following criteria (Chang et al., 1999):

- *Role of component, essential versus support.* A component is essential if equipment accomplishes a main function; otherwise, it is support.
- *Impact.* Severe means that the situation threatens operability, middle means that it degrades performance, slight means that performance is minimally affected.
- *Response time.* Urgent means that the situation requires immediate operator attention, imminent threat; intermediate means that action is required but not immediately.

An established alarm priority scheme avoids problems with inconsistent interpretation and assignment of high-priority alarms. One example of how a team can develop an alarm philosophy and standardize alarm categorization is to use a matrix such as the one shown in Figure 5-16, based on potential severity and urgency to classify alarms consistently.

This way, there is no doubt by the operator in interpreting the relative priority of the alarm. The use of this method also helps prevent too many alarm states from being identified as high-priority alarms. Another aspect of the alarm philosophy is how an active alarm is consistently presented to and acknowledged by the operator.

Each alarm should be assigned a consistent code, such as color, symbol, and tone, with codes selected to convey the priority of the alarm to the operator.

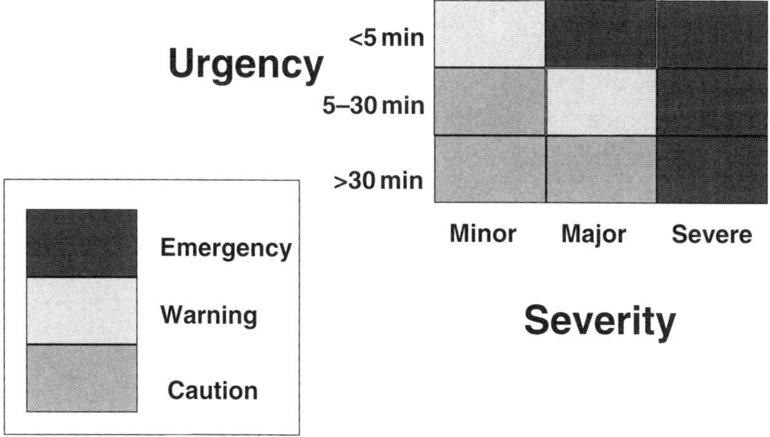

Figure 5-16 Example alarm priority matrix.

The following are some alarm guidelines:

- Rationalize the alarm system to an acceptable relative number of alarms.
- Use a minimum number of priority levels; for example, three (high, medium, and low).
- Do not assign alarms to status information that requires no operator action.
- Minimize nuisance alarms.
- Ensure alarm priority is consistently assigned.
- Ensure that the alarm coding scheme is meaningful to the operators and conveys appropriate priority.
- Ensure that alarms are coded by combinations of schemes, such as both color and symbol, and audible tone.
- For control rooms, avoid alarm tones and loudness that startle the operator.

Technologies are being developed to aid the operator in identifying a potential abnormal process condition, tracking its development, and intervening to resolve the problem. One concept of this type of tool is a series of windows that bring up relevant process diagrams, alarm trends, and process parameter trends for a particular problem (Kawai, 1997). Another concept integrates online emergency procedures linked to process variables and relevant displays to confirm steps of the procedure (Chang et al., 1999).

5.7.1.6 Text Messages

Text messages in displays should be simple, use mixed upper- and lowercase letters, and be understandable to the operator.

5.7.1.7 Lines and Arrows

In displays, lines are used to represent process flow lines, and arrows represent the direction that the contents are flowing. Distinguish between main and secondary flows by line thickness. The standard flow direction is from left to right. Label all entry and exit path lines. Avoid crossing lines, clearly distinguish them. The preference is that, if lines cross, the main line should remain intact and the crossing line can be represented by an arc across the line or by leaving space on either side of the main line. Ensure arrowheads look different than valves.

5.7.1.8 Numeric Values

Display numbers only to the significant decimal point needed by the operator for accuracy (e.g., does the operator need to know that the pressure is 75.046, 75.05, 75.1, or 75 psi?). Suppress leading zeroes except for use in decimals, where one leading zero should be used. Display all numbers upright and include units of measurement next to the number. For values in columns or imbedded tabled values, line up the decimal points. The operator may need only a qualitative indication of a data point, in which case it is better to provide that than a numeric value. Do not require the operator to hold figures or numbers in memory to perform mental arithmetic, calculations, or conversions between units.

5.7.1.9 Use of Color

The effective use of color can decrease the time required to search for salient signals, such as alarms (Van Laar and Deshe, 2002). When developing a display, a suggested way to prevent the overuse of color is to develop the layout and design of the display before introducing the color-coded elements. Color is typically used for coding priority of alarms, indicating if valves are open or closed, on/off status of equipment, identifying materials in the process flow, or distinguishing between plant areas, such as two identical process trains. The use of color on process displays should be meaningful, and the color on the display should not be used just for aesthetics. Where color is used as an identification code, ensure it is used consistently within each display, between displays, and with other coding systems the operator encounters. The use of color coding should be redundant with another coding scheme, as shown in Table 5-2.

TABLE 5-2

Example of a Color Coding Scheme for Process Displays

Display Feature	*Color Code*	*Additional (Redundant) Codes*
Emergency, high-priority alarm	Red	Letter H for high priority, flashing symbol
Caution, medium-priority alarm	Yellow	Letter M for medium priority, flashing symbol
Warning, low-priority alarm	Cyan	Letter L for low priority, flashing symbol
Utilities feeder/returns, reagent feeds, drains	White	Line style-thin line, text labels
Crude oil, processed oil	Green	Line style: medium thickness; text labels
Condensate, produced water	Orange	Line style: medium thickness; text labels
Gas, vents, flare lines	Magenta	Line style: medium thickness; text labels
Background detail, control logic connections	Blue	Line style; dashed

Source: Modified from Danz-Reece and Noerager, 1998.

Attention-getting features such as bright colors (red, yellow, cyan) and flashing or blinking should be limited to portions of the display that demand operator attention. For example, flashing or blinking should not be used just to indicate that a piece of equipment (e.g., a fan) is under normal operation. Good uses for flashing are to indicate unacknowledged alarms and indicate valves changing position; that is, in the process of moving.

5.7.1.10 Display Access

The operator should be able to readily locate each display as well as critical controls from within the system. The operator should have multiple approaches available for calling up the displays. Examples of these approaches include

- Menu list of available displays by title and area covered.
- Process overview overlaid with navigation targets for access to more detail on an area schematic.
- Process lines labeled with targets that the operator can use to view the next schematic downstream or upstream of the current view.
- Alarm summary list with access to the page of each alarm.
- Text input of title name.
- Search tool by phrase or name of display.
- Next and Previous navigation buttons.

Active or "hot" areas on the display that support navigation, such as arrow symbols or named rectangular targets, should be obvious to the operator. One method of doing this is to change the appearance of the target if the mouse cursor floats over it (e.g., from an arrow to a hand).

5.7.1.11 Symbols

In process displays, equipment is represented by symbols. The symbols used should be standard, used consistently, and meaningful to the operators. In addition, the symbols should be clearly distinct from one another.

The Instrument Society of America (ISA) provides a standard for graphic symbols in process displays (1986).

To summarize some guidelines for process control displays, we recommend

- Displays support the operators' tasks.
- Navigation through the displays be compatible with how people operate.
- Displays allow selecting directly and moving backward and forward.
- Displays not be cluttered.
- Display layouts be consistent.
- Screens be labeled and similar screens for different processes not be confused.
- Display information be presented in the proper format.
- Symbols be standardized.
- Color be used as a redundant code to clarify information.
- Information be grouped for clarity.
- Information not require mental conversion to be usable.
- Text messages be simplified to aid understanding.

5.7.2 Approach for Developing Process Control Displays

Optimal design of the operator interface begins with the conceptual design. Although the application of human factors techniques can significantly improve an existing interface, it is preferable to apply these techniques during the initial design of the control system. Human factors principles apply to all stages of design, including new acquisition, modification, upgrade, or near-term enhancement of equipment. Human factors principles are integrated into the design process in steps that correspond to increasing levels of detail design. The steps to design the interface aspects of the process control system are best approached by a team consisting, at least, of a process engineer, control system technician, experienced board operator, inexperienced board operator, and a human factors specialist. Human factors should be thoroughly considered throughout the design process, beginning with the initial scoping of the project and selection of equipment to the conceptual and preliminary design of the screens, as well as at the detailed design level. The activities to integrate human factors into the control system design are (Danz-Reece and Noerager, 1998).

- Conduct an initial survey.
- Scope the design or improvement effort.
- Prepare the design team.
- Brief the board operators.
- Execute the design.
- Obtain user feedback.
- Transfer to the new system.

5.7.2.1 Initial Survey

Conduct an initial survey of the existing system to identify aspects that can be improved. A human factors specialist is trained to evaluate the system interface according to specific design criteria. Examples of tools that specialists use include guidelines to perform a tabletop review of the existing displays, interviews, questionnaires administered to users of the system, and objective assessments such as mental workload while performing a task on the system. If a complete transfer from analog to digital control will take place, the initial survey is more detailed and emphasizes the identification of the performance requirements and goals of the new system rather than evaluation of the existing system.

The design process differs slightly depending on whether the control system is being implemented for the first time, converted from an analog to a digital system, or upgraded to an improved digital system. In all cases, provisions must be made for involving the users (board operators, system technicians, and system maintainers) in the design process and the interface evaluation and developing training and providing structured user practice on the new system.

5.7.2.2 Scope the Improvements

Based on the results of the initial survey, define the scope of improvements. These can include near-term improvements such as simple quick fixes and long-term improvements such as modifications or upgrades. If a new system is to be acquired, human factors considerations should be included in the equipment selection. Human factors considerations vary according to whether the system is analog or digitally controlled, with control panels or computer screens as interfaces to the operator.

For example, a number of near-term fixes for existing analog systems and corresponding control panels can be applied to improve the organization of information to the operator. Some examples include adding demarcation lines to group panels of displays or controls, adding labels to indicate sequence of control use or display checking, identifying the matrix of annunciator tiles with row and column codes, and organizing annunciator tiles in process groups. These enhancements to the organization of information are more complex in a computerized digital system because screens and the items on them may have to be reprogrammed and reinstalled on the system.

Planned milestones and resources may be revised after the design team has been trained and further examines the system. At that time, needed evaluations and actions can be better defined, and the plan may need refinement.

5.7.2.3 Prepare the Interface Design Team

The operators are the key players in the interface design process and provide an abundance of useful information for design decisions. While operator preference and opinion is important, it must be drawn out and applied judiciously. Research studies have shown that situations preferred by users do not necessarily result in the best performance outcome. Rather, the situation designed with human factors principles results in the best performance and fewest errors while performing the task. The approach described by Eason (1994) is to "design for the users with the users," instead of "design for the users" or "design by the users." This is accomplished by operator participation on the interface design team and training the team in basic human factors principles and techniques applied to process control design.

5.7.2.4 Brief the Board Operators

Successful introduction of a new system depends on board operator buy-in. Earn the buy-in of the operators who use the system by providing them awareness about human factors and process control design. Operators aware of human factors examples are well equipped to add value to the design process and survey activities.

Use operator input constructively to improve the design throughout the design process. Invite operator input on the control system setup and displays by posting paper mock-ups in a special area of the control room bulletin board easily accessed by operators; invite volunteers to review the mock-ups and lead informal discussions. Additional review and design evaluation techniques, such as usability tests, are covered in the section "Obtain Operator Feedback."

Advise operators of the design team's progress. Notify them about the progress by posting a special project newsletter and announcing milestones at shift meetings.

5.7.2.5 Execute the Interface Design

Design of the interface occurs over three phases: conceptual, preliminary, and detailed design (Stultz and Schroeder, 1988). The conceptual design phase is the most important phase from a human factors point of view because critical decisions are made about how the interface will present information to the operator. Once these basic issues are addressed, the preliminary and detailed design phases are used to assign specific content and format to the information presented to the operator.

Conceptual design involves the development of the relationships between the groups of displays within the system and their organization and hierarchy according to the process. This phase is a checkpoint for determining if the features of the selected hardware accommodate the conceptual design criteria, such as the number of available areas and groups and the format of alarms. Other issues addressed during conceptual design include the number of keystrokes and the type of input/output device used to access each level of the hierarchy, response requirements of the hardware based on results of task analysis, control strategies (such as access to displays and changing set points and controller states), and feedback from the system to the user. Availability of menus, overviews, help, function keys, and custom keys should also be planned during the conceptual design.

Whether in an initial design or a retrofit, the organization of information determines the criteria for the remaining design decisions for the control system. Analysis of operator functions, such as start-up or shutdown of a particular system, is used to determine several fundamental design decisions, including the minimum number of monitors required, the number and identification of areas, and the number and identification of groups and graphics within the areas. The relationship of the groups and areas to each other is also important; for example, in chemical operations it may be necessary for the operator to clearly distinguish one train from another. The use of functional analysis also aids the development of procedures quite well, so the effort invested is effectively used (Pikaar, 1986).

To make these decisions, the design team can be trained to use a technique called *task analysis* to define and analyze operator functions in managing the process. Examples of operator functions include start-up, shutdown, changing batches, handling various upsets, and monitoring normal operations. Task analysis techniques are discussed in Chapter 7 of this book. A recommended practice to analyze the operator control room tasks is available from the Instrument Society of America (1993). The tasks required to successfully complete each function are defined and analyzed for the following information (Staples, 1993):

- The cue that prompts the operator to perform the task.
- The information that tells the operator the computer has understood the task request.
- The communication contacts required to perform the task.
- The criteria the operator evaluates to determine if the task was performed correctly or incorrectly.
- The factors that affect the likelihood of successful task completion.

- The information the operator needs to know to take an action (e.g., procedures).
- The physical actions required to complete the task.
- The potential errors.
- The likely causes of each potential error.
- The consequences of each potential error,
- The ways in which the potential errors can be recovered.
- The priorities of the consequences.
- Rating the task itself on time criticality and importance.

The information identified in the task analysis is used in the preliminary and detailed design phases to select the contents of displays, identify necessary special displays, analyze mental workload and staffing issues, identify training needs, and perhaps develop operating procedures.

The tasks and functions are analyzed to determine whether the operator or the computerized system is better equipped to perform particular tasks and actions. This, called *functional allocation*, assigns tasks and functions to a person or computer depending on the type of task, its complexity, frequency of execution, and the like (Boff and Lincoln, 1988). More advanced analysis techniques can be used to develop displays that can improve operator understanding of the technical complexity and dynamics of specific multivariable processes (Olsson and Lee, 1994; Vicente and Rasmussen, 1990).

Information from the operators who will use the system supplements the task analysis and helps tailor the control system interface to the specific site needs. For example, the board operators provide input on how an existing interface could be improved by answering questions about their most difficult or confusing tasks, features of the system they find difficult to use, features of the system that they prefer, the colors that they associate with the priority of alarms or parts of the process. Interviews with open-ended questions can be used as well as questionnaires (Pennycook and Danz-Reece, 1994). In addition, objective methods can be used to measure mental workload while using the existing system (see more in the "Obtain Operator Feedback" section).

Another source of information used to supplement the task analysis is the lessons learned from industry and site incidents, near misses, and operator logs. Site records may indicate particularly difficult functions or upset situations and those functions or tasks prone to error in production, quality, or safety. Lessons learned from the industry are examples of complex scenarios that resulted in severe consequences, and recommendations in reports can be reviewed and included in the design if appropriate.

The preliminary design phase involves the determination of the total number of displays needed, development of the alarm philosophy, identification of special displays, type of information required on each display, and development of the coding philosophy, such as identification of transmitters, controllers, areas of the process, materials, and so forth.

During this phase, the team develops and agrees on the interface design standards that will be documented as a "style guide" for the project. Interface standards should be used consistently for all displays.

During the detailed design phase, each display is planned and entered into the system. During this phase decisions are reached about the format of the information on each display; for example, the information could be displayed as columnar data, in a plot, in a chart, as a figure, written as text, or as a combination of these on a single page. For retrofit situations, each display is evaluated and, if necessary, targeted for improvement. The priority of the information is assigned, and the format of the information is selected so that the priority and content of the information are best conveyed to the operator. Guidelines for detailed interface design are discussed in Section 5.7.1.

5.7.2.6 Obtain Operator Feedback

Evaluation of the design is not actually a final step in a series but a step that should occur in parallel with the previously described steps. Design evaluation can be in the form of informal and formal surveys, mini-tests or trials with groups of operators, situation walk-throughs using paper or computerized mock-ups, and full evaluation of a single new control station (van der Schaaf and Kragt, 1992). Evaluations can also include objective measurement of an operator's mental workload under various situations and observation of how the operator locates data and what data are important to the operator's current decision.

The difference between user preference and performance is an important distinction when evaluating user feedback. Feedback from users should be structured so that areas of improvement are indicated by the outcome of using the system while performing a trial (van der Schaaf, 1989, 1991). For example, some observations might be "could not locate screen for controller P-141," "schematic on the A unit is confusing," "operator did not know where to find the tag description," or "locating the controllers that indirectly lower the plate #5 temperature took longer than we can allow under actual circumstances." Observations like these are based on the results of an operator using the system to perform a task (or a simulated task) rather than a judgment of a screen's appearance, such as "the colors look nice—in fact, add more of them—I like lots of color."

Incorporating operator feedback and experience into the system should be an ongoing process. Operator feedback on new individual screens can be obtained from paper or computer screen mock-ups. The paper mock-ups can be examined by each operator for accuracy and consistency in layout. In addition, a group of paper mock-ups can be used to simulate the series of screens that would be used to handle a particular situation, to determine if the information required by the operator is actually available on the screen.

A software prototyping tool can be used to mock up a series of screens with dynamic features to more closely mimic the actual system. This tool can be used to evaluate different color coding and alarm coding strategies. Also, operators could keep a log book to record their difficulties with using the system to help identify aspects of the system that need improvement.

5.7.2.7 Transfer to the New System

It is recommended that transfers to a new system occur gradually (Bollen and van der Schaaf, 1993, recommend 9 months). The first transfer step actually begins during the design process, as operators evaluate the proposed mock-ups. The mock-ups can be made available to the operators by installing a single read-only unit in the control room early in the design. It is suggested that the completed screens in view-only format be provided to the operators for an extended period of time (Bollen and van der Schaaf recommend 6 months) before the old screens are dismantled. During this period, formal and structured training should be provided to orient the operators to the new system and its features. After the formal familiarization period, the new units can be made operational in parallel with the old system (Bollen and van der Schaaf recommend 3 months). After this period to ensure familiarization and confidence with the new system, the old system can be dismantled.

The training and adjustment of the operator to the new system is critical, particularly in the case of converting from an analog to digital control system. In the analog setup, the operator was accustomed to the information available at a single hierarchy—a glance around the room. The operator had a mental picture of the room and the process, and there was probably a "blackboard" concept in effect, meaning that, if the unit was operating within limits, the displays were not lit or lit green (Rankin et al., 1985). However, introducing a digital control system requires that the information be layered and accessed in steps through the computer interface. The operator now has to modify the mental image of the process from a single hierarchy to a layered hierarchy, which steps from area to groups by increasing levels of detail. Also, the glance around the room is no longer available to the operator as a status check; this is replaced with a walk-through of selected computer displays.

The effectiveness of operator training on the new system not only involves familiarization with the organization of the screens but also with the hardware itself. This includes a thorough understanding of codes and symbols as well as the functions of all keys and available display options. These concepts are best learned with a formal introduction to the system followed by structured practice.

5.7.2.8 Summary

Integration of human factors/ergonomics considerations into the design of the process control interface requires coordinating and executing team activities to optimize usability of the interface, resulting in numerous benefits:

- Reduced rework of the system after it has been installed. Some modifications and adjustments are expected, just not as many and not as drastic as they might have been without human factors input to the design.

- Concurrent development of sound and well-documented operating procedures based on human factors principles, preventing a duplicated effort to produce procedures after the fact.

- Identification of specific training needs for board operators.

- Identification of specific tasks and functions that are documented and can be incorporated into the description of the board operator's responsibilities.

- Selection of control hardware and software features that accommodate the process goals and the operators' capabilities.

- Thoroughly considered potential for errors, consequences, and points of recovery for the tasks. This information is used for the human factors considerations in human reliability assessments and is available for further analysis.

- Operator and personnel buy-in and understanding of the new system.

- Members of the design team trained in human factors and ergonomics principles have lasting practical knowledge that can be applied to other projects.

- Most important, a control system interface configured to help the operator manage the normal operations and control the upsets, to reduce the potential for error in operation.

5.8 CASE STUDY

This case study shows how an emergency control panel was improved solely by improving the labeling (Swensen, personal communication, 2000). A review of the original design, shown in Figure 5-17, indicated the following deficiencies:

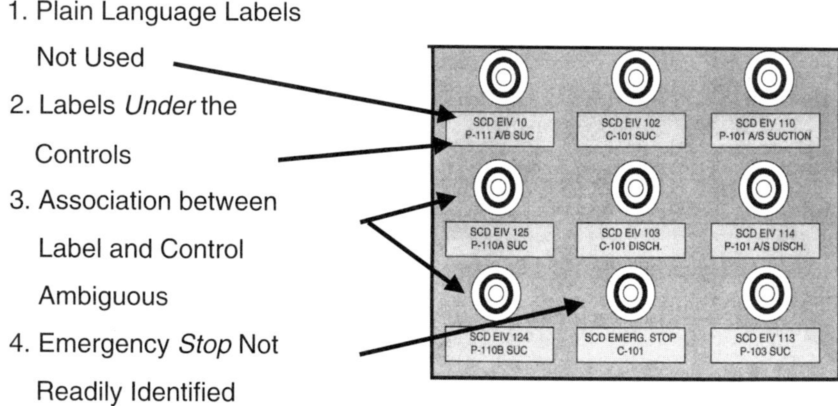

Figure 5-17 Original design of an emergency isolation valve control panel.

- Lack of color coding to indicate that the panel contains controls to isolate an area in the event of an emergency.
- Labels placed below the controls, so that they could be obscured by the hands while activating the controls.
- Lack of clarifying instructions on how the controls on the panels should be operated.
- Placement of labels that does not make it obvious to which control each label corresponds.

The improved labels on the control panel had the following features (see Figure 5-18):

- Large color-coded sign placed above the control panel to indicate controls to perform emergency isolation.
- Revised the labels to use the full names of the equipment where possible.
- Labels placed above the controls.
- Improved the association between the labels and push buttons.
- Uscd mixed-case letters to improve readability.
- Clearly identified the emergency stop control button.

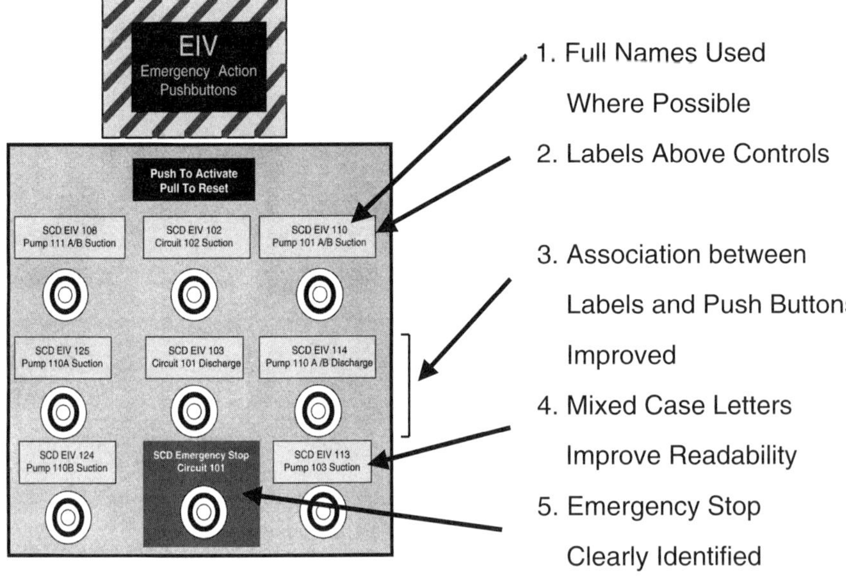

Figure 5-18 Improved design of an emergency isolation valve control panel.

APPENDIX 5-1. CHECKLIST FOR EQUIPMENT DESIGN

Displays

1. The type of display selected must be appropriate for its use.
2. LED and LCD displays must be capable of being read under all possible lighting conditions.
3. All numbers on displays must increase in a clockwise, upward, or left-to-right direction.
4. Multiple pointers and multiple scales on the same display must be avoided.
5. Color-coded zone markings must be clear and understandable (e.g., red for danger, yellow for caution, green for acceptable).
6. Displays must be mounted where each can be seen from the operator's normal working position.
7. Safety critical displays must be mounted in the primary field of view.

8. Similar functions must have the same layout of graduation marks and characters.
9. Displays must be free of parallax and glare.
10. All displays must have labels.

Controls

1. Controls must be selected to be appropriate for their use.
2. Correct colors must be used for each control.
3. It must be easy to understand the position of each control.
4. All controls must operate as expected.
5. Similar controls must operate the same way.
6. Controls must be large enough to be easily grasped.
7. Adequate room around each control avoids the chance of inadvertent activation of the wrong control.
8. Safeguards on emergency controls must avoid accidental operation.
9. All controls must be labeled.

Control Panel Layout

1. Related functions at each control panel must be grouped together.
2. Frequently used controls and displays must be located in primary areas.
3. "Mirror image" layouts must be avoided.
4. Displays must be located above their associated controls.
5. Each display must be viewable while its associated control is being operated.
6. Emergency controls must be easily identified, readily accessible allow quick activation.

7. Distinct subunits of panels must be grouped together for safer, easier operation.
8. Seated control panels must have the controls within 65–113 cm (26–45 in.) and the displays within 75–150 cm (30–60 in.) above the floor.
9. Standing control panels must have the controls within 50–133 cm (20–55 in.) and the displays within 50–160 cm (20–64 in.) above the floor.
10. At least 79 cm (28 in.) of free access space must be allowed in front of standing control panels.
11. Control panels must be viewable from normal operator work locations (e.g., from a walkway when on rounds).
12. Adequate space for maintenance must be provided.
13. Controls and displays operated sequentially must be oriented left to right.
14. Controls and displays must be located for use by the user population, or from top to bottom.
15. Field graphic displays (LCD, LED) must be readable under all lighting conditions.
16. Controls must be mounted near the displays they affect.
17. Safety-critical displays must be mounted in the primary field of view.
18. Safety critical controls must be mounted in the primary area for hand use.
19. All displays and controls must be labeled.

Process Control Displays

1. Displays must support the operators task.
2. Navigation through the displays must be compatible with how people operate.

3. Operators have access to multiple ways to call up specific display screens.
4. Displays must not be complex or cluttered.
5. Display layouts must be consistent.
6. Screens must be labeled and similar screens for different processes must not be confused.
7. Display information must be presented in the proper format.
8. Symbols must be standardized.
9. Color must be used as a redundant code to clarify information.
10. Information must be grouped for clarity.
11. Information must not require mental conversion to be usable.
12. Text messages must be simplified to aid understanding.

REVIEW QUESTIONS

Test your understanding of the material in this chapter.

1. In what three ways can controls be easily improved?
2. What is redundant coding? Give an example of redundant coding for process control displays.
3. You are selecting new field temperature instruments that will have digital readouts. What features of the digital temperature display would you evaluate to determine if the instruments are suitable for check reading by operators during their plant walk-throughs?
4. When you assess the process control interface, what aspects of the alarm system do you evaluate?

REFERENCES

Amell, T. K., and Kumar, S. (2001) "Industrial Handwheel Actuation and the Human Operator: A Review." *International Journal of Industrial Ergonomics* 28, pp. 291–302.

Attwood, D. A., Nicolich, M. J., Pritchard, K. K., Smolar, T. J., and Swensen, E. E. (1999) "Valve Wheel Rim Force Capabilities of Process Operators." Human Factors and Ergonomics Society 43rd Annual Meeting, Houston TX.

Boff, K. R., and Lincoln, J. E. (1988) *Engineering Data Compendium: Human Perception and Performance.* Wright-Patterson Air Force Base, Fairborn, OH: AAMRL.

Bollen, L. A., and van der Schaaf, T. W. (1993) "How to Survive Process Control Automation." Fifth International Conference on Human Computer Interaction, vol. 1. Orlando, FL: Harcourt.

Chang, S. H., Choi, S. S., Park, J. K., Heo, G., and Kim, H. G. (1999) "Development of an Advanced Human-Machine Interface for Next Generation Nuclear Power Plants." *Reliability Engineering and System Safety* 64, pp. 109–126.

Connelly, C. S. (1995) "Toward an Understanding of DCS Control Operator Workload." *Instrument Society of America Transactions* 34, pp. 175–184.

Danz-Reece, M. E., and Noerager, J. A. (1998) "Process Control Interface Issues in the Upstream." Paper No. SPE 46425. Society of Petroleum Engineers International Conference on Health, Safety and Environment. Caracas, Venezuela.

Eason, K. D. "User Centered Design: For Users or by Users?" Inaugural lecture at the IEA Annual Meeting, Toronto Canada, 1994.

Eastman Kodak Company. (1983) *Ergonomic Design for People at Work.* New York: Van Nostrand Reinhold Company.

Electric Power Research Institute. (1981) *Human Factors Review of Electric Power Dispatch Control Centers*, vol. 2. Palo Alto, CA: Electric Power Research Institute.

Gilmore, W. E., Gertman, D. I., and Blackman, H. S. (1989) *The User-Computer Interface in Process Control.* San Diego: Academic Press.

Instrument Society of America. (1985) *Human Engineering for Control Centers.* Recommended practice ISA-RP60.3. Research Triangle Park, NC: Instrument Society of America.

Instrument Society of America. (1986) *Graphic Symbols for Process Displays.* American National Standard ANSI/ISA-S5.5-1985. Research Triangle Park, NC: Instrument Society of America.

Instrument Society of America. (1992) *Instrumentation Symbols and Identification.* American National Standard ANSI/ISA-S5.1-1984(R1992). Research Triangle Park, NC: Instrument Society of America.

Instrument Society of America. (1993) *Human-Machine Interface: Task Analysis.* Draft recommended practice ISA-dRP77.65. Research Triangle Park, NC: Instrument Society of America.

Instrument Society of America. (1995a) Human-Machine Interface: CRT Displays. Draft technical report ISA-dTR77.64.01. Research Triangle Park, NC: Instrument Society of America.

Instrument Society of America. (1995b) *Human-Machine Interface*: Alarms. Draft recommended practice ISA-dRP77.62. Research Triangle Park, NC: Instrument Society of America.

International Instrument Users' Associations (WIB) (1998a) *Ergonomics in Process Control Rooms. Part 1. Engineering Guidelines.* Report M2655X98. The Hague.

International Instrument Users' Associations (WIB) (1998b) *Ergonomics in Process Control Rooms. Part 2. Design Guidelines.* Report M2656X98. The Hague.

International Instrument Users' Associations (WIB) (1998c) *Ergonomics in Process Control Rooms. Part 3. Analysis.* Report M2657X97. The Hague.

ISO. (1999a) *Ergonomic Requirements for the Design of Displays and Control Actuators. Part 1. Human Interactions with Displays and Control Actuators.* ISO 9355–1:1999. Geneva: International Standards Organization.

ISO. (1999) *Ergonomic Requirements for the Design of Displays and Control Actuators. Part 2. Displays.* Geneva: International Standards Organization.

Kawai, K. (1997) "An Intelligent Mulitmedia Human Interface for Highly Automated Combined-Cycle Plants." *Control Engineering Practice* 3, pp. 401–406.

Kletz, T. (1998) *What Went Wrong? Case Histories of Process Plant Disasters,* Fourth Ed. Houston, TX: Gulf.

Mulley, R. (1994) *Control System Documentation.* Research Triangle Park, NC: Instrument Society of America.

O'Hara, J. (1993) *Advanced Human-System Interface Design Review Guideline.* NUREG/CR-5908, Washington DC: U.S. Nuclear Regulatory Commission.

O'Hara, J., Brown, W., Higgins, J., and Stubler, W. (1994) *Human Factors Engineering Guidance for the Review of Advanced Alarm Systems.* NUREG/CR-6105, Washington DC: U.S. Nuclear Regulatory Commission.

Olsson, G., and Lee, P. L. (1994) "Effective Interfaces for Process Operators—A Prototype." *Journal of Process Control* 4, no. 2.

Pennycook, W. A., and Danz-Reece, M. E. (1994) "Practical Examples of Human Error Analysis in Operations." Paper SPE27262. Proceedings of the 1994 Society of Petroleum Engineer's Conference on Health, Safety and Environment in Oil and Gas Exploration and Production in Jakarta, Indonesia.

Pikaar, R. N. (1986) "Man-Machine Interaction in Process Control." In H. P. Willumeit (ed.), *Human Decision Making and Manual Control.* Amsterdam: North Holland: Elsevier Science Publishers B.V.

Rankin, W. L., Rideout, T., Trigs, T., and Ames, K. (1985) *Computerized Annunciator Systems.* NUREG/CR-3987. Washington DC: U.S. Nuclear Regulatory Commission.

Sanders, M. S., and McCormick, E. J. (1987) *Human Factors in Engineering and Design,* Sixth Ed. New York: McGraw-Hill.

Shih, Y., Wang, M. J., and Chang, C. (1997) "The Effect of Valve Hand Wheel Type, Operating Plane and Grasping Posture on Peak Torque Strength of Young Men And Women." *Human Factors* 39, no. 3, pp. 489–496.

Staples, L. J. (1993) "The Task Analysis Process for a New Reactor." In *Proceedings of the Human Factors and Ergonomics Society 37th Annual Meeting.* Santa Monica, CA: Human Factors and Ergonomics Society.

Stultz, S., and Schroeder, L. R. (1988) "Incorporating human engineering principles in distributed controls upgrades." Paper 88-0421, 1988. Research Triangle Park, NC: Instrument Society of America.

Swensen, E. (2000) Personal Communication.

U.S. Nuclear Regulatory Commission. (1981) *Guidelines for Control Room Reviews.* NUREG-0700. Washington, DC: U.S. Nuclear Regulatory Commission, 1981.

U.S. Nuclear Regulatory Commission. (1982) *Human Engineering Design Considerations for Cathode-Ray-Tube Generated Displays.* NUREG/CR-2496. Washington, DC: U.S. Nuclear Regulatory Commission.

van der Schaaf, T. W. (1989) *Interface Design for Process Control Tasks.* Report EUT/BDK/45, Eindhoven, the Netherlands: Eindhoven University of Technology.

van der Schaaf, T. W., and Kragt, H. (1992) "Redesigning a Control Room from an Ergonomic Point of View: A Case Study of User Participation in a Chemical Plant." In *Enhancing Industrial Performance.* H. Kragt, Ed. London: Taylor and Francis.

Van Laar, D., and Deshe, O. (2002) Evaluation of a Visual Layering Methodology for Colour Coding Control Room Displays. *Applied Ergonomics* 33, pp. 371–377.

Vicente, K. J., and Rasmussen, J. (1990) "The Ecology of Human-Machine Systems II: Mediating 'Direct Perception' in Complex Work Domains." *Ecological Psychology* 2, no. 3, pp. 207–249.

6

Workplace Design

6.1 INTRODUCTION

At this point in the book, you have been exposed to a number of concepts in human factors and ergonomics, from the effects of light on human behavior to the design of local control panels. A good workplace design is based on the biomechanical, physiological, and psychological requirements of the user. For the human factors practitioner, workplace design is often a balancing act—balancing these requirements, when one can be offset by the other. In addition, you must consider the cost of the solution. The site must survive financially, so solutions need to be practical and cost effective.

This chapter deals with two concepts in workplace design: the workplace and the workstation. In this chapter, the *workplace* is defined as "the physical area where a person performs tasks." The workplace may include physical fixtures such as furniture, equipment, hallways, stairs, vehicles, and displays and is affected by environmental variables such as lighting, temperature, and noise (discussed in Chapter 4). A *workstation* is defined as "a location where the operator may spend only a portion of the working shift." Clearly, workstations are a subset of the workplace. An operator may travel between and work at several different workstations in the workplace.

This chapter also deals with two design concepts: existing and new (nonexisting) workplaces. We approach the analysis and design of each type of workplace by first acquiring an understanding of the design process involved, then illustrating the application with a case study. Finally, when you feel comfortable with the process, example design problems are included in an independent study section of the chapter.

The design of the physical workplace has been addressed in the HF literature for over 60 years. Many excellent books provide the details of workplace design. The purpose of this chapter is to ensure that workplaces are designed according to widely accepted human factors principles. Consequently, this chapter has three objectives:

1. To identify the principles involved in the design, installation, operation, and maintenance of workplaces. The literature that supports each of these principles is provided.

2. To review the techniques in place to analyze the work situations that determine workstation design.
3. To provide guidance (models) on the evaluation and redesign of existing and new (grassroots) workstations. A case study is presented on the use of these models for the redesign of a control room.

6.2 WORKPLACE DESIGN PRINCIPLES

6.2.1 Introduction

In keeping with the how-to theme of this book, recommendations for workplace design are addressed in this chapter by a series of basic design principles. The principles are presented in a hierarchical structure that consists of principles and subprinciples. The approach is to provide enough theory to help the reader understand why the principle is important, then explain the principle with actual examples. A table that summarizes all the principles is provided at the end of this section.

Two principles that affect the design of the workplace have already been addressed in previous chapters. But, they need to be repeated as statements in this chapter to ensure that the reader has a complete list.

6.2.2 Controls and Displays Must Be Optimally Located (Refer to Chapter 5 for this Discussion and Examples)

6.2.3 Equipment Must Be Visually Accessible (Refer to Chapter 2 for this Discussion and Examples)

6.2.4 The Workplace Must Be Designed for the User Population

6.2.4.1 People Differ in the Characteristics Necessary to Perform within the Workplace

The capabilities, limitations and characteristics of people are discussed in Chapter 2. The information shows that people can be classified in two ways:

1. Every person is a member of a distinct population. Populations could include
   Nationality, populations of different nationalities differ on many characteristics such as body size, language, stereotypic behavior.
   Gender.
   Process specialty, the population of process specialists includes men and women with specific skills, experiences, and expectations about how to perform the job.

2. Every person is a data point within the distinct population. Figure 6-1, for example, demonstrates how "rim force strength capability" varies among a group of North American, male operators (distribution is not accurate). In this case the characteristic rim force strength is for operators attempting to crack open a valve positioned at about waist height. As the valve position changes, the distribution of rim force strength changes as well (Attwood et al., 2002). The data illustrate that the force capability of the population varies widely.

To properly design a workstation, the designer needs to know the characteristics of the user population and the human characteristics of interest and how they vary within the user population.

6.2.4.2 *Workplaces Must Be Designed to Accommodate the Extremes of the User Population*

Humans vary widely on each of their characteristics. The variation typically follows a "normal" statistical distribution as shown in Figure 6-1.

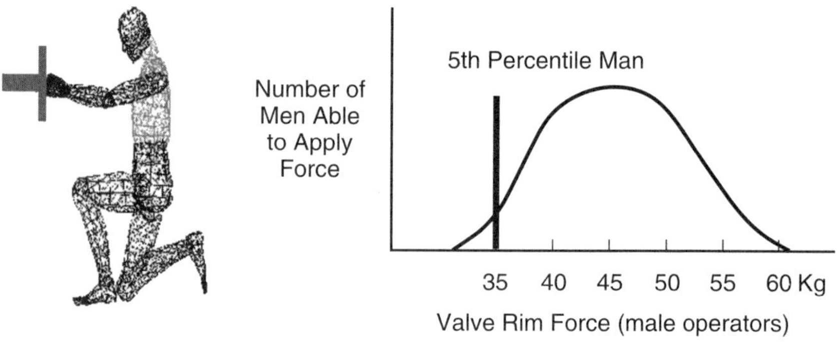

Figure 6-1 Normal distribution of rim force capability of male operators (Data are for demonstration purposes only).

(It is well known that force distributions are positively skewed. For the purposes of this demonstration, the distribution is assumed to be symmetrical.) The normal distribution is asymptotic, which means that the *y* values never reach zero (except in the case of force). So, referring to Figure 6-1, it is possible, although unlikely, that some operators could apply 135 kg (300 lb) of force to the rim of the valve.

When designing the valve, designers need to know the minimum rim force that accommodates the capabilities of most of the users. The lower is the force that opens the valve actuator, the more difficult the valve is to design and the more expensive the valve becomes. Clearly, the minimum actuator force should not be based on the force capabilities of a few weak people in the user population. It is generally agreed that a design should accommodate most of the population but not everybody. Depending on the importance of not being excluded, the accommodation limit is set to accommodate 90, 95, or 99% of the user population.

Figure 6-2 is a copy of the distribution shown in Figure 6-1 that illustrates the force that would accommodate all but 10, 5, or 1% of the user population. The area under the curve shows the percentage of the user population *not* accommodated to the left of each point. These points are referred to as *percentiles*. So, the point that delimits 10% of the area under the curve is referred to as the *10th percentile* of the population.

Note that, in this example, the percentiles are at the lower extreme of the "rim force" distribution. If, for example, we are installing a door in a

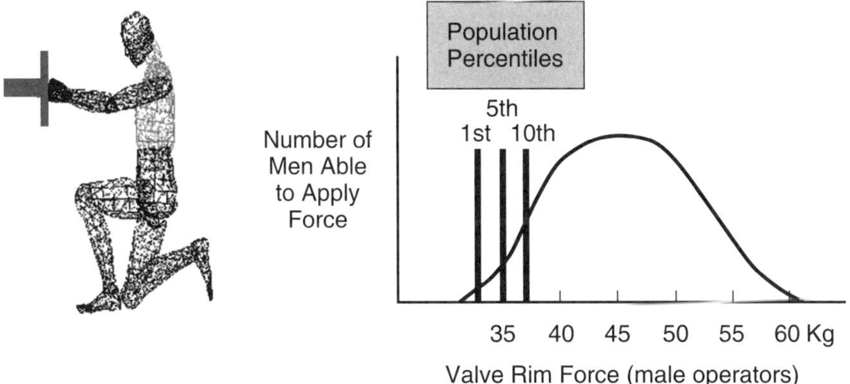

Figure 6-2 Rim force distribution indicating percentiles.

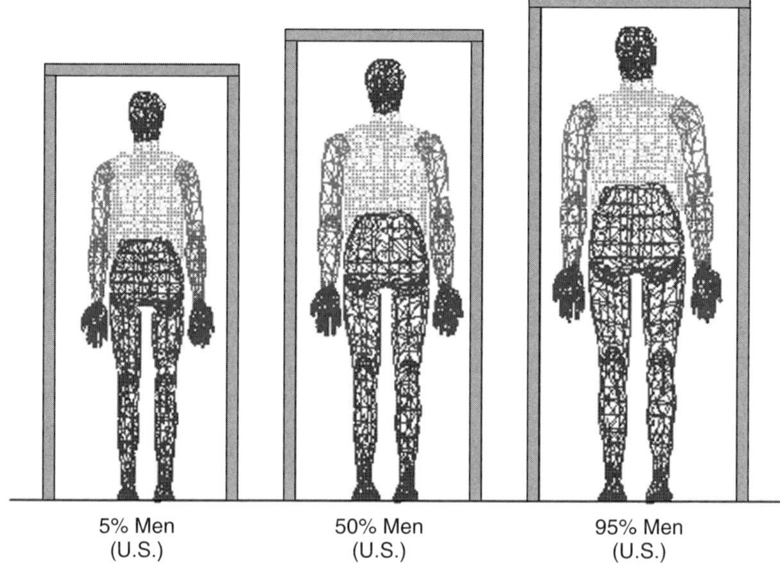

Figure 6-3 The need to design to *fit* taller operators.

wall (Figure 6-3) and interested in the minimum height of the door that would accommodate the majority of users, the characteristic of interest would be the stature of the user population. In this case, we want to identify the taller people in the population, since if they can walk under the doorframe without hitting their heads, shorter people would also have clearance. Considering the extremes, this time we would want to select a door height that, say 95% of the user population could walk through without ducking their heads.

These two examples illustrate the use of population extremes in design: Extremes at the lowest and extremes at the highest ends of the populations. The lower extreme could represent maximum reach to a control lever, maximum force, or maximum test score to enter a program. The upper extreme could represent fit, such as minimum height of an opening or minimum diameter of a manway necessary to accommodate 95% of the users.

Figure 6-4 Office workstation showing range of adjustable heights.

6.2.4.3 Workplaces Must Be Adjusted to the Characteristics of the User Population

In many cases, the same piece of equipment can be designed to be adjustable to the characteristics of the user population. Figure 6-4, for example, illustrates an office workstation in which the height of the keyboard surface adjusts over a range of about 20 cm (8 in.) to accommodate those who sit lower or higher in their chairs.

In the workplace, many pieces of equipment adjust to the characteristics of the worker. Figure 6-5, for example, illustrates a platform on the back of a well-servicing rig that adjusts the user to the height of the wellhead.

When possible, the workplace should be designed to adjust to the range of users; that is, from one extreme of the user population to the other, say, the 5th to the 95th percentile.

Figure 6-5 Height adjustable platforms on a well-servicing rig.

6.2.5 Equipment Must Be Physically Accessible

6.2.5.1 Aisles and Corridors

The minimum dimensions of aisles and corridors are determined by how they are used and the "size" of the user population. If the aisle or corridor is used to accommodate only people, the widths and heights are determined by how many people pass through simultaneously or walk side by side. Minimum aisle widths are intended for a single file of people, determined by the body width of the largest users. Figure 6-6 shows the widths and heights of 95th percentile men from two very different populations: United States and Japan (Pheasant, 1994). ABS (1998) recommends a 508-mm (20-in.) width for travel in areas with limited access.

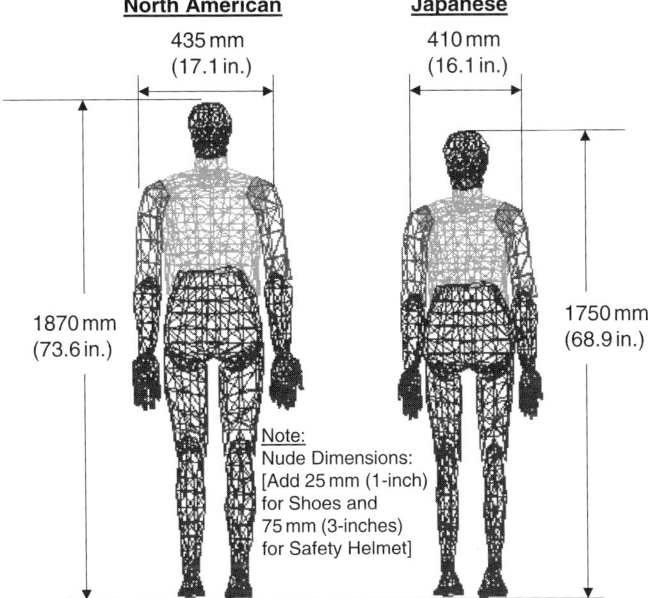

Figure 6-6 Widths and heights of 95th percentile male workers from North America and Japan. (Data from Pheasant, 1994.)

If the aisle is designed to permit two people to pass or walk alongside each other without touching, the minimum width is twice that shown in Figure 6-6 for North American or Japanese populations. There is a wide variation of recommendations of aisle widths among published guidelines. ABS (1998) suggests that the minimum width for normal two-way traffic is 36 in. (914 mm), while Eastman Kodak (1983) recommends 1372 mm (54 in.) as the minimum width.

If the aisle is used to move carts or equipment, its width may depend on the width of the largest pieces of equipment moved and consideration of how they will be moved rather than the people moving them.

In summary, the design of aisles and corridors is determined by how they are used. Nothing substitutes for a thorough analysis of the needs of the facility before it is designed.

6.2.5.2 Distances between Adjacent Pieces of Equipment

The distance between adjacent pieces of equipment again depends on how they are used. If the separation is merely to permit a person to pass safely between equipment, the minimum distance is determined by the width (or depth) of the largest user. In most cases, the minimum distance is specified to be 60 cm (24 in.). Clearly, if the equipment is hot or sharp or if footing is uncertain, a safety factor must be added. Normally, an additional 8–13 cm (3–5 in.) is sufficient.

If, however, the distance between equipment is determined by the task that the operator must perform, it depends on the job performed. Figure 6-7, for example, illustrates a technician squeezing between two pieces of equipment to access an instrument. In this case, merely fitting between the equipment is insufficient.

Figure 6-8 illustrates the minimum dimensions that are necessary to permit a maintenance technician to work on a pump in a squatting position. The minimum clearance is again determined by the largest technician (95th percentile) in a squatting position to perform the job.

Figure 6-7 Operator passing between two pieces of equipment.

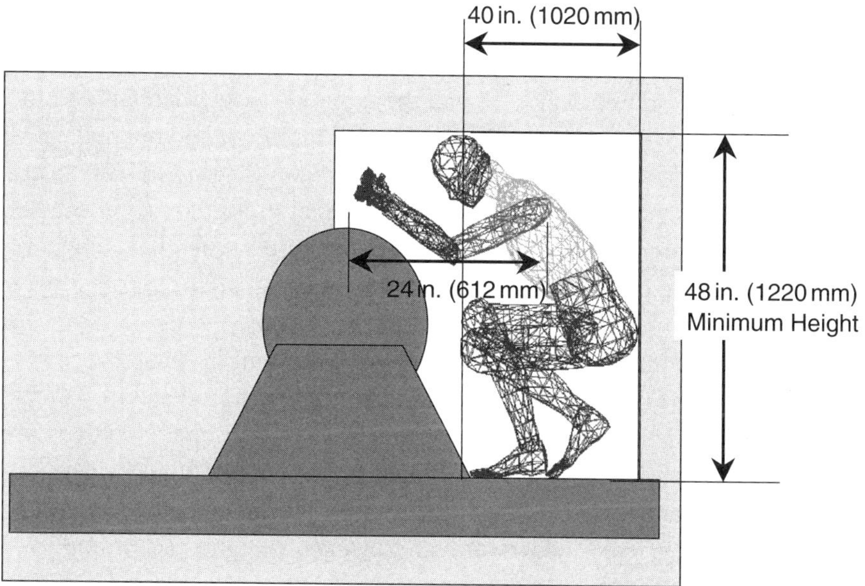

Figure 6-8 Minimum clearance required for a technician to work on a pump from a squatting position.

6.2.5.3 Ladders, Stairs, Ramps, Walkways, and Platforms

Ladders, stairs, ramps, walkways, and platforms are fundamentally important in the design of process plants. Process plants require workers to operate and maintain equipment above grade. Above-grade access can be accomplished in many ways, but in most plants, operators climb to the equipment using ramps, stairs, and ladders. And, once at elevation, workers access equipment using walkways and platforms.

The proper design of stairs, ramps, ladders, walkways, and platforms can affect the safety and efficiency of work above grade. Most government agencies responsible for the safety of process plants have enacted legislation (United States: OSHA, 1997; Norway: NPD, 1977; Great Britain: HSE, 2001) that lays out the minimum requirements for the design of ladders, stairways, ramps, walkways, and platforms. Many standards organizations have also addressed the issue (ABS, 1998) and several textbooks have reported design specifications for this equipment (Eastman Kodak, 1983; Grandjean, 1988). The purpose of this section is to advise the reader what design issues are important and guide the user to appropriate standards.

Stairs, Ladders, and Ramps

It is generally agreed that the design and use of a ramp, stairway, or a ladder depends on two major issues: the angle of ascent/descent and the nature of the task. Most research suggests that ramps should be used when the angle of ascent/descent is between 0° and 20°. Angles above 15° are difficult to negotiate if material is being manually pushed or pulled. So, power equipment is required. Stairs are installed when the angle of ascent/descent is between 30° and 50° to the horizontal (OSHA, 1997), and ladders are used from 50° to 90°.

Variations from traditional designs occur often in process plants. Eastman Kodak (1983), for example, describes a stair ladder, with a slope of between 50° and 75° (Figure 6-9), which is used for operators to move "between several levels on an occasional basis." In addition, ladders for floating tanks can assume angles between 0° and 90° (Figure 6-10). So, they can be considered ladders or stairways and designed accordingly.

Figure 6-9 Stair ladder installed on a conveyor line.

Figure 6-10 Stair ladder installed on a floating roof tank.

When facilities permit a wide range of options, the optimum slope of a stairway is determined by the energy required to use it. Grandjean (1988) cites research that found that the least energy consumed when climbing stairs was with a gradient of 25–30° and tread height (*h*) and depth (*d*) that met the formula

$$2h + d = 630 \text{ mm}$$

The design characteristics of a stairway are shown in Figure 6-11. Data from Eastman Kodak (1983) suggests that, as the slope of the stairway increases, the riser height should increase and the tread depth should decrease. Table 6-1 provides data from three sources that illustrate the wide range of recommendations on step height, tread depth, and tread overlap for the North American or European user populations. For stairways installed for use by other populations, the dimensions in Table 6-1 would vary. Designers should consult the regulations in each jurisdiction to ensure that stairways meet the requirements of the user population.

Quite apart from the angle of ascent and descent, how a person ascends or descends determines, to a large extent, what equipment is used. Ladders are cheaper to install than stairs. So, in most plants, ladders, not stairs, are the default method of getting from one level to another. But, ladders are

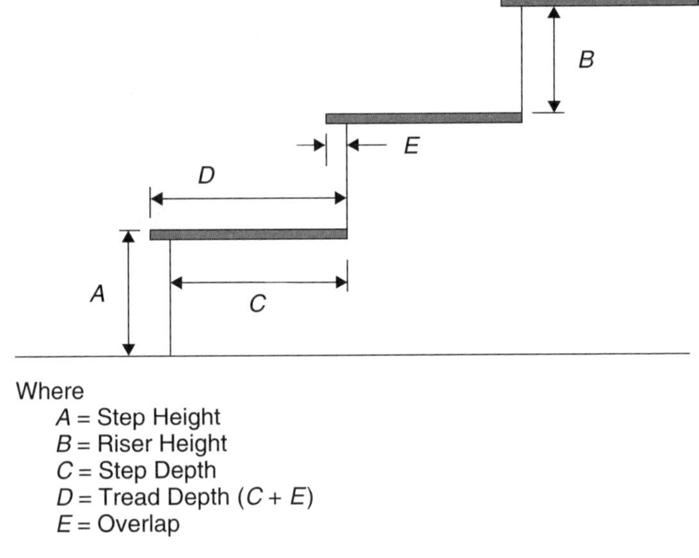

Figure 6-11 Characteristics of stair design.

TABLE 6-1

Recommended Stairway Dimensions from Three Sources

Reference	*Step Height mm (in.)*	*Tread Depth mm (in.)*	*Overlap mm (in.)*
Eastman Kodak (1986)	160–190 (6.5–7.5)	280–300 (11–12)	25–40 (1–1.5)
Grandjean (1988)	170 (6.7)	290 (11.4)	
ABS (1998)	203*–229 (8–9)	229–276* (9–10.9)	22 (0.9)

* = recommended.

generally harder to climb than stairs. So stairs, not ladders, should be installed, if

- Personnel are required to carry large tools or pieces of equipment up or down the structure.
- Equipment must be accessed or personnel evacuated during emergencies (e.g., battery limit stations).

- Hazardous material must be carried manually between levels.
- Equipment is frequently accessed (at least once per shift on the average).

While climbing stairs is considered a good form of exercise, stair climbing can also promote wear and tear to the body if stairs are used excessively or not designed properly. Field data suggest, for example, that workers complain more about leg and knee discomfort when stairs have steeper angles and many back and forth transitions. Depending on the way plant areas are accessed on a day-to-day basis (e.g., surveillance rounds, preventive maintenance), a decision to provide elevators or power lifts should be made. Based on the authors' experience, the facility is a candidate for a powered lift of some type if all of the following conditions apply.

1. The height of the climb is more than 10 m (32.5 ft).
2. The climb is performed by the same person more than four times per shift (on average).
3. Equipment and tools are often carried up and down.

Other combination of climbing heights and conditions would have to be analyzed to determine whether elevators are indicated using a combination of energy expenditure and musculoskeletal analyses.

Another way to reduce the degree of discomfort from using stairs is to find an alternate way to move tools and equipment up and down the structure. When available, cranes can be used to lift and lower tools, materials, and equipment. Small elevators, sometimes referred to as *dumbwaiters*, can be installed for routine use. Finally, one of the best ways to reduce the wear and tear from ladders and stairs on workers is to reduce their use. Often, moving equipment to grade (e.g., sampling stations or field panels) or changing the work processes (e.g., rounds protocols) can reduce worker exposure to stairs and ladders.

As a final note, stairways and ladders, if not properly designed, can contribute to slips and falls. It is important that the stairway and ladder covering is designed to maximize friction in all types of weather and for the uses intended.

Walkways and Platforms

Walkways and platforms provide access to equipment that is above grade. Platforms are required to provide access to the following facilities and equipment when access from grade is not possible:

- Instruments including transmitters and rod-out ports, which must be monitored, adjusted, or serviced while the plant is in operation.
- Service and inspection openings (includes observation points on fired heaters).
- Critical valves, including control valves, emergency block valves, safety valves, and motor- or air-operated valves.
- Sample points.

The physical specifications of platforms are shown in Table 6-2.

Portable platforms are acceptable for use as long as

- They are made of lightweight material.
- No one person must lift more than 23 kg (51 lb) to move them.
- They are equipped with wheels on at least one end to help them move.
- Wheels can be locked when the platform is stationary.
- They are equipped with railings, as in Table 6-2, if they are more than 610 mm (24 in.) high.

TABLE 6-2

Physical Specifications of Platforms

Specification	*Value*
Maximum horizontal reach from a platform or ladder	508 mm (20 in.)
Maximum overhead grip reach	1800 mm (72 in.)
Minimum walkway clearance between vessels and protruding equipment and platform railings	610 mm (24 in.)
Platform railing heights	915–1067 mm (36–42 in.)
Height of mid-rails	457–533 mm (18–21 in.)

6.2.5.4 Pathway Obstructions

Pathway obstructions, such as the one shown in Figure 6-12, can be a potential hazard to workers. Often, they are not easy to identify because of location and blending in with the surroundings. Lines and valves that create the most concern are located below the knees or above the eyes. Since our normal line of site is about 10° below the horizon, those obstructions in the head area can be especially vicious.

It is important to identify pathway obstructions and either eliminate them or mark them so workers can avoid them or pad them so contact has little effect. Moving a valve or a line during normal operations often is expensive, but they could be tagged for elimination during a turnaround when the unit is not operating. If the obstruction is marked, it is often possible not only to paint or sign it but also reroute the pathway around it.

Figure 6-12 Pathway obstructions are often difficult to recognize.

6.2.6 Work Must Be Positioned as Best for the Operator

Work should be positioned where it causes the least stress on the limbs, joints, and soft tissue of the body.

6.2.6.1 Position Work within the Range of Motion of the Body

The body is designed to move. So, the dimensions of the workplace should consider the range of flexibility of the limbs and the trunk of the body. Figure 6-13 illustrates an operator leaning forward and constrained by railings of different heights. As the height of the railing is reduced from chest height to waist height, the operator can extend his reach by bending at the waist. Note that, in the figure on the far left of Figure 6-13, much more force is placed on the lower spine for the same pulling force. So, increased range comes at the price of reduced force capability.

6.2.6.2 Place Frequently Used Materials and Tools within Easy Reach

Placing materials and tools within easy reach of workers saves time and energy. In addition, easily reached tools may prevent an accident when time and accessibility are critical.

An associated principle is to ensure that the location of tools and materials is the same from one workstation to another. This minimizes errors among operators who transfer between locations. They can always

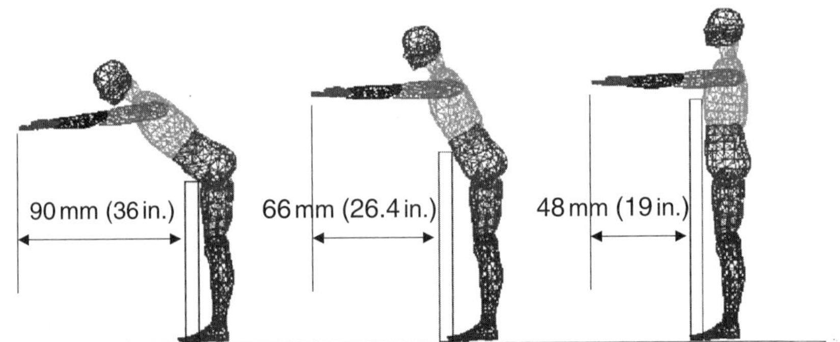

Figure 6-13 Range of forward motion, data are in cm (inches), is affected by the design of railings.

count on the tools being in the same location from one installation to the next.

6.2.6.3 Avoid Static Loads and Fixed Work Postures

Static postures require the muscles to work without movement. Muscle movement promotes blood flow. As static muscular effort increases above about 20% of maximum voluntary contraction (MVC), blood flow through the muscles is restricted, and at about 60% of MVC, blood flow is almost completely interrupted (Grandjean, 1988). Restricting the flow of blood to the muscle results in muscular fatigue, discomfort, and eventually, muscle failure. It is estimated that static work that requires exerting 50% of a person's MVC can last no more than 1 minute. But, if the force is less than 20% of maximum, it can last for some time. Static effort can also lead in the short term to higher energy consumption, raised heart rate, and longer rest periods. In the long term, continued static effort can lead to the deterioration of joints and soft tissue (Grandjean, 1988).

Static postures can include holding items in the arms with the arms extended, standing for long periods of time, and pushing or pulling objects for considerable time.

6.2.6.4 Design to Encourage Frequent Changes in Body Posture

To counteract the effects of static muscular effort, workstations should be designed to encourage frequent changes in body posture. The effects of static posture can lead to constant movement, fidgeting, crossing and uncrossing legs, and moving weight from one leg to the other (Pheasant, 1994).

Figure 6-14, illustrates an adjustable-height workstation that encourages the user to go from a sitting to a standing work posture. The same result can be achieved with a static workstation positioned at standing height if the user is provided with a "stool-chair," as shown in Figure 6-27.

6.2.6.5 Avoid Causing the Upper Limbs to Work above the Shoulder

The ideal location for arm work is in the range that begins at 75 mm (3 inches) below the elbow and ends at a position between the elbow and

Figure 6-14 Adjustable height workstation. (Photograph courtesy Evans Consoles Inc, Calgary, AB, Canada.)

shoulder (Figure 6-15). For light loads requiring high levels of dexterity, the upper limit is recommended. Heavy loads should be performed at the lower end of the range.

If the arms must work beyond these ranges for prolonged periods of time, support is required. Figure 6-16 illustrates the use of a computer, mouse where the arm is supported.

6.2.6.6 Avoid Work That Causes the Spine to Be Twisted

The spine should not be twisted while it is subjected to forces that pull the body forward. All lifts should be performed with a straight (untwisted) spine.

6.2.6.7 Ensure That Forces on the Limbs and Joints Are within Human Capabilities

Each joint can sustain a certain level of force without injury. The force that can be safely sustained depends on the design of the joint, the direction of the force, and the duration of the force. The maximum forces that

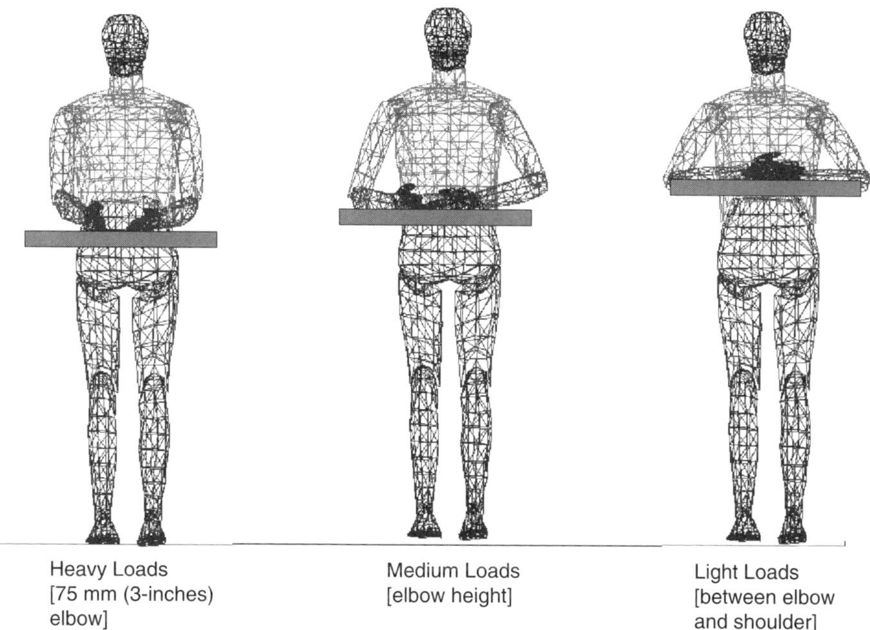

Figure 6-15 Recommended heights of work surfaces for the standing operator based on the weights being handled.

can be sustained by joints are generally expressed by the characteristics of the task being performed. Figure 6-17, for example, illustrates an operator carrying a part with two hands. The maximum force that can be maintained in this posture by female and male operators for a 2.1 m carrying distance and a rate of 10 carries per minute is given by Mital, Nicholson, and Ayoub (1993) in Table 6-3. The data show that only 10% of the male population are able to safely carry 21 kg (46 lb) continuously under the conditions specified over an 8-hour shift. The reader is referred to Chapter 3 for more detailed information.

6.2.6.8 Minimize Manual Handling

By now, it should be evident that the more material operators handle, the higher the potential is for injury and the more operators are fatigued. One objective of workplace design should be to reduce the frequency and weight of manual handling. Reducing the frequency of manual handling

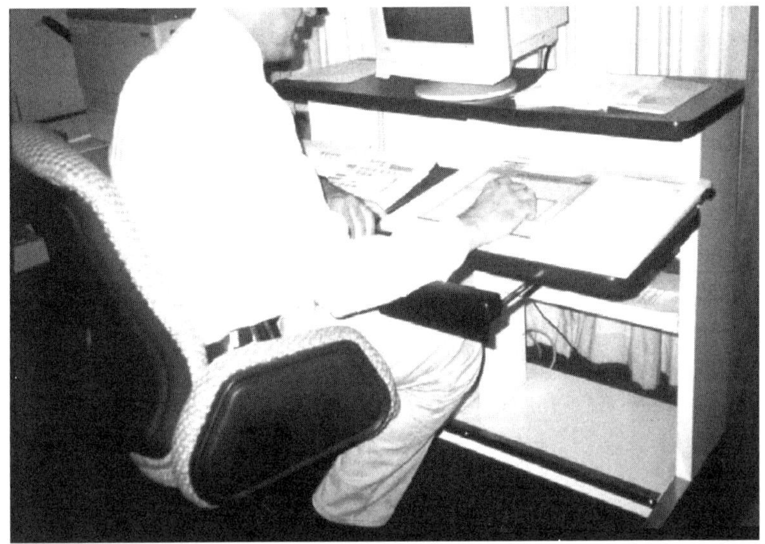

Figure 6-16 Arm support installed on a CAD workstation.

Figure 6-17 Operator carrying a machine part with two hands.

TABLE 6-3

Maximum Recommended Force for a Two-Handed Carry (2.1 m carrying distance, 10 carries per minute for an 8-hour shift)

Male Operator			*Female Operator*		
Carrying Height (cm)	*Population Percentile*	*Weight (kg)*	*Carrying Height (cm)*	*Population Percentile*	*Weight (kg)*
111	90	10	107	90	11
111	50	15	107	50	13
111	10	21	107	10	16

Source: Adapted from Mital, Nicholson, and Ayoub (1993).

means providing equipment to eliminate the need for operators to lift and carry material. Hoists, jacks, and carts provide this support. Reducing the weight of manual handling means designing the workstation to lower the amount of effort as much as possible. For example, positioning shelves to store heavy items at waist height rather than on the floor meet this requirement. Figure 6-18(a) illustrates an office worker reaching to a wall storage unit to remove a binder. Figure 6-18(b) illustrates the wall-storage relocated to the work surface. Now, binders can be slid into position rather than lowered.

6.2.6.9 Provide Specialized Tools to Reduce Body Stress

Tools provide the assistance necessary to allow the operator to perform tasks he or she could not perform without assistance. Tools amplify the force of the body or provide all the force necessary to accomplish the task. The operator in Figure 6-19 pulls on an extension bar that is about 240 cm (96 in.) long. The extension is fitted over the handle of a pipe wrench attached to a block on a wellhead. If the pipe wrench is 120 cm (48 in.) long, the extension allows the operator to put twice as much torque on the block as he could using only the pipe wrench. This is an example of force amplification. In many cases, amplification is all that is required to provide the necessary force.

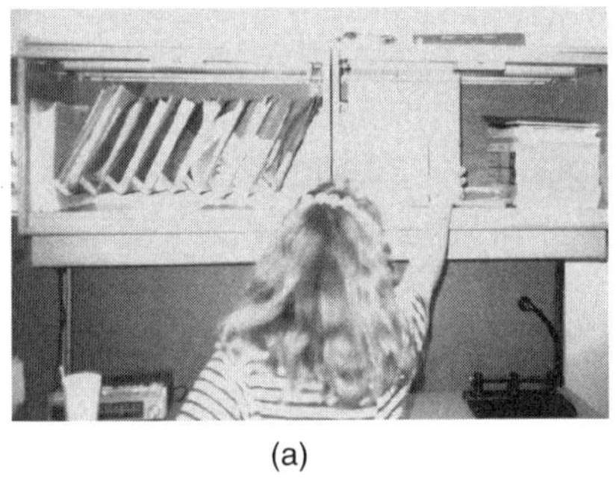

(a)

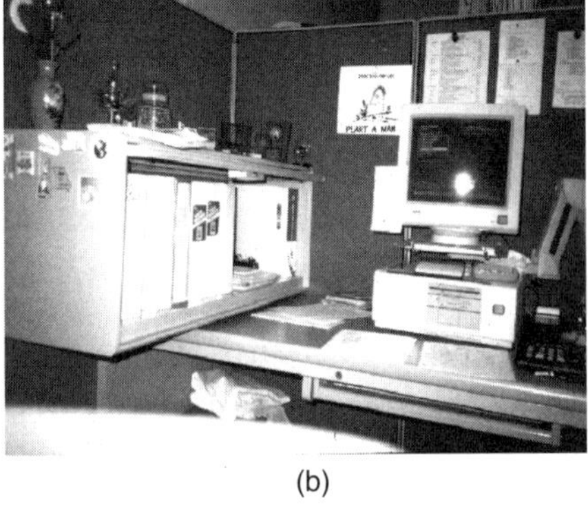

(b)

Figure 6-18 Lowering binder storage (a) to the desk work surface (b) reduces the stress of handling binders.

Figure 6-19 Operator using an 8-ft extension to increase the force applied to a wellhead block.

Figure 6-20 Hydraulic wrench designed to reduce the requirement for operator force. (Courtesy Greg Reynolds, Cold Lake, AB, Canada.)

However, in some cases, amplification is not practical. Either the force required is so excessive that the tools necessary become unwieldy—imagine a 360-cm (144-in.) extension—or the forces on the tool become excessive (the jaws of the pipe wrench eventually fail). In this case, another tool that provides all of the force necessary is required. Figure 6-20 illustrates a wrench powered by hydraulic pressure.

6.2.7 Workstations and Seating Must Be Designed According to Accepted Human Factors Standards

A workstation was defined earlier as "a location where an operator may spend only a portion of his or her working shift." In the process industry, a workstation may be as diverse as the control board in a control room, the rod basket on a service rig, or the fume hood work area in a process laboratory. Workstations can be divided into three major types: seated, standing, and seated/standing. Each has characteristics that make it more acceptable for some types of tasks and less for others. This section examines the characteristics of these major types of workstation from three perspectives:

1. The design characteristics of each.
2. The criteria that make one category more or less acceptable than the others for a given set of tasks.
3. The design specifications of each.

6.2.7.1 Major Categories of Workstations in the Process Workplace

Seated Workstations

By definition, seated workstations (Figure 6-21) require the operator to be seated at a panel or at a desk to perform his or her task. The major issues with seated workstations are

- Lack of adjustability to the wide range of body sizes that occupy the workstation.

Figure 6-21 A seated workstation.

- Access to the areas that support seated work such as filing, storage, binders, communications, and computer equipment.
- Communication with other seated operators.
- The design of the operator's chair.
- Difficult to lift objects or apply force.

Standing Workstations

Standing workstations (Figure 6-22) require the operator to stand at a workstation, display panel, or bench to perform a task. The major issues with standing workstations are

- Fatigue caused by long periods of standing.
- Access to tools, materials, and equipment needed to perform the task.
- Height adjustability of the work surface with regard to both the stature of the user and the type of task being performed.

Figure 6-22 A standing workstation.

Sit/Stand Workstations

Sit/stand workstations (Figure 6-23) are used when some of the tasks are done standing and others done sitting. The work surface is elevated to a standing height and the operator sits at the workstation on a chair that is also elevated. The base of the chair is typically fitted with a footrest in the form of a ring. This category of workstation can offer users the best of all worlds since they do not suffer the fatigue found in standing workstations, yet have more range of motion than seated workstations would offer. The only disadvantage to the elevated work surface is that all other equipment, storage, and filing facilities must be similarly elevated to be accessed effectively.

6.2.7.2 Selecting the Optimal Workstation Design

Each of the workstations just described is more useful for some task conditions than others. In many instances, the selection of workstation

Figure 6-23 A sit/stand workstation. (Courtesy Brad Adams Walker Architects, Denver, Colorado.)

category is obvious (e.g., working at a compressor control panel), but in some situations, the selection is not entirely obvious. And, if the wrong choice is made (for example, a standing workstation rather than a sit/stand one), the resulting design may be difficult to retrofit when the mistake is identified (lack of knee room in the standing workstation for the seated operator). The choice of workstation is made easier by considering the characteristics of the task(s) performed. Table 6-4 is provided as an aid in selecting the correct workstation configuration based on the tasks performed. If, for example, the user is in one position over extended periods of time, a seated or sit/stand workstation is indicated.

TABLE 6-4

Optimizing the Selection of a Style of Workstation

	Recommended Style of Workstation		
Questions	*S*	*ST*	*SS*
Is a high degree of body stability required?	*		*
Does the user perform the same task over long work periods?	*		*
Does the user need to use his or her feet for control?	*		*
Does the user need to be continually mobile to perform the work?		*	
Does the user work over a counter with standing clients?		*	*
Does the user work with heavy loads?		*	
Does the user need to monitor displays that are widely separated?		*	*
Does the user need to monitor displays whose lines of sight are obstructed by other equipment?		*	*
Does the work surface need to vary widely in height from one task to another?		*	

S = seated workstation.
ST = standing workstation.
SS = sit/stand workstation.

6.2.7.3 Workstation Design Standards

Design standards ensure that equipment and facilities meet minimum engineering design requirements. They also provide a benchmark for measurement, they are used to set priorities on issues, they help guide the selection of equipment, and they measure the effectiveness of the implementation.

According to Attwood (1996), the design specifications of a workstation are influenced by four factors:

1. Science, in the form of construction and materials; for example, the process control board shown in Figure 6-24 is made from highly reflective stainless steel, consequently, it is an indirect source of glare for the board operator.

2. Company goals are usually expressed in terms of procurement specifications.

3. Task requirements, as demonstrated in the previous section.

4. Worker needs.

Figure 6-24 Highly reflective stainless steel console is the source of specular glare.

The last factor refers to the ability of the workstation to adjust to the characteristics of each user. These characteristics may be only physical, such as arm reach or stature, or sensory, such as adjustable lighting systems to compensate for the vision of the older operator. Keep in mind that workstation adjustability puts pressure on the design of associated systems. For example, one disadvantage of continuously adjustable work surface heights is the difficulty of mounting storage cabinets over the work surface. Cabinet height might have to adjust to the work surface.

Finally, do not forget that the user is the customer for the workstation. No design should be completed without considering the requirements of the group of users who work at the workstation. Having said this, the specification of every workstation should be determined through the use of a comprehensive task analysis or a survey such as the one in Table 6-5 for process control room equipment.

The physical dimensions of seated, standing, and sit/stand workstations are well known and available in several texts (Grandjean, 1995; Sanders and McCormick, 1993; Eastman Kodak, 1983; VanCott and Kinkade, 1972; Joyce and Wallersteiner, 1989; Attwood, 1996) and standards (ANSI/HFS, 1988; CSA, 1989). The following paragraphs provide the basic physical design considerations for each workstation category. The ranges of adjustment cover the 5th percentile North American/European woman to the 95th percentile North American/European man. A wider range of adjustment is necessary to account for the dimensions of the non-North American/European populations.

Seated Workstations

Figure 6-25 summarizes the dimensions the literature generally recommends for a seated workstation. The major dimensions to key on include

- The height of the top of the work surface.
- The clearance between the floor and the bottom of the work surface.
- Clearance for the knees and feet.

Work surface depth depends on that task to be conducted, such as whether the person is using a computer monitor.

TABLE 6-5

Control Room Needs Assessment

Completed by:______________________ Room No: ___________ Date:_______

Do you share your workstation? (Y/N) ______. If yes, with whom?

____

If yes, how many days per week do you use it? __________

A. Principal Activities

Please choose from the following list, the principal activities you perform and the percentage of time you spend on each over a year. Then, answer the questions in each section that corresponds to the activities you perform.

Principal Activity	*% Time*	*Go to Section*
1. Computer use		B
2. Deskwork (reading, writing, etc.)		C
3. Filing/storage		D
4. Telephone use		E
5. Fax/copy use		F
6. Meetings in CR		G

B. Computer Use

1. What type of computers do you use in your control space?

   a. DCS _______________
   b. Desk Top for Local Area Network _______________
   c. Others? (explain) ____

2. What tasks do you perform on your computer and for what percentage of time?

Task	*Computer System from B1*	*% Time*
a. Process control?	_________	______
b. Data/text input?	_________	______
c. E-mail?	_________	______
d. Receive data/reports?	_________	______

continued

3. What type of support materials do you use while operating your computer?

Material	Computer System from B1	% Time
a. Documents (letter size)	________	______
b. Books (28 × 43 cm, 11 × 17 in.)	________	______
c. Binders (28 × 43 cm, 11 × 17 in.)	________	______
d. Printed output (55 × 35 cm, 22 × 14 in.)	________	______
e. Large drawings (60 × 90 cm, 24 × 36 in.)	________	______

4. In addition to monitors, keyboards, and pointing devices, do you use any specialized computer devices? (Y/N)

   a. External disk drives ________________________
   b. Printer/plotters ________________________
   c. Other devices (specify) ________________________

5. What type of pointing devices do you use for each of the systems identified in Question 1 (e.g., mouse, trackball, touch screen)?

Computer System from B1	Pointing Device
__________	______________
__________	______________
__________	______________

6. What type of input devices do you use for each of the systems identified in Question 1 (e.g., standard keyboard, specialized keypad, other [specify])?

Computer System from B1	Input Device
__________	______________
__________	______________
__________	______________

continued

TABLE 6-5

Control Room Needs Assessment *Continued*

C. Deskwork

1. What type of support materials do you use?

Material	*Yes/No*
a. Reports (letter size)	_____
b. Binders (opening to 28 × 43 cm, 11 × 17 in.)	_____
c. Computer printer output (opening to 55 × 35 cm, 24 × 14 in.)	_____
d. Drawings (up to 90 cm, 36 in. square)	_____
e. Other (specify) ______________________________	

2. What other equipment or material is routinely on your work surface while you perform deskwork?

Equipment/Material	*Yes/No*
a. Telephone	_____
b. Plant radios	_____
c. Stacks of paper in temporary storage	_____
If yes, how many stacks?	_____
d. Other items of material (specify)	_____

D. Filing and Storage	cm	in.
1. Hanging files can be measured in cm (in.)		
a. How many cm (in.) of hanging files are provided?	____	____
b. How many cm (in.) of those do you have to access		
i. Frequently	____	____
ii. Occasionally	____	____
iii. Infrequently	____	____
c. How many cm (in.) of hanging files could you discard or put into records retention?	____	____
2. Books and binder storage can be measured in cm (in.)		
a. How many cm (in.) of storage are provided?	____	____
b. How many cm (inches) of those do you access		
i. Frequently	____	____
ii. Occasionally	____	____
iii. Infrequently	____	____

continued

c. How many cm (inches) of storage could you discard or put into records retention? ____ ____

3. Personal storage can be measured by drawers (10 cm, 4 in. high) or by cm (in.) of shelf space.

   a. How many drawers of personal storage do you have? ____
   b. How many cm (inches) of shelf space do you have? ____ cm? or in.?
   c. By how much could you reduce your personal storage?
      i. Drawers? ____
      ii. Shelf space? ____ cm? or in.?

4. Special requirements for the storage of office supplies? (Y/N). If yes, please explain.

   ____.

E. Telephone Use

1. Which activities do you perform while using the telephone?
   a. Writing? (Y/N) ____
   b. Accessing the computer? (Y/N) ____
   c. Accessing files/drawers? (Y/N) ____
   d. Other (explain) ________________________

F. Fax/Copy Use

1. Do you have a fax machine on any part of your workstation? (Y/N) ____

2. Do you have a copy machine on any part of your workstation? (Y/N) ____

3. Are you required to add paper or change ribbons or cartridges? (Y/N) ____

G. Meetings

1. How many meetings do you typically hold in the control room in a week? ____

2. How many people typically attend the meetings? ____

3. What types of materials would you and your attendees bring to the meetings?
   a. Reports? ____________
   b. Open reports (23 × 43 cm, 11 × 17 in.) ____________
   c. Other materials? (please explain) ____________

continued

TABLE 6-5

Control Room Needs Assessment *Continued*

4. How long do your meetings last (on average) ____
5. Do you have enough seating for the attendees? (Y/N) ____
6. Do you have enough work surfaces for them to take notes? (Y/N) ____

H. Control Space Traffic

1. Who is permitted in your control room?
   a. Forepersons and supervisors? (Y/N) ____
   b. Outside operators? (Y/N) ____
   c. Contractors? (Y/N) ____
   d. Others (explain) ____________________
2. How many "guest" seats do you have available? ____
3. Does the location of exits and entrances promote unnecessary traffic? (Y/N) ____
4. Are tasks done inside the control room that encourage unnecessary people? ____
   If so, explain ____________________

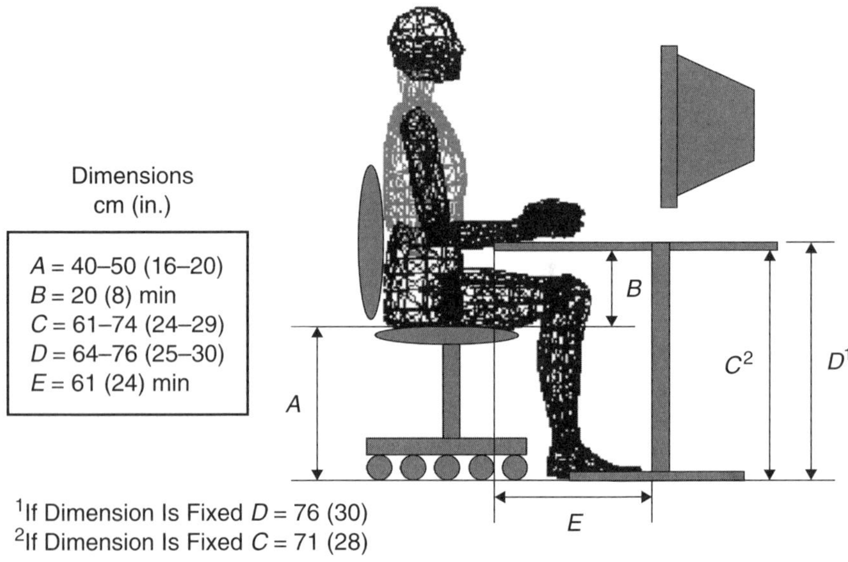

Figure 6-25 Specifications for a seated workstation.

Standing Workstations

Figure 6-26 summarizes the dimensions the literature generally recommends for a standing workstation. The major dimensions to key on include

- The height of the top of the work surface, which varies according to the range of heights of the user population and the task being performed (see Section 6.2.4.4).
- Clearance for the feet.
- Work surface depth, which depends on that task to be conducted.
- Height of the footrest.

Sit/Stand Workstations

Figure 6-27 summarizes the dimensions the literature generally recommends for a sit/stand workstation. The major dimensions to key on include

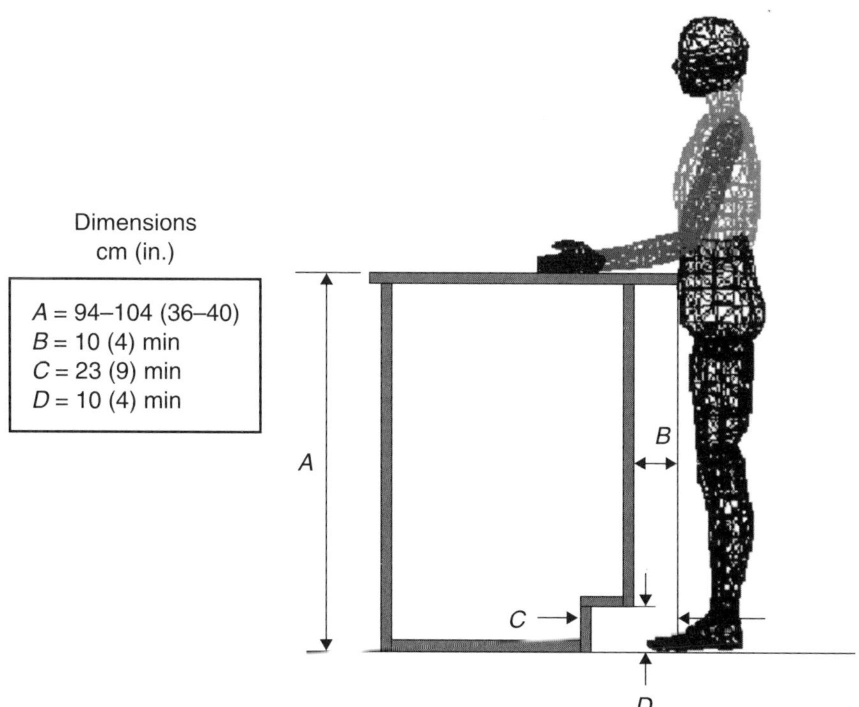

Figure 6-26 Specifications for a standing workstation.

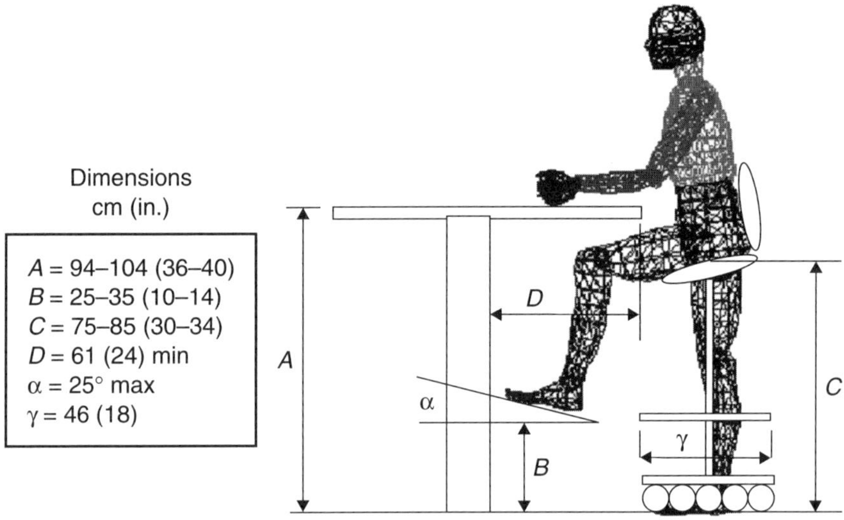

Figure 6-27 Specifications for a sit/stand workstation.

- The height of the top of the work surface, which varies according to the range of heights of the user population and the task being performed (see Section 6.2.4.4).
- Clearance for the knees and feet.
- Work surface depth, which depends on that task conducted.
- Height of the footrest on the workstation or the one attached to the seat or stool.

6.2.7.4 Seating

The worker's chair may be his or her most important piece of personal furniture. Certainly, it is the piece of furniture with which the sedentary worker has the most contact.

When we change from a standing to a seated position, the curvature of our spine changes and we put more pressure on the intervertebral discs (Grandjean, 1988). The upright seated posture is difficult to maintain for long periods of time; hence, we slump forward, sometimes leaning on our elbows to do our work. This posture relaxes our back muscles but increases the pressure on our intervertebral discs.

The more we can increase the angle between our body and legs, the less pressure we put on our discs. If we tilt the seat pan downward in the front to increase the angle, more body weight is taken by the chair, thus reducing muscle stress. So, a seated position that allows an angle of 110–120° between trunk and legs reduces the pressure squeezing our discs and the stress on our back muscles.

Unfortunately, in the process plant working environment, we cannot always work in a reclined position. To do so, the work surface would have to be angled.

So, for the foreseeable future, the seated worker will continue to work at a horizontal work surface and vary his or her posture between reclined, erect, and slumped forward, thus shifting the stress between different muscle groups. With this in mind, the job of the chair is to move with the worker and to support the way he or she works.

Seating Guidelines

Figure 6-28 illustrates a well-designed workstation chair. At a minimum, the chair requires

- Five casters for stability.
- A "gas lift" height adjustment that will raise and lower the seat pan at least 10 cm (4 inches) and preferably 13 cm (6 inches).
- A seat pan that is
  - — no more than 43 cm (17 inches) deep
  - — firm, yet well cushioned
  - — covered with a non-sticking fabric, and
  - — designed with a rounded front that puts no pressure on the underside of the thighs.
- Good lumbar support in the backrest.
- The ability to recline (when reclined, the seat pan should remain horizontal)
- Armrests that are adjustable in height, distance apart (from each other) and can rotate to provide support for the arm when using a pointing device.

Figure 6-28 The "TAG" chair from SMED meets required ergonomic specifications (photograph courtesy SMED Manufacturing Calgary, AB, Canada).

An addition, premium chairs optionally provide

- Lumbar supports that are adjustable in height or thickness.
- An adjustable seat pan angle.
- High backrests to provide head support for employees to lean back from time to time to relax the back muscles.

The dimensions and adjustment ranges for

- Seat pan height.
- Seat pan breadth.
- Seat pan angle.
- Back rest height and angle.
- Lumbar support adjustment.
- Armrest lengths and height adjustments.

are available in the ANSI/HFS (1988) and the CSA (1989) guidelines.

The quality of seating varies widely from one manufacturer to the other. It is recommended that workplace seating meet the safety and performance standards set out by the Business and Institutional Furniture Manufacturers Association (BIFMA).

6.2.8 Maintenance and Maintainability

Workstation equipment is designed for ease of maintenance and maintainability. Malfunctioning equipment affects productivity and results in costs related to lost opportunities, idle time, and equipment replacement or repair. So, equipment should be selected and installed according to ease of maintenance.

The proper design and installation of equipment has a greater effect on maintenance efficiency than the skill or training of maintenance technicians. This section, therefore, provides guidelines on equipment design and installation as they relate to maintainability and maintenance.

Basic distinctions are made in this section between maintenance and maintainability to differentiate between the focus of different people on the site. Maintainability is concerned with the design of equipment and systems for ease of maintenance. Consequently, engineers and maintenance staff usually make maintainability decisions for procurement purposes. Maintenance, on the other hand, is concerned with the technical problems of installing the equipment after purchase so it can be maintained properly.

6.2.8.1 Design Considerations

Equipment design should be considered when selecting equipment for maintainability and maintenance. Equipment must be designed with regard to the safety and health of the personnel who service it as well as those who use it. These items should be looked at during the selection process:

- Maintenance access, including doors and panels and labels.
- Shelves and drawers.

- Adjustment and lubrication points.
- Access to components that frequently wear out and must be replaced (e.g. belts, bushings).

Maintenance Access

Maintenance access should be located on the same face of the equipment as are the related controls, displays, instructions, and the like. Accesses to high-voltage areas should be equipped with safety interlocks.

Hinged access doors with mechanical devices for holding them open are better than cover plates and panels. Lift-off covers are acceptable if they can be removed and put back easily. Access covers, cases, handles should have rounded corners and edges.

Fasteners for covers and access panels should be kept to a minimum and hand operated whenever possible. (Avoid equipment where special purpose tools are required.)

Check that labels provide information on how access is to be made. Warning labels must be provided on accesses leading to hazardous equipment. Labels should

- Identify the controls, displays, control panels, and so forth.
- Alert the user to hazards.
- Be large enough to be read by users with corrected vision.
- Be located so users with bifocal glasses can read them.

The reader is referred to Chapter 8 for more information on labels

Shelves and Drawers

Pullout shelves and drawers are better because they make components accessible from several sides. If more than one drawer can be pulled out at a time, check that equipment does not tip.

Adjustment and Lubrication Points

Equipment that must be adjusted, inspected, or lubricated frequently should be equipped with easily accessible adjustment, inspection, and lubrication points.

6.2.8.2 Maintenance Considerations

Several things must be addressed when selecting equipment for ease of maintenance, including access by operators and maintenance personnel, space and clearance considerations, procedures, and lighting systems.

Access

Operators and maintenance personnel must be able to access the equipment safely and easily to prevent injury and to enhance efficiency of routine inspections or preventative maintenance. Therefore,

- Locate access openings so that other equipment need not be moved to gain access.
- Provide platforms, scaffolds, stairs or ladders for technicians when accesses must be made at heights that exceed their reach.
- Ensure that equipment that requires more frequent service or maintenance is in the most accessible locations.

Space and Clearance

Equipment that cannot be adequately positioned for access must be easily moved. Sufficient clearance must be allowed within and around the outside of the equipment to

- Use the tools necessary.
- Permit convenient removal and replacement of components.
- Permit convenient connections to utilities such as nitrogen, air or water.
- Ensure adequate visual access and lighting on the work surfaces.

The equipment designer must pay special attention to

- Postures dictated by the maintenance task, such as stooping, kneeling, or reaching overhead.
- Body sizes of operators and repair technicians.

- Clothing factors, such as bulky clothes for winter use or protection.
- Care and handling of toxic substances.

Traffic areas should be designed to permit the necessary support equipment, such as portable cranes or portable stairs, to be transported to the maintenance site.

Equipment with hinged access doors or pullout shelves or drawers require outside space that is equal to their openings.

Space should be available for the planned growth of equipment.

When possible, systems should be designed so that their functions are not disrupted if components must be removed from the site for repair.

Procedures

Every maintenance operation should be covered by written procedures, prepared and evaluated during the development of the workstation.

Lighting Systems

The performance of lighting systems, although adequate when new, may deteriorate over time, thereby creating visual problems in the workplace, especially when dirt accumulates on fixtures, lamps burn out, power sources malfunction, or lamp output decreases (with age).

6.2.9 Summary of Design Principles

Table 6-6 summarizes the principles in this section in a checklist format. The table can be used to help identify issues. Of course, like most checklists, this one should only be used by someone who is knowledgeable with the subject.

6.3 ANALYTICAL TECHNIQUES IN WORKPLACE DESIGN

The following techniques are more general and deal with the work system rather than the worker. For a complete description of each of the

TABLE 6-6

Workplace Design Checklist

Design Principle and Chapter or Section	*Meets Design Specs? (✓)*	*Comments*
Controls and displays are optimally located (Chapter 5)		
Equipment is visually accessible (Chapter 2)		
The workplace is designed for the user population (6.2.4)		
Meets the capabilities and limitations of the user population (6.2.4.1)		
Accommodates the extremes of the user population (6.2.4.2)		
Adjusts to the characteristics of the user population (6.2.4.3)		
Equipment is physically accessible (6.2.5)		
Aisles and corridors are properly designed (6.2.5.1)		
Distances between adjacent pieces of equipment are optimal (6.2.5.2)		
Ladders, stairs, ramps, walkways and platforms are properly designed (6.2.5.3)		
Pathway obstructions are eliminated or marked to increase recognition (6.2.5.4)		
Work is positioned as best for the operator (6.2.6)		
Position work within the range of motion of the body (6.2.6.1)		
Place frequently used materials and tools within easy reach (6.2.6.2)		
Avoid static loads and fixed work postures (6.2.6.3)		
Design to encourage frequent changes in body posture (6.2.6.4)		
Avoid causing the upper limbs to work above the shoulder (6.2.6.5)		
Avoid work that causes the spine to be twisted (6.2.6.6)		
Ensure that the forces on the limbs and joints are within human capabilities (6.2.6.7)		
Minimize manual handling (6.2.6.8)		
Provide specialized tools to reduce body stress (6.2.6.9)		
Workstations and seating are designed according to accepted ergonomic standards (6.2.7)		
The optimal workstation design has been selected for the task (6.2.7.2)		

continued

TABLE 6-6

Workplace Design Checklist *Continued*

Design Principle and Chapter or Section	*Meets Design Specs? (✓)*	*Comments*
Workstation meets accepted design standards (6.2.7.3)		
Seated workstations		
Standing workstations		
Sit/stand workstations		
Seating meets accepted design standards (6.2.7.4)		
Maintenance and maintainability are proper (6.2.8)		
Design considerations are met (6.2.8.1)		
Equipment is accessible for maintenance		
Shelves and drawers are used to improve access		
Lubrication points are accessible		
Components that frequently wear out are accessible		
Maintenance considerations are met (6.2.8.2)		
Openings are positioned to improve access		
Work stands and ladders are provided to improve reach to equipment		
Problem equipment is most accessible		
Space and clearance is provided to improve access		
Maintenance procedures are provided		
Lighting is designed to facilitate maintenance		

following analyses, see Sanders and McCormick (1993), VanCott and Kinkade (1972), and Eastman Kodak (1983).

6.3.1 Activity Analysis

The purpose of activity analysis is to determine how long each activity is performed over defined time period. Before beginning the analysis, it necessary to know what activities are performed over the time period of interest. The activities are listed, then the work is observed at predetermined intervals and the activity being performed at each interval is recorded.

For example, in one study, it was necessary to know how the activities performed by control room operators were distributed over their shifts. Five categories of activities were identified prior to the analysis:

1. Use of the phone or radio.
2. Report preparation, paperwork.
3. Breaks, social discussions.
4. Passive monitoring.
5. Active monitoring.

The control room operator was observed at sample intervals of 2 minutes, the activity was noted, and the observer waited for another 2 minutes to take the next observation. Over a 12-hour shift, 360 samples (observations) were taken. The results of the analysis are shown in Figure 6-29.

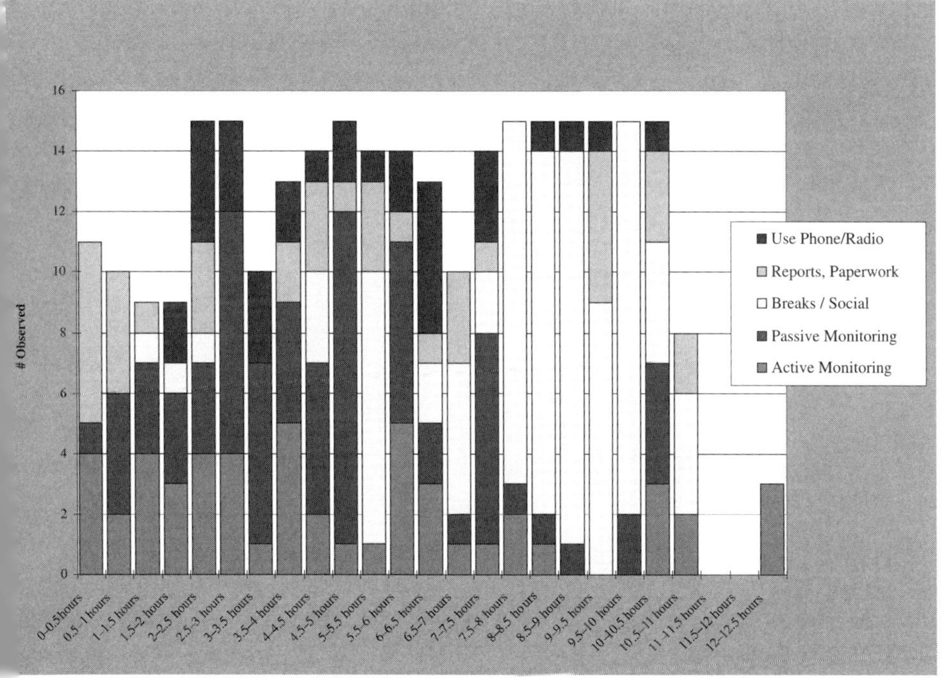

Figure 6-29 Results of an activity analysis of a control room operator over a 12-hour shift.

The data indicate that, during the first part of the shift, the operator is busy setting up maintenance permits, adjusting the process, and working with outside operators. Further into the shift, assuming no upsets, the operator performs more passive monitoring and spends more time in social activities.

6.3.2 Task Analysis

Task analysis is the basic analytical technique of human factors specialists. The technique is described in Chapters 5, 7, and 9.

6.3.3 Link Analysis

Link analysis is one of the most used analytical techniques in the process industry. Link analysis is a technique that optimizes the location of people and equipment in a workplace, or a workstation.

The fundamental considerations when laying out a workplace or workstation include

- Services needed by several people should be placed in a central location.
- People and equipment should be located to maximize the communication between them.
- Distances between components that are important to each other or frequently communicate should be minimized.
- The probability of unnecessary physical contact between personnel or equipment should be reduced.

Link analyses can be applied over a wide range of layout applications, such as layout of controls and displays on a panel, a workstation, or an office, plant, loading area or warehouse. Let us look at an example.

Figure 6-30 shows five displays on a control panel. Let us assume that the importance of one display to another is determined by how often the

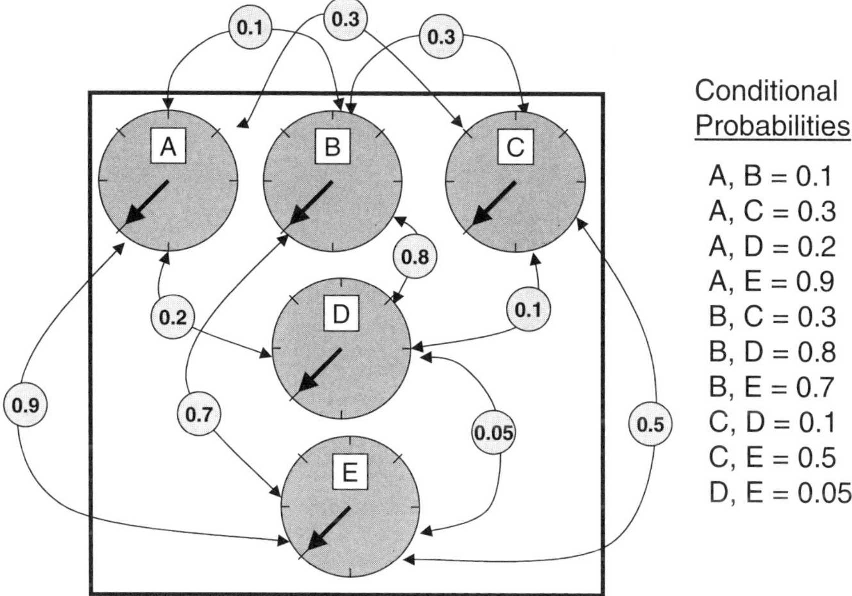

Figure 6-30 Probability of moving the eyes from one display to another.

operator looks from one display to another. Techniques are available to measure where the eyes are looking, and these techniques can be used to tell which display the operator is observing at any time. The data from this analysis can be used to determine the probability that the operator will look at display *x*, given that he or she is now looking at display *y*. The data in Figure 6-30 illustrate these probabilities. If we assume that the displays most important to each other are those that have the highest transitional probability, it is possible to rank the importance of one display to another. This is illustrated in the table in Figure 6-30.

The display layout can now be optimized based on the order shown in the table. Clearly, it is most important to put displays A and E as close together as possible, then displays B and D, and so forth. The resulting layout is shown in Figure 6-31.

Link analysis is revisited in Chapter 9, at which time, a demonstration of how the technique can be used in the design of a new project will be given.

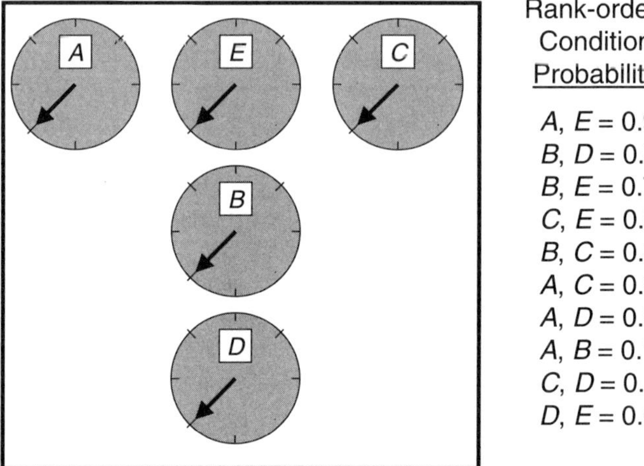

Figure 6-31 Rearrangement of displays to reduce the cumulative distance the eyes have to move when using the display panel.

6.4 HUMAN FACTORS DESIGN PROCESSES FOR EXISTING AND NEW WORKSTATIONS

6.4.1 Existing Workstations

There are several reasons why an existing workstation may have to be redesigned (Attwood, 1985):

1. *To increase efficiency.* For example, production figures may show that the throughput of a particular workstation is so low that it affects the output of the whole plant.
2. *To improve health and safety.* The employees that share a particular workstation may be overrepresented in safety incidents or using toxic substances.
3. *To adapt to a new work process.* A change in an associated industrial, commercial, or business process may require a corresponding change in the design of the workstation.
4. *To adapt to a change in equipment.* The workstation may have to be redesigned to accommodate a new piece of equipment.

The redesign of an existing workstation is often much easier to accomplish than designing a new workstation. An existing workstation can be analyzed, a nonexistent workstation cannot. The steps to redesign an existing workstation are outlined in Table 6-7. Let us work in detail with each of the steps shown in Table 6-7.

6.4.1.1 Familiarization

First, become familiar with the physical workstation (the equipment and equipment layout), the task, the working environment, and the work performance criteria. It is essential that you observe the operation and discuss it with the operator and first line supervisor. This is a good opportunity to meet the people affected by your recommendations and learn the relationship between the particular workstation and the rest of the operation. At this point, also become familiar with any new processes or equipment the workstation accommodates.

6.4.1.2 Problem Identification

Identify the extent of the design deficiencies with the existing workstation. These indicate whether and to what extent a redesign is required and often give you a clue about which data to collect to develop the new design. Deficiencies can be identified in a number of ways:

1. Performance records may indicate a sudden increase in the scrap material produced.

TABLE 6-7

Activities in the Design Process for an Existing Workstation

1. Familiarization
2. Problem identification
3. Background
4. Data collection and analysis
5. Evaluation
6. Recommendations
7. Redesign
8. Follow-up

2. Employee records may show a higher than normal level of absenteeism at certain times or on particular days. Find out what is so different about these periods.
3. Employees may complain about discomfort.
4. Employees may modify their own workstation to make them more comfortable.
5. Employees may have suffered an injury or experienced a near miss.
6. Safety inspections may have uncovered a deficiency.

Any of these indicators may signal that a redesign is required.

Specific indicators may provide clues to the cause of the deficiency. Shoulder complaints for example, may indicate that the work surface is too high. High scrap levels or spills may be caused by excessive material handling.

6.4.1.3 Background

Obtain background information on these issues:

- What are the physical differences between this workstation and others like it that perform adequately?
- Have other studies examined the discomfort produced from similar workstations?
- Do individual components operate the way they should?

6.4.1.4 Data Collection and Analysis

Analyze the workstation. The type of analysis to do depends on the situation. Many of the techniques for musculoskeletal and physiological analyses have been discussed in the previous chapters and facilities analyses are discussed in this chapter.

6.4.1.5 Evaluation

Compare the results of your analyses with the performance criteria that have been developed for the workstation. The criteria may take many forms, such as injury-free operation or output must exceed X bales per minute.

6.4.1.6 Recommendations

Modifications to the existing workstation are recommended based on the results of the analyses. For example, a link analysis might recommend relocation of equipment.

6.4.1.7 Follow-Up

The follow-up process should repeat the same steps followed when originally evaluating the need for workstation redesign, comparing these results with original results and making sure that you achieved the desired improvements.

In the case study, these steps are applied to the redesign of a process control room.

6.4.2 New Workstations

New workstations are designed and built everyday. The requirement for a new workstation may be to support a new process in an existing facility or as part of the requirement from a new plant.

We approach the design of a new workstation in a way that is similar to that for the redesign of an existing workstation. The main difference is that, now, we cannot take advantage of history. That is, there is nothing to improve on. We cannot become familiar with the workstation and the task, because neither exists. We can become familiar only with the current objectives. We cannot analyze a workstation that never existed, although we can gain background on similar workstations that are now in operation. With a new workstation, the analysis and design process proposed for existing workstations changes slightly to the following steps:

1. *Develop objectives.* Understand what the workstation must accomplish and identify the performance criteria it must meet.

2. *Familiarization.* Become familiar with the design constraints (such as workstation dimensions, the equipment specified, the work processes) as best you know them and the interaction with associated processes that may as yet be undefined. It is important at this point to have as good an idea as possible of the tasks that the operator must perform. Write an initial version of the procedures from what you know of the process and improve them as more information becomes available.
3. *Prototype (model).* Develop a prototype of the proposed workstation that meets the established objectives and work processes. Models can take many forms, and several different models may be developed as more information comes available. The first may be a sketch of the workstation, with the various components laid out in relation to each other. The next version may be a scale drawing in plan and elevation. For complex designs, it may be useful to render the design in a 3-dimensional drawing, such as 3D-CADD. Finally, miniature cardboard or full-size, static, foam-core mockups may be developed.
4. *Final design.* Base the final design on the models just developed.
5. *Follow-up.* The follow-up process should evaluate the final design against the objectives.

6.5 CASE STUDY

The design of a process control room is critical to the safety and efficiency of a process plant. For existing plants, a control room may have to be updated for a number of reasons, including design and construction of new control building or the consolidation of buildings, consolidation of process control systems, introduction of new equipment, or reduction in the potential for human error.

This case study deals with the redesign of a process control room. It will not address operator workload, training, or the design of the process control system. For more information on these topics, see CCPS (1994), ASM (2002), EEMUA (2002), Nimmo (2002), and Strobhar (2002).

The case study follows the process introduced in Section 6.4.1 for the redesign of existing workstations up to the recommendations stage. The activities are listed in Table 6-7.

6.5.1 Familiarization

A plan layout of the hypothetical control space is illustrated in Figure 6-32. It is modeled on a space typically designed to control a gas processing facility. The space contains the areas, equipment, and facilities listed in Table 6-8.

6.5.2 Problem Identification

In this hypothetical example, the design of the control room has been identified as a causal factor in missed alarms and two instances with spills

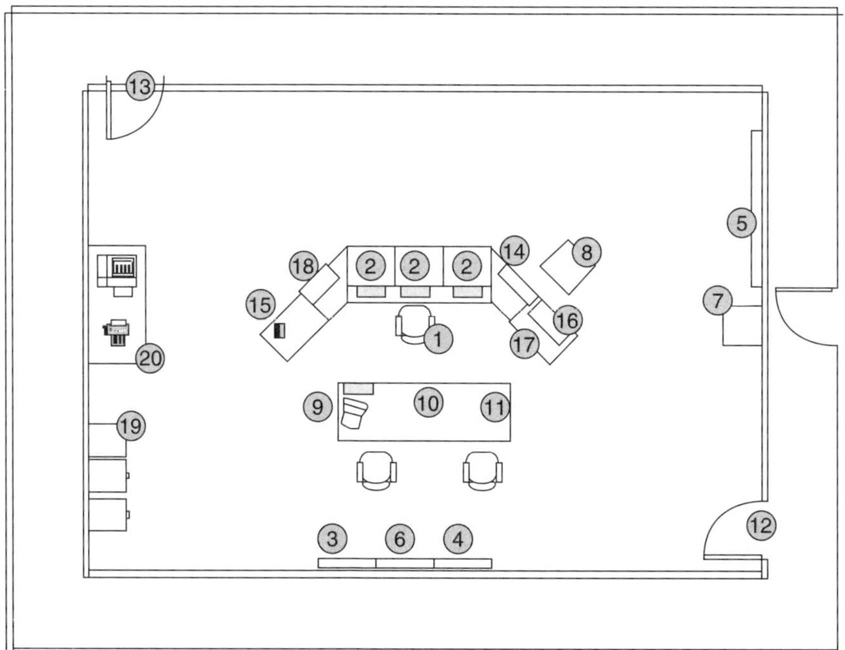

Figure 6-32 Original process control room layout.

TABLE 6-8

Case Study: Control Room Affinity Matrix Components

1. Board operator position
2. Process control system monitors and keyboards
3. Fire alarm panel
4. Emergency shutdown (ESD) panel
5. Defeat board/daily orders
6. H_2S/LeL panel (hydrogen suphide/lower explosive limit) alarm panels
7. Security camera (an elevated television screen used to monitor access gates and loading racks)
8. Flare monitor (an elevated television screen used to monitor the operation of the plant flare)
9. Desktop computer (PC) used to attach to the local area network (LAN)
10. Work table (a general-use work surface used by the board operator, outside operators, applications engineers, and the like)
11. Work permits (a place on the work table that is reserved for issuing work permits)
12. Exit to office and toilets
13. Exit to locker room and plant
14. Plant radio panel
15. Plant telephones
16. Binder (procedure) storage
17. Log book work area
18. Security camera controls
19. Storage and files
20. Printer and Fax

on the loading rack that were not detected in time to prevent a major event. It was felt that a reevaluation of the design of the room would help prevent similar losses in the future. So, the objective of this redesign is to make the control room more usable and efficient, by understanding the way people use the room and designing it for the way they work.

6.5.3 Background

The process control room should be designed to ensure that

- Traffic is properly routed through the control room.
- Space is used efficiently.

- Controls and displays are located optimally (to be seen and manipulated by the operator).
- Work areas and equipment are located for optimal interaction.
- Entries and exits are properly located.
- Ergonomic standards are met.

In addition, the control space design should comply with the general workstation design principles introduced in Section 6.2.

6.5.4 Data Collection and Analysis

This section introduces human factors tools to identify the proper layout of the control space.

6.5.4.1 Affinity Analysis

Affinity analysis is a technique to identify relationships among various spaces and determine how close these spaces should be placed to each other. The analysis starts by identifying each of the items in the control space that could have a relationship with others (Table 6-8). In some cases, items are physical areas such as "17. Log book work area." While in others, the items are pieces of equipment such as "4. Emergency shut down (ESD) panel." The items listed in Table 6-8 are located on the plan view of the control space in Figure 6-32. Each is relisted as row headings in Table 6-9, the affinity matrix. Note that each of the 20 items identified in our example is uniquely numbered. Each number is also a column heading in Table 6-9.

The affinity analysis process consists of estimating the subjective strength of the relationships between each item. These are the cell values in Table 6-9. Typically, the opinions of several board operators are obtained and averaged in each cell in the table. The strengths of the relationships are repeated here, as follows:

1 = Avoid closeness
2 = Closeness unimportant
3 = Ordinary closeness
4 = Closeness important
5 = Closeness absolutely important

TABLE 6-9

Affinity Matrix

Area	*1*	*2*	*3*	*4*	*5*	*6*	*7*	*8*	*9*	*10*	*11*	*12*	*13*	*14*	*15*	*16*	*17*	*18*	*19*	*20*
1. Operator position	—	5 AC	3 H	5 BD	3	3 H	4 DH	4 DH	4 B	4 B	2	3	2	4 B	4 B	2	3	3 B	2	3
2. PCS monitors and keyboards		—	3	3	2	3	2	4	2	2	2	2	2	2	4 F	2	2	2	2	2
3. Fire alarm panel			—	4 F	2	2	2	2	2	2	2	2	2	4 F	4 F	2	2	2	2	2
4. ESD panel				—	2	4 F	2	2	2	2	2	2	2	4 F	4 F	2	2	2	2	2
5. Defeat board/daily orders					—	2	2	2	2	2	4 E	2	4 E	2	2	2	2	2	2	2
6. H_2S/LeL panel						—	2	2	2	2	2	2	2	4 F	4 F	2	2	2	2	2
7. Security monitor (elevated)							—	2	2	2	2	2	2	4 F	2	2	2	3 E	2	2
8. Flare monitor (elevated)								—	2	4 D	2	2	2	3 B	2	2	2	2	2	2
9. PC (LAN)									—	3	3	2	2	2	2	2	4	2	2	3
10. Work table										—	2	2	3	2	4 B	3	4 A	2	3	3
11. Work permits											—	3	4 G	4 G	4 G	4 G	2	2	2	4 G
12. Exit to office and toilets												—	1	2	2	3	3	2	4 G	2

13. Exit to locker room and plant	—	2	2	4 G	2	2	3	4 G
14. Plant radio		—	5 F	3	2	2	2	2
15. Plant telephones			—	3	3	2	2	2
16. Binder (procedure) storage				—	2	2	2	2
17. Log book work area					—	2	3	3
18. Security camera controls						—	2	2
19. Storage/files							—	3
20. Printer/Fax								—

Note: Rate affinity (need for items to be close) of "items" from 1 to 5, where

1 = Avoid closeness
2 = Closeness unimportant
3 = Ordinary closeness
4 = Closeness important
5 = Closeness absolutely important

A = Within hand reach
B = Reachable from a seated position
C = Within the optimal visual field
D = Within visual field defined by head movement (±60°)
E = Visually adjacent
F = Common issues
G = Reduce through traffic
H = Close enough to be read or understood from operator position

Note that this technique is similar to a link analysis. An objective method of rating the strength of the relationships between areas could be determined by observing operator activities. However, observation looks at only physical activities and does not take into account the nonphysical characteristics such as viewing distances.

6.5.4.2 Space Requirements

The space requirements within the control room can be determined by completing the survey in Table 6-5, "Control Room Needs Assessment."

6.5.4.3 Ergonomic Analysis

An analysis of the environmental and ergonomic status of the control room can be completed by answering the questions posed in Table 6-10.

6.5.5 Evaluation

In this hypothetical example, data are collected from the affinity analysis only. The space requirements and the ergonomic surveys are not completed. The first indication of whether the items in the control room are positioned correctly comes from identifying the cells in Table 6-9 that contain high numbers. Table 6-11 lists the strong affinities identified by the control room operators and, when given, the associated comments identified in the affinity matrix. The table indicates four basic reasons for rearranging the items identified in the control room:

1. To make them easier to monitor.
2. To make them easier to reach.
3. To bring together items with common uses (e.g., communication equipment, emergency panels).
4. To reduce traffic flow through the control room.

TABLE 6-10

Control Room Environmental and Ergonomics Survey

Site ———————————— Date ————————————

Item	*Human Factors Criteria*	✓	*Comments*
HVAC			
Ambient temperature	21–27°C or 70–80°F		
Relative humidity	20–60% (45% at 21°C/70°F)		
Thermal variation	No more than 10°F difference between head and floor level		
Ventilation	Air is introduced at minimum rate of 15 cfm per occupant, approximately two-thirds of which is outside air filtered to remove dust, particles, etc.		
Air velocity	Maximum 45 fpm measured at head level and no noticeable drafts		
Lighting			
Glare	No glare on control panels and displays No glare on CRTs		
Emergency lighting level	10–50 lux (1–5 FC).		
Overall illumination	92–927 lux (9–93 FC), with mean of 240 lux (24 FC) during the day and 184 lux (18 FC) during the night shift		
Task lighting	Available in areas where paperwork (and other tasks) done		
Illumination level	Adjustable by the operator		
Noise			
Background noise	Level below 55 dBA		
Additional sources of noise	No excessive noise besides equipment (e.g., radio, personnel, startling alarms)		
Ambient noise	Free from high-frequency tones (8000 Hz)		
Noise interference	Noise does not interfere with communication		

continued

TABLE 6-10

Control Room Environment and Ergonomics Survey *Continued*

Site ______________________ Date ______________________

Item	*Human Factors Criteria*	✓	*Comments*
Alarms	10 dBA over background		
Vibration			
Noticeable vibration	No noticeable vibration from either standing or seated position		
Control Room Layout			
Minimum clearance around control panels	54 in. minimum in front of all main consoles 36 in. minimum for maintenance Maintenance clearance around equipment.		
Traffic	Traffic does not flow through main operating area		
Space	Available for ______ personnel needed to handle upsets		
Distance between keyboards of consoles	30 in. minimum (42 in. preferred)		
Visual angle	Equipment within normal reach and ± 30°		
Placement of equipment	Consistent with importance and frequency of use Dedicated to portions of the process consistent with process flow		
VDT Ergonomics			
Chairs	Meet ANSI standards and in good repair; include circular footrest of 18 in. diameter		
Level of midmonitor	15–117 cm (6–46 in.) with 99 cm (39 in.) preferred; viewing angle within 35° of horizontal line of sight (LOS) with 15° below LOS preferred		
Viewing distance	33–80 cm (13–30 in.) with 46–61 cm (18–24 in.) preferred		

continued

Site ______________________ Date ______________________

Item	*Human Factors Criteria*	✓	*Comments*
Screen orientation	Adjustable		
Height to home row of keyboard	66–78 cm (26–30.5 in.)		
Working level height	66–81 cm (26–32 in.)		
Working level width	61–76.5 cm (24.4–30.1 in.), 76.5 cm preferred		
Clearance available	Knee depth 46–51 cm (18–20 in.)		
	Leg depth 100 cm (39 in.)		
	Leg width 51 cm (20 in.)		
Frequently accessed items	Within functional reach of 64–88 cm (25.2–34.6 in.)		
Labels on keyboard?	Labels provided on keyboard		
Quality of monitor	For example, flicker, resolution, focus		
Glare	Monitor free from glare		
Printer			
Supplies	Readily available and instructions on how to load attached to printer and clearly written		
Printer location	Primary operating area		
Printer operation	Does not add significant ambient noise		
Printer	Capable of printing alarm data, trends, and plant status data		
	Can provide copy of any page/screen at operator request without altering screen content		
	Has take-up device to collect outgoing paper		
	Possible to write or highlight on the paper while still in the machine		
	Operator can read the most recently printed line		
Printed material	Has adequate contrast		
Printer	If printer is down or being loaded, information that should have printed is not lost		
Print confirmation signal	If copy is printed at a remote location, the operator is provided a print confirmation signal		
Printer speed	Minimum speed of 300 lines per minute		

TABLE 6-11

Item Pairs from the Affinity Matrix with Strong Associations

Item Pair	*Affinity*	*Comment*
1. Operator position (1)–PC monitors/keyboards (2)	5	Within hand reach Within optimal visual field
2. Operator position (1)–ESD panel (4)	5	Reachable from a seated position Within visual field defined by head movement Close enough to be read and understood from the operator position
3. Plant radio (14)–plant telephones (15)	5	Common issues
4. Operator position (1)–security monitor (7)	4	Within visual field defined by head movement Close enough to be read or understood from the operator position
5. Operator position (1)–flare monitor (8)	4	Within visual field defined by head movement Close enough to be read or understood from the operator position
6. Operator position (1)–LAN system (9)	4	Reachable from a seated position
7. Operator position (1)–work table (10)	4	Reachable from a seated position
8. Operator position (1)–plant radio (14)	4	Reachable from a seated position
9. Operator position (1)–plant telephones (15)	4	Reachable from a seated position
10. PCS monitors and keyboards (2)–flare monitor (8)	4	
11. PCS monitors and keyboards (2)–plant telephones (15)	4	Common issues
12. Fire alarm panel (3)–ESD panel (4)	4	Common issues
13. Fire alarm panel (3)–plant radio (14)	4	Common issues
14. Fire alarm panel (3)–plant telephones (15)	4	Common issues
15. ESD panel (4)–H_2S/LeL panel (6)	4	Common issues
16. ESD panel (4)–plant radio (14)	4	Common issues
17. ESD panel (4)–plant telephones (15)	4	Common issues
18. Defeat board/daily orders (5)–work permit board (11)	4	Visually adjacent

continued

Item Pair	*Affinity*	*Comment*
19. Defeat board/daily orders (5)–exit to locker room and plant (13)	4	Visually adjacent
20. H_2S/LeL panel (6)–plant radio (14)	4	Common issues
21. H_2S/LeL panel (6)–plant telephones (15)	4	Common issues
22. Security monitor (7)–plant radio (14)	4	Common issues
23. Flare monitor (8)–work table (10)	4	Within the visual field defined by head movement
24. PC LAN (9)–log book work area (17)	4	Common issues
25. Work table (10)–plant telephones (15)	4	Reachable from a seated position
26. Work table (10)–binder (procedure) storage(16)	4	Reachable from a seated position
27. Work permits (11)–exit to locker room and plant (13)	4	Reduce through traffic
28. Work permits (11)–plant radio (14)	4	Reduce through traffic
29. Work permits (11)–plant telephones (15)	4	Reduce through traffic
30. Work permits (11)–binder (procedure storage) (16)	4	Reduce through traffic
31. Work permits (11)–printer/Fax (20)	4	Reduce through traffic
32. Exit to office and toilet (12)–storage and files (19)	4	Reduce through traffic
33. Exit to locker room and plant (13)–Binder (procedure) storage (16)	4	Reduce through traffic
34. Exit to locker room and plant (13)–printer/Fax (20)	4	Reduce through traffic

6.5.6 Recommendations and Redesign

The affinities noted and explained in Table 6-11 direct the designer to make a number of changes in the layout of the control space. These are listed in Table 6-12. The resulting layout is shown in Figure 6-33.

In summary, the case study suggests how the design of the control space can be systematically analyzed and improved.

Models drawn with ManneQuin PRO Software, courtesy of NexGen Ergonomics, Montreal, Quebec, Canada.

TABLE 6-12

Explanation of Major Changes to Control Room Layout and the Source of the Change

Major Change	*Lines in Table 6–11 Driving the Change*
1. ESD (4), fire alarm (3) and H_2S/LeL (6) panels relocated to a position behind and just above the process control monitors (2)	2, 12, 13, 14, 15, 16, 17, 20, 21
2. Telephone system (15) repeated in three additional locations	3, 25, 31
3. Both security (7) and flare (8) monitors relocated to a position on the ceiling in front of the board just above the emergency panels (3), (6), (2);	4, 5, 10, 22
the security monitor located on the west side of the board so it can be seen from the area of the board near the radios (14); the camera controls (18) incorporated into the radio panel (14)	14
4. Binder storage and log book work area (16) moved to the west side of the board to make more room for the radios (14) and security monitor controls (18) on the east side; the LAN system (9) moved to the west end of the board to be near the binder storage (16) and log book work area—if the LAN did not have to be on the board, the binder storage could have been moved back to the work table (10)	6, 24
5. A permit counter (11) built into the wall next to the exit to the plant (13) and the defeat board (5) relocated to the north wall of the control room, so it can be seen by the board operator and the person issuing permits—contractor traffic into the control room eliminated; in addition, the permit counter now closer to the	18, 19, 27
printer/FAX table (20)	31
6. A second flare monitor (8) hung from the ceiling south of the work table, so the board operator can monitor the flare while working at the table	23
7. Plant radio (14) repeated at the work permit counter (11)	28

continued

Item	Figures
8. An additional set of procedures (16) located at the work permit counter (11), which also is close to the plant exit and available to outsiders without creating traffic into the control space	30, 33, 34
9. Storage and files (19) relocated to the east wall, nearer to the exit to offices; this location reduces office traffic into the control room	32

Note: Numbers in parentheses ( ) identify the item from the affinity matrix.

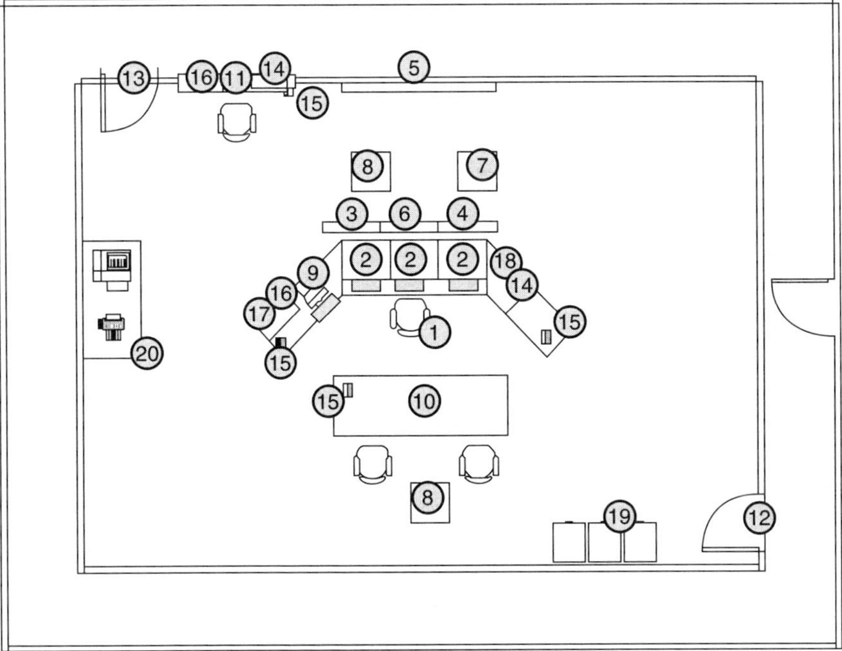

Figure 6-33 Revised process control room layout.

REVIEW QUESTIONS

Test your understanding of the material in this chapter.

1. For standing workplaces, as the work becomes less strenuous, the work surface height (floor to work surface) should decrease: true or false?

2. The following questions deal with the design of industrial seating:
   (a) List five design criteria for operators chairs (dimensions are not necessary) that chair purchases should consider.
   (b) Explain briefly the working conditions in which a "gas lift" chair would be recommended.
   (c) Explain briefly why five-caster chairs are preferred over four-caster chairs.
   (d) Under what working conditions would locking casters (casters that resist turning unless there is weight on the chair) be recommended?
   (e) Explain, in 10 words or less, the population extreme (e.g., 50th percentile male) that governs seat pan depth.
3. Fill in the blank with a word from the following list to complete this sentence: "A workstation is defined as the space, facilities, and ________________ allocated to a person at work."
   (a) Furniture.
   (b) Workers.
   (c) Air quality.
   (d) Environment.
   (e) Working surface.
4. Which work does not belong in the following list of workstation prototypes?
   (a) Dynamic mock-up.
   (b) Scale drawing.
   (c) Miniature model.
   (d) Static mock-up.
   (e) Dimensional graphic.
5. Which term in the following list does not fit with this sentence? "Operators working at improperly designed workstations frequently complain of ________________."
   (a) Sore eyes.
   (b) Aching in the lower back.
   (c) Pain in the upper arms and shoulders.
   (d) Shortness of breath.
   (e) Headaches.

6. Which of the following techniques would be best to analyze the layout of a kitchen in a process control building?
   (a) Link analysis.
   (b) Decision analysis.
   (c) Activity analysis.
   (d) Functional analysis.
   (e) Discomfort rating analysis.

7. Your company is undergoing a total quality program to improve customer service and reduce operation expenses. Management has decided to eliminate office space for the marketing sales force and have them work out of their leased vehicles. You have the job of designing a workplace within each salesperson's vehicle. The objectives of the workplace are
   (a) Interference with driving must be minimal. Safety is of paramount importance. So, for example, while you would not be expected to initiate cell phone calls while driving, you would be expected to receive calls safely.
   (b) Full voice communication between the vehicle and the office support staff and between the vehicle computer and the company's LAN is vital.
   (c) On-board computing facilities are required to access customer databases, prepare invoices, and so forth.
   (d) The vehicle must be able to be returned to resale shape for under $500.
   (e) All equipment must be secure in the event of a sudden deceleration.
   (f) Enough space must be available in the vehicle to carry one passenger.

   The major pieces of equipment used in the vehicle are a cellular phone, a laptop computer, a portable printer, a Fax machine, and a handheld portable computer. Your task is as follows:
   (a) Briefly discuss the principles or assumptions you developed to determine the priorities and the locations relative to the driver (assume that the steering wheel column is the centerline for the driver).

(b) Sketch the workplace, identifying the layout of each piece of equipment and the priority attached to controls and displays.
(c) In what type of vehicle would you choose to install your workstation? Why?

8. (a) Explain, briefly, the difference between designing for maintainability and designing for maintenance.
(b) Name three considerations when designing a piece of equipment for maintainability.
(c) From the point of view of a person who must service the equipment, what is a major consideration in the design and installation of a motor-operated valve? Explain briefly.

9. The following population extremes can influence the designs of facilities and equipment.
i. 5th percentile woman.
ii. 95th percentile woman.
iii. 5th percentile man.
iv. 95th percentile man.
v. 50th percentile member of the user population.
What is the number of the preceding population extreme (e.g., i) that most influences the designs listed that follows:
(a) Maximum shelf height for women's lockers.
(b) Men's locker room door height.
(c) Bench height for equipment assembly while standing.
(d) Height of a supermarket checkout counter.
(e) Height of a desk work surface.

10. Within the work environment, the placement of people and equipment should reduce the distance between components that are important to or frequently communicate with each other, and the probability of unnecessary physical interaction between personnel. Figure 6-34 illustrates the arrangement of personnel and equipment in a typical maintenance shop. The personnel are illustrated by triangles, the equipment by squares. The components are connected by two types of "links":
(a) Solid lines illustrate the degree of importance between components. The higher is the value, the more important is the link.

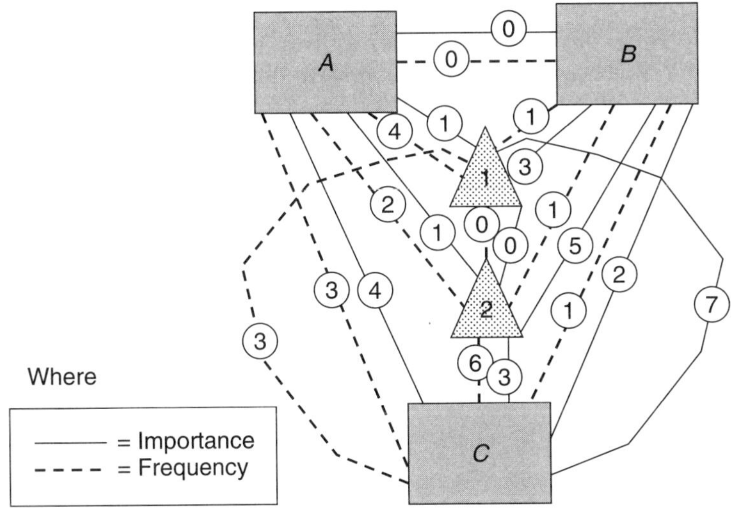

Figure 6-34 Link values between the elements in Review Question 10.

TABLE 6-13

Link Analysis Data for Figure 6–34 and Review Question 10

	1	*2*	*A*	*B*	*C*
Importance data					
1	—	0	1	1	7
2		—	1	5	3
A			—	0	4
B				—	2
C					—
Frequency data					
1	—	0	4	3	3
2		—	2	1	6
A			—	0	3
B				—	1
C					—

(b) Dotted lines illustrate the frequency with which each of the components interacts over a given period of time. The higher is the value, the more frequently the components interact.

A summary of the importance and frequency links between components and their values is listed in Table 6-13.

Using the information given,

(a) Generate a single criterion for component placement.
(b) Sketch the optimal relative location of each component, drawing in the appropriate links and their values.
State all your assumptions.

REFERENCES

ABS. (1998) *Guidance Notes on the Application of Ergonomics to Marine Systems.* New York: American Bureau of Shipping.

ANSI/HFES. (1988) *American National Standard for Human Factors Engineering Design of Visual Display Terminal Workstations.* ANSI/HFS Standard 100-1988. Santa Monica, CA: Human Factors Society.

ASM. (2002) *ASM Consortium Guidelines: Effective Operator Display Design.* Minneapolis, MN: Abnormal Situation Management Consortium, Honeywell Laboratories.

Attwood, D. A. (1985) "Applied Workstation Design." Distance Education in Ergonomics, Section 12. Post-diploma course in Occupational Safety and Occupational Hygiene. Community College Network of Ontario, Canada.

Attwood, D. A. (1996) *Office Relocation Sourcebook. A Guide to Managing Staff throughout the Move.* New York: John Wiley & Sons.

Attwood, D. A., Nicolich, M. J., Doney, K. P., Smolar, T. J., and Swensen, E. E. (2002) "Valve Wheel Rim Force Capabilities of Process Operators." *Journal of Loss Prevention in the Process Industries* 15, pp. 233–239.

CCPS. (1994) *Guidelines for Preventing Human Error in Process Safety.* New York: Center for Chemical Process Safety of the American Institute of Chemical Engineers.

CSA. (1989) *Office Ergonomics: A National Standard of Canada.* CAN/CSA—Z412M89. Toronto, ON, Canada: Canadian Standards Association.

Eastman Kodak. (1983) *Ergonomic Design for People at Work.* New York: Van Nostrand Reinhold.

EEMUA. (2002) *Process Plant Control Desks Utilising Human Computer Interfaces: A Guide to Design, Operational and Human Interface Issues.* Technical Publication 201. London: Engineering Equipment and Materials Users Association.

Grandjean, E. (1988) *Fitting the Task to the Man.* London: Taylor and Francis.

Grandjean, E. (1995) *Ergonomics in Computerized Offices.* New York: Taylor and Francis.

HSE (prepared by Bornel Limited). (2001) D*ecks, Stairways, Gangways and Their Associated Handrails.* Report 2001/69. London: Health and Safety Executive.

Joyce, M., and Wallersteiner, U. (1989) *Ergonomics: Humanizing the Automated Office.* Cincinnati, OH: South-Western Publishing.

Mital, A., Nicholson, A. S., and Ayoub, M. M. (1993) *A Guide to Manual Materials Handling.* London: Taylor and Francis.

Nimmo, I. (2002) *Determining Operator Workload and Console Loading Is More Than a Simple Loop Count.* Anthem, AZ: User Centered Design Services, LLC.

NPD. (1977) *Regulations Relating to the Systematic Follow-up of the Working Environment in Petroleum Activities.* Oslo: Norwegian Petroleum Directorate.

OSHA. (1997) *Stairways and Ladders.* Regulation 3124. Washington, DC: US Department of Labor, Occupational Safety and Health Administration.

Pheasant, S. (1994) *Bodyspace: Anthropometry, Ergonomics, and the Design of Work.* Second Ed. New York: Taylor and Francis.

Sanders, M. S., and McCormick, E. J. (1993) *Human Factors in Engineering and Design.* Seventh Ed. New York: McGraw-Hill.

Strobhar, D. A. (2002) *Controller Workload: What Are They Doing and Why Are They Doing So Much of It?* Dayton, OH: Beville Engineering.

VanCott, H. P., and Kinkade, R. G. (1972) *Human Engineering Guide to Equipment Design.* Sponsored by the Joint US Army-Navy-Air Force Steering Committee. Washington, DC: American Institutes for Research.

7

Job Factors

7.1 INTRODUCTION

This chapter discusses work-related issues that can affect the process operator. Shift work and work scheduling are reviewed in terms of design, their effects on performance, health, psychosocial life, safety, and coping. Fatigue and stress as a product of shift work are also reviewed. In addition, techniques on how to analyze tasks, create a working team approach, and using behavior-based safety to improve efficiency, productivity, and safety are discussed.

At the end of this chapter is a case study on ergonomic assessment of tasks at a loading rack using the task analysis technique. In addition, a set of review questions are included to help you check your understanding of the material covered in this chapter.

7.2 SHIFT WORK AND WORK SCHEDULES

For decades now, shift work has been instituted in hospitals, police and fire departments, in the military, and in petrochemical and other process industries, on a 24-hour basis (Presser and Cain, 1983). In addition, looking at the recent workplace environment, the expectation of more access to service around the clock, and the stiff competition among the service sectors, we find that, for example, supermarkets and drug stores operate seven days a week, 24 hours per day. This further increases the percentage of the working population involved in shift work.

To date, no longitudinal studies have been reported that explore the effects of a given work schedule on the business goals of increased productivity and workers' health, safety and quality of life. In fact, the work schedules practiced today have an unknown origin and no references may be cited from the literature. However, given past experiences and the wide recommendations found in the literature, we can design work schedules to better meet these goals.

The goal of this section is to provide general information on shift work and work schedules and on the effects of the work schedules on sleep, fatigue, performance, and the health of the human operator. It provides guidance on coping strategies that can be implemented by the shift worker, the family, and the organization. This section is *not* a book on sleep patterns and sleep disorders or shift schedules. This section provides a summarized overview of these topics. People interested in more detail can consult Tepas and Monk (1987); Monk and Folkard (1992); and Tepas, Paley, and Popkin (1997).

7.2.1 Sleep and Sleep Disorders

Normally, adults sleep between 7 and 8 hours per night. However, some may sleep less or more, creating large individual differences that may be related to age and health, stress and working hours, or social activities and drug or alcohol use.

The 7–8 hours per night sleep is regulated by the two daily cycles the human organism experiences. These are "the ready to perform cycle" (ergotropic phase) during the daylight or daytime and the "relaxed phase for recovery and replenishment of energy" (trophotropic phase) during the night (Grandjean, 1988). Given these two phases, working at night carries with it physiological and medical problems, not to mention family and social issues, since the person is trying to perform during the relaxed phase. Therefore, work schedules and shift work should be designed to produce the fewest negative effects on the health, family, and social life of the individual. This topic is elaborated on later in this chapter.

The human body experiences a biological rhythm, identified by a periodic change in body physical measurements, such as body temperature, heart rate, blood pressure, and chemical response (like the production of catecholamines such as epinephrine, norepinephrine, and dopamine, and adrenalin that function as hormones, neurotransmitters, or both). This biological rhythm fluctuates in a 24-hour cycle, referred to as *circadian rhythm*. *Circadian* is from the Latin word of *circa* ("about") and *dies* ("day"). More specifically, this can be referred to as *endogenous* (internal clock) *circadian rhythm* and operates on a cycle of between 22 and 25 hours and varies among individuals.

7.2.1.1 Normal Sleep

Obtaining the proper amount of sleep (i.e., total hours and quality) has been associated with health, well being, and efficient performance. In determining if an individual is sleeping, two methods are employed. The first measures brain electrical activity using an EEG (electroencephalogram). The second measures eye movements or muscle activity using an EOG (electrooculogram). We can identify two main categories, describing the quality and depth of sleep, REM (rapid eye movement) and NREM (nonrapid eye movement). Table 7-1 provides a description of the two categories and their cyclical stages.

The quantity (length) of sleep and its quality are not uniform. In terms of quality, NREM and REM sleep alternate cyclically. For example, slow-wave sleep, or light sleep, dominates during the first one third of the night. However, REM, or maximum relaxation of the muscles and resistance to being awakened, dominates during the last one third of the night.

7.2.1.2 Sleep Behavior and Disorders

The age of the worker plays a role in sleeping patterns and disorders. As people get older, they have less sound sleep and are easier to arouse than younger people. This is experienced as more and longer nighttime awakenings, more shifts from one sleep stage to another, and more body movement during sleep. This fluctuation leads to a decrease in the restorative power of sleep (normal sleep) and an increase in daytime drowsiness and "sleepy" feelings. The buildup of sleep deprivation (younger or older people) can lead to irritability, fatigue, inability to concentrate, and other psychological problems. When people are able or allowed to sleep after prolonged sleep deprivation, they experience increased sleep duration as well as an increase in the percentage of recuperative (NREM, Stages 3 and 4) sleep.

Sleep disorders are divided mainly into three categories: inability to get to sleep or stay asleep, disrupted timing of sleep and waking as occurs with jet lag or shift work, and feeling too sleepy during daytime.

Shift work can aggravate sleep disorders by leading to a difficulty in initiating and maintaining sleep (insomnia), cessation of breathing during sleep (sleep apnea syndrome), irresistible sleepiness occurring unpredictably throughout the day (narcolepsy), and gastrointestinal reflux issues. In addition, the consumption of caffeine and alcohol can generally lead to a decrease in length and quality of sleep, where sleep becomes

TABLE 7-1

REM and NREM Stages of Sleep

Stage	*Description*	*EEG*	*Muscle Tone*	*Eye Movement*
REM	Most dreams occur in REM sleep Physiological systems are active: heart rate, blood pressure, breathing Sleep paralysis: muscle tone absent, protects person from possible injury in acting out a dream	Resembles Theta (4–9 Hz) in NREM Stage 1	Absent	Rapid and irregular
NREM	Transitional stage between awake and sleep	Theta (4–9 Hz)	Moderate	Slow rolling
Stage 1. Drowsiness	Lasts only a few minutes Physiological stable: heart rate, breathing slower and more relaxed			
Stage 2. Sleep	First true sleep Constitutes 45–55% total sleep time	Sleep spindles peaks (12–14 Hz)	Relaxed	Slow rolling
Stages 3 and 4.	Constitutes 10–20% total sleep time	Delta less than 6 Hz	Low	Absent
Slow-wave sleep	Most restorative functions occur during this stage of sleep			

Notes: Theta wave, or theta rhythm, is a relatively high-amplitude brain wave pattern between approximately 4 and 9 Hertz (1 Hz = 1 cycle per second). Delta wave, or delta rhythm, is a relatively slow brain wave, having a frequency of fewer than 6 Hertz, that is associated with deep sleep in normal adults.

lighter and more fragmented. Common sleep disorders that can be aggravated by shift work arc:

- Insomnia, difficulty in initiating and/or maintaining sleep:
  Most common adult sleep complaint.
  Can be caused by disease or chronic pain.
  Most common cause is psychological, because of extreme stress, depression, or anxiety.
- Sleep apnea syndrome, cessation of breathing during sleep:
  More common in working-age men.
  About 1% of men over 30 may suffer from sleep apnea.
  Commonly associated with obesity.
  Associated with airway disorders, causing snoring.
  Potentially fatal.
  Treatment is positive airway pressure devices worn during sleep.
- Gastrointestinal reflux, acidic stomach juices coming back up into the esophagus when a person lies down to sleep at night:
  Causes burning sensation and sour taste.
  Awakens sleepier.
  Common in rotating shift workers, who often have irregular sleeping patterns and digestive problems.
- Narcolepsy, irresistible sleepiness occurring unpredictably throughout the day:
  Believed to be a disorder affecting the brain's mechanism for controlling sleep.
  People with narcolepsy fall asleep suddenly in any situation.

7.2.1.3 Fatigue

From the preceding discussion of work schedules, it becomes clear that fatigue is a component that cannot be ignored. *Fatigue* is generally defined as physical or mental weariness resulting from exertion. Regardless of its definition, in the context of this chapter, fatigue plays a key role in affecting productivity, error rates, and accident rates. For example, alertness and performance change as a function of time of day or night that relate to the body's circadian rhythms. This variation undoubtedly includes a fatigue component. In addition, the inadequate quality and quantity of sleep manifest themselves in increased fatigue during work hours. Fatigue also can be related to two additional issues that companies, and especially management, always struggle with—the length of the shift and the breaks built into a shift. Extending the shift time beyond 8 hours (12 hours is a common extended shift work) for jobs requiring heavy physical work or sustained attention (vigilance) may decrease performance and productivity and increase error rates and accident rates. Indeed, the literature predicts an increase of 80–180% in error rates for such jobs when extending the shift work time from 8 to 12 hours (Kelly and Schneider, 1982; Tepas, 1985). Shift breaks are a way of breaking up the shift length to overcome the effects of acute fatigue. However, for an 8-hour shift, for example, shift breaks extend the length of the time period beyond those 8 hours. Bhatia and Murrell (1969) report that appropriate break scheduling prevents performance decrement and can increase productivity throughout the whole shift.

Major industrial accidents where shift work and fatigue played a role in leading to human error and accidents are the Three Mile Island nuclear power plant, Bhopal chemical plant, and Chernobyl nuclear power plant. The data show that these accidents occurred between the hours of midnight and 4 AM. Human error is likely to occur at a rate twice the average during this time period as contrasted with the rest of the day. Indeed, the data from the literature also indicate that, for every daytime accident (daytime accidents normalized at 1), there are 1.15 evening (4:00 PM–12:00 midnight) and 1.2 night (12:00 midnight–8:00 AM) accidents. Other statistics related to road accidents find fatigue responsible for up to 20% of fatal truck accidents (Haworth and Heffernan, 1977) and 31% of fatal-to-the-driver accidents (NTSB, 1995). The factors that may affect fatigue can be related to one or more (usually a combination) of the following:

- Inadequate rest.
- Sleep loss or disrupted sleep, affecting sleep quantity and quality.
- Changes in biological rhythm (i.e., jet lag).
- Excessive physical activity.
- Excessive mental or cognitive work.

- Individual physical fitness.
- Environmental conditions, such as temperature, humidity, noise, and light.
- Time of day, such as at 2:00 PM we feel more tired and sleepy than the morning hours.
- Stress.
- Substance abuse.

The fatigue state caused by these factors can be classified into acute or chronic. Acute fatigue is short term, where a short break or a good night's sleep usually can provide relief; for example, after driving for several hours or attending to an emergency maintenance task that needs to be accomplished quickly. Chronic fatigue is relatively longer term, resulting from repeated and cumulative effect of these factors. This state requires an extended break, relatively longer than the acute fatigue state. Table 7-2 presents the characteristics of mental and physical fatigue.

TABLE 7-2

Characteristics of Mental and Physical Fatigue

Mental Fatigue	*Physical Fatigue*
Feeling tired after extended or repeated performance of nonphysical tasks	Temporary loss of muscle strength to respond to demands
Feeling of monotony or boredom created by lack of novel stimulation	Feeling of muscular tiredness, soreness, or other discomfort
Slower reaction time affecting decision making and response	Decrease in physical performance
Significant decrease in attention, vigilance, and alertness leading to impaired judgment	
Subjective feeling of loss of motivation, tiredness, and the desire for rest and sleep	

7.2.2 Effects of Shift Work on Performance

Rotating shifts are a fact of life for many workers within the petrochemical and other process and manufacturing industries. Most refineries, chemical plants, off-shore platforms, and upstream (development) operations run 24 hours a day. As we have seen earlier in this chapter, the combination of both sleep loss (quality and quantity) and low-arousal circadian rhythm phase (time of day or night) leads the individual into a sleepy state and thus affects all aspects of performance. When we are in a sleepy state, we experience an increase in the number of eye blinks and closures and short duration nodding off. Performing tasks that depend on visual input are particularly affected. In addition, performing tasks requiring vigilance (see Chapter 8) and other monotonous tasks are also affected, since the individual is unable to sustain alertness and attention (Hildebrandt, Rohmert, and Rutenfranz, 1974; Rutenfranz and Knauth, 1976; Horne, Anderson, and Wilkinson, 1983). Other situations affected by the sleepy state involve tasks requiring (1) decision making and judgment; (2) information processing and memory, such as learning or storing new information; and (3) self-initiated cognitive activity, like maintaining situation awareness and planning.

It is important to note here that we cannot assume that all tasks show the same performance decrement as alertness. For example, for a particular work schedule and same time of day (i.e., early morning hours), performance is best for one type of task and poorest for another task (Folkard and Monk, 1979). Tasks with high arousal states involving interesting and exciting materials and tasks requiring physical activities are not affected by the sleepy state. Performance is generally measured in terms of three variables:

1. *Productivity.* General findings in the literature conclude that there is no reduction in productivity as a function of shift work (day versus night).

2. *Number of errors.* The highest number of error rates is detected in the 1:00–4:00 AM time period. In addition, the literature reports an increased peak in error rates occurring at about 3:00 AM and a lower peak at about 3:00 PM.

3. *Number of accidents.* The literature reports that driving accidents show two distinct peaks. The first peak occurs in young drivers 18–20 years old between midnight and 6:00 AM. The second peak occurs in drivers 56 years old and older during the late afternoon hours (Summala and Mikkola, 1994).

7.2.3 Effects of Shift Work on Health

The actual cause of sickness among night workers is related to two main reasons: the circadian rhythm is "out of synchrony" (desynchronization of the internal body timekeeping mechanism that tells us when to sleep, eat, and be active) and the disturbance (eating habits, social, and level of activities) experienced in the change from day to night work.

Most research and surveys show a clear, direct association between night-shift workers and health. For example, studies made between 1948 and 1959 in Norway (Aanonsen, 1964; Thiis-Evenson, 1958) show that the sickness rate among night-shift workers was three times higher than that of day workers. The ailments are due mainly to unhealthy eating habits and chronic fatigue and its associated symptoms of psychosomatic disorders (Table 7-3). However, it is important to add that other studies (Barnes, 1936) also found a tendency on the part of night-shift workers to misuse drugs, such as taking stimulants during the night and sleeping

TABLE 7-3

Symptoms and Psychosomatic Disorders Associated with Chronic Fatigue.

Symptoms	*Psychosomatic Disorders*
Mental and physical weariness	Poor quality and quantity of sleep
Mental irritability and confusion	Digestive problems
Loss of motivation	Loss of appetite
Episodes of mood changes and depression	Stomach and intestinal problems
	Ulcers
	Nervous disorders

Source: Grandjean (1988).

pills during the day. This adds to the severity of the sickness rate (Grandjean, 1988).

Most companies have no medical programs for night-shift workers. In addition, with the trend toward cost cutting, since the middle 1990s, many companies either eliminated or did not initiate a regular physical examination program for their employees. Yet, and ironically, almost all companies require applicants to undergo a physical exam to determine if their health and fitness meet a required standard to gain initial employment. It is recommended that

- A specific medical examination be in place to select and place individuals for rotating shift work.
- A program be in place to ensure regular medical examination for rotating shift workers.
- Physicians be well trained and up-to-date on recent medical literature related to rotating shift work. Physicians should concentrate particularly on issues such as sleep disorders, chronic fatigue, and those listed in Table 7-3.
- Educate and counsel workers and supervisors on how to cope with rotating shifts.
- Use synchronizers, such as light exposure (for example, bright lights).

7.2.4 Effects of Shift Work on the Psychosocial Life

Studies over the years have shown that the main issues and disruptions arising from shift work are related to the psychosocial (psychological and social) factors. These psychosocial factors can, in turn, affect work performance and satisfaction. The issues and disruptions are generally related to three factors: the shift schedule worked, individual differences, and the worker's personal and social life.

7.2.4.1 Shift Schedule Worked

The type of shift schedule worked can affect what the individuals can do when they are off the job. For example, workers in the first shift have between the middle to late afternoon and the middle or late evening hours to attend to their own needs. The workers in the second shift have the morning to midafternoon hours. Third-shift workers usually have from the middle afternoon until late evening hours. In reality the off-time activities are mainly spent not on leisure or free time but everyday needs, such as commuting to and from work, buying and preparing meals, doing housework, children's activities, and sleeping. This can directly affect the personal, recreational, and social events, such as family, friends, and leisure time.

7.2.4.2 Individual Differences

Individual differences have significant impact on many dimensions of the shift worker's life. For example, individuals who cannot tolerate night shift either request a change to the day shift or quit the job when they find a more-acceptable work schedule. The following is a list of some of the individual differences:

- *Age.* Older people, over 40 years old (Barnes, 1949; Grandjean, 1988), are less tolerant of changes in schedule, take longer to adapt to shift changes, tire more easily, are more prone to disturbed sleep, have less resistance to circadian disruption, and complain of their ill-health.
- *Individual susceptibility.* The symptom levels, ill health, or ailments affect different people differently. Even with age, there are people who are fairly flexible and resistant to changes in work/rest schedule and "shift lag" or "jet lag," while others show the effects dramatically and require more time to recover.
- *Gender.* In general, female shift workers, compared with male workers, have pressures to provide same level of home care as a day worker, require more sleep per night, and require limited working hours during pregnancy.
- *Morning versus evening types.* Some research suggests there are some differences between morning type and evening type individuals. In

general, morning types do not adapt to shift work as well a evening types, transfer out of shift work more often, and have difficulty in adjusting their circadian rhythms to night schedules.

- *Others.* Other factors that may contribute to individual differences in adjusting to shift work and shift changes are heavy domestic workload, history of addiction to alcohol or drugs, history of sleep disorders, psychiatric illness, epilepsy, diabetes, and heart disease.

7.2.4.3 Personal and Social Life

A number of studies clearly demonstrated the influence of shift work on a person's social and leisure time, including family and friends as well as participating in recreational, educational, and community activities. For example, the second- and third-shift workers are less satisfied with the amount of time they are able to spend with family and friends and often pursue solitary hobbies.

The social effect of shift work is seen as a major source of dissatisfaction, especially when weekend work is scheduled. In general, weekends are identified as time for religious and family activities, recreation as well as work around the house. For single parents, shift work schedules carries with it difficulties in terms of raising a family. Special arrangements for the children have to be made when the parent is at work (Wedderhum, 1981). Since most human society is day oriented and most recreational and social events are scheduled for the day worker, shift workers find it very difficult, if not impossible, to participate. In addition, when workers are part of a rotating shift schedule, they find it difficult to develop and maintain social interaction with friends who happen to be on different shifts because of the rotation process. Therefore, such workers may experience social isolation (Sergean, 1971).

7.2.5 Shift Work Schedule Design

The type of work schedule is identified by two major characteristics: length of shift and rotation pattern.

7.2.5.1 Length of Shift

Rotating shifts are typically 8 or 12 hours long. Each has its advantages and disadvantages. One is not generally considered better than the other. In fact, some operations schedule both 8- and 12-hour shifts. The advantages and disadvantages of 12-hour shifts compared to 8-hour shifts are listed in Table 7-4. Because 12-hour shifts already occupy one-half of the day, shift workers should be discouraged from working overtime or double shifts. Extra hours should be worked by those working 8 hours, day workers, or off-duty workers.

7.2.5.2 Rotation Pattern

Rotation pattern refers to the cycle of consecutive shifts worked and days off work. The major variables in each cycle are the direction of rotation, speed of rotation, and number of consecutive days off. Each of these

TABLE 7-4

Advantages and Disadvantages of 12-Hour versus 8-Hour Shifts

Disadvantages	*Advantages*
Allows opportunity for moonlighting, which may reduce the quantity of sleep	Increased worker satisfaction, since it allows more days and weekends off
Women may find it difficult to arrange home schedules around 12-hour shifts	Fewer consecutive days spent on night shifts
Replacement of absent workers may be difficult, or if workers are held over for a second 12-hour shift, dangerous.	12-hour shifts lessen the number of shift changes (least productive time of an operation)
Administration could be more difficult, as hours of work and exposure to noise and chemicals are based on a 8-hour workday	Absenteeism may be reduced
Shift may be too long to perform physically demanding tasks, especially for older workers	Data on accidents and productivity rates for 8- and 12-hour shifts are equivocal

variables has an impact on the quality of life of the shift worker. For example, many family and social activities occur on weekends, so shift schedules that give no weekends off are not recommended.

Direction of Rotation

For 8-hour shifts, the direction of rotation should be in a forward; that is, days followed by evenings followed by nights.

For 12-hour shifts with a fast return, the days-to-nights and the nights-to-days rotations allow sufficient time (a 24-hour break) between rotations. For the days-to-nights rotation, the worker should obtain a "normal" night's sleep before the shift, then a short nap before the shift. For nights to days, the worker should take a short nap after completing the night shift, then obtain a normal night's sleep before going on the day shift.

Speed of Rotation

Speed of rotation, defined by the number of consecutive shifts before a day off, can be classified into three categories: fast rotations have four or fewer consecutive shifts, medium rotations have five to seven consecutive shifts, and slow rotations have seven or more consecutive shifts. The advantages and disadvantages of each category are shown in Table 7-5.

Number of Consecutive Days Off

For more complex jobs, or those requiring up-to-date situation knowledge, long periods of consecutive days off can make it more difficult to bring the worker up to speed on the status of the operation and contribute to the deterioration of job skills and knowledge. In supervisory control tasks, for example, a rule of thumb is to limit the operator to 4 consecutive days off. Otherwise, the operator requires an extensive briefing on return to work. On a related issue, the work schedule should be designed to provide each worker with at least one weekend off in four.

TABLE 7-5

Advantages and Disadvantages of Various Speeds of Rotation

Speed of Rotation	*Advantages*	*Disadvantages*
Fast (four or fewer consecutive shifts)	Avoids problems associated with body clock adjustment Does not allow sleep debt to accumulate Social contacts are maintained	Body clock of night shift workers does not adapt, so performance could be affected Greater impact on family life
Medium (five to seven consecutive shifts)	Typically accompanied by 7 consecutive days off; can be desirable in remote locations or on rigs	Can result in excessive buildup of sleep debt Body clock does not fully adapt during night shift
Slow (seven or more consecutive shifts)	More chance of body clock adjustment during night shift	Can result in excessive buildup of sleep debt Most likely to produce physiological problems

7.2.6 Coping Strategies with Shift Work

Both shift workers and their families and friends can help reduce the adverse affects of shift work. Studies have shown that an individual shift worker can have a profound effect on his or her own health and safety by observing a few commonsense lifestyle principles. These principles are listed next.

7.2.6.1 Sleep

- Give sleep the attention that it deserves. Do not take the sleep process for granted.
- Do not let social or domestic pressures take away sleep time.
- Preserve sleep opportunities during the work week. Do not spend days off catching up.

- Regulate sleeping times ensuring that you sleep at the same time during each shift.
- Do not try to sleep on the couch in front of the TV.
- Advise family and friends of your schedule so they respect your need for privacy.
- Develop techniques for relaxing before bedtime or work. These might include reading, listening to music, or light exercise.
- Develop a regular routine of bathroom/shower use and darken the room when sleeping during the day. This acts as a stimulus for sleep and reinforces the importance of sleep to the family.
- Prepare the house: Advise the family to be quiet and, if possible, turn off the TV.

7.2.6.2 Diet

- Maintain regular, predictable eating patterns as much as possible.
- Pay attention to foods eaten at night. Heavy, fatty foods should be avoided.
- Drink plenty of fluids on the night shift.
- Avoid the use of stimulant drugs and medicines without medical advice. Do not use caffeine for 5 hours prior to your expected bedtime.
- Alcohol is not a sleep aid. In fact, it may cause you to awaken within several hours.
- Some workers find that a light snack or milk before bedtime promotes sleep.

7.2.6.3 Keeping the Body Clock in Synch

As mentioned previously, each of the three categories of shift rotation speed (fast, medium, and slow) has a different effect on the internal body clock that regulates body functions during a 24-hour cycle. For each class of rotation speed, the following strategies can help the shift worker to reduce the effects on health and well-being.

Fast Rotations

Try to maintain a daytime routine:

- Avoid heavy night meals.
- Spend some time in the daylight each day.
- Try to grab a quick nap during the "lunch" break on nights.

Medium to Slow Rotations

Try to reset the body clock from a daytime to a nighttime schedule:

- Have a light meal before sleep.
- Go to bed as soon as possible after the night shift ends. The sooner to bed, the less adjustment the body clock needs to make.
- Try to avoid daylight in the drive home after the shift (wear dark glasses) and in the house (wear eye shades to bed).

7.2.6.4 Personal and Mental Hygiene

- Take leisure seriously.
- Minimize the opportunity for stress.
- Physical fitness is important. Try to obtain a half-hour of physical exercise (i.e., take a walk, garden) a day.
- Keep frustration out of the bedroom. Pay bills, watch TV, and discuss family problems outside of the bedroom.

7.2.6.5 Strategies for Night Work

Plan your work schedule:

- Schedule activities to keep active and alert.
- Take frequent short rest breaks to walk and get fresh air.
- Socialize with colleagues to the extent possible.

Diet:

- If caffeine works to increase alertness, drink coffee. But restrict caffeine intake after about 3 AM so it will not affect the morning sleep.
- Have lunch halfway through the night shift.

7.2.6.6 Organizational Strategies

In addition to promoting awareness and optimizing work schedules, the organization can contribute to a worker's shift tolerance by the way it designs its work systems and work environment. The following points are provided as examples of what other organizations have implemented.

Education

It is essential that shift workers and their families and friends be educated on how to support the shift worker. Families and friends can be made aware of shift work issues and coping strategies that shift workers develop with the use of videos and pamphlets.

Facilities Design

- Maintain proper air quality. Maintain temperature, relative humidity, air quality, and air movement within recommended limits.
- Preserve the best noise levels. Provide a varying-sound environment to stimulate alertness. Avoid high-noise environments over prolonged periods of time, they can induce fatigue.
- Design night jobs to maintain alertness, such as frequent changes in job tasks.
- Provide rest facilities for "on-call" or extended hours.

- Provide exercise facilities for night-shift workers. They can maintain body tone and stamina and promote alertness.
- Ensure that the food and beverage service at night is equivalent to that provided during the day.
- Design and lay out facilities to increase alertness and facilitate social interchange.
- Provide night shift workers with bright lights (Table 7-6). Bright lights have been shown to aid in significant adjustment to night work and the "no-days-off" effect (i.e., rapid reversion to daily rhythm), improve sleep quality and quantity, aid rapid adjustment of the internal clock, and enhance performance (reduced latency in responding to alarms and improved problem solving).
- In control rooms, computer rooms, and offices, lighting can induce fatigue if not properly designed. Provide for adjustment of illumination, reduce glare, and provide task lighting.
- Do not require night worker to perform tasks requiring, for example, complex mental calculations, and vigilance.

Career Opportunities

Studies show that shift workers do not have the same opportunity for career advancement as nonshift workers because management does not

TABLE 7-6

The Advantages and Disadvantages of Bright Lights

Advantages	*Disadvantages*
Works to suppress melatonin, the hormone that promotes sleep	Can be a glare source in a VDT environment
Acts to stimulate the worker to stay alert on the night shift	Does not consistently improve performance on the night shift
	Does not prevent dozing on night shift
	Employee acceptance is problematic

see them as often and they are not available for projects that can enhance their careers. Consequently, effort should be made to provide shift workers with equal opportunities by providing them the same training opportunities as nonshift workers, increasing the number of day rotations, and limiting the maximum age for working shifts.

Planned Maintenance Napping

Planned maintenance napping occurs on-the-job during the work shift. Naps of 30–60 minutes can compensate for daytime sleep loss or serve as a bridge over the low point in daily alertness (e.g., 1–3 AM). Maintenance naps have been tested in Japan: 35% of manufacturing and 64% of service industries use nap facilities (Matsumoto et al., 1982). Current evidence suggests that napping during the first one or two night shifts could ease the transition from days to nights. More important, it can increase the level of alertness of the worker and decrease unplanned microsleeps (dozing).

7.2.7 Process for Creating or Changing Shift Schedules

The process employed for creating or changing shift schedules is critical to the acceptance of the workforce. The following is a list of recommended changes (Tepas, 1985, 1993):

- Educate management and workers and their families.
- Clarify why a change is required in their shift schedule.
- Consult and plan with all parties involved:
  Enlist experts to help site workers develop the requirements for the work schedule.
  Develop preliminary schedules tailored to the needs of the site (different parts of the organization may need different schedules).
  Put proposals to the workforce and outline the agreed-on criteria.
  Allow as much time for discussion as is needed.
  Have the workforce vote on the final shift schedule.
- Coordinate reorganization of work practices and rosters.

- Plan evaluation techniques before the schedule is implemented.
- The trial period should be at least 6 months.
- Evaluate the work schedule and adjust it.

7.3 STRESS

The general definition of *stress* is the reaction of an organism (human or animal) to a threatening situation. More specifically, stress is a biological reaction to any adverse stimulus (physical, chemical, mental, emotional, internal, or external) that disturbs the organism's homeostasis (stable or balanced state). Since stress is a necessary condition in our life, it can be classified into two main types:

- Positive stress is the desire to continue doing things in life and solve daily and difficult problems. Generally speaking, positive stress is good for us, and life without it can be unnatural and perhaps boring.
- Negative stress is basically the stress that we wish to live without. It puts us in an environment or state of anger, frustration, depression, confusion, and tiredness. Continued, long-term negative stress can lead to emotional problems and physical illnesses, such as cardiovascular disease and musculoskeletal disorders.

This section discusses, very briefly, the factors that can cause and lead to stress and how to cope with it.

7.3.1 Sources and Causes of Stress

It is almost impossible to point to only one factor as the cause of stress. Given the complexity of modern life and the dynamic changes in the work environment, there is no wonder why. Three main categories classify the sources of stress. Based on the literature (Karasek, 1979; Lindstrom, 1991; Buunk and Janssen, 1992; Smith and Carayon, 1995; Yang and Carayon, 1995), Table 7-7 provides summary examples of the sources of stress.

Stress can be measured qualitatively (through questionnaires, interviews, observation) or quantitatively (by measuring heart rate, heart rate

TABLE 7-7

Sources of Stress

Category	*Factors*	*Leads to*
Workplace environment	Lack of control	Job dissatisfaction Emotional and physiological strain
	Lack of management and peer support	Increased stress
	Work overload in terms of quantity (too much to do or under a strict, short deadline) and quality (task demands exceed individual capabilities or individual feels incapable)	Increased stress and vulnerability to stomach ulcers and high blood pressure
	Work underload: individual's knowledge and capabilities far exceed job demands	Increased stress, boredom Decreased feelings of competence and self-esteem
	Job uncertainty, task ambiguity, technology changes, and job insecurity	Increased stress
	Lack of organizational commitment to employees, in terms of loyalty, career development (i.e., training, skills, security)	Increased stress, health complaints, job dissatisfaction, poor performance, turnover, absence
	Job insecurity related to streamlining businesses, mergers, downsizing, bankruptcy, and unethical corporate practices	Increased stress
Physical environment	Noise, poor lighting, vibration, heat/cold	Contribute to increased stress
Individual differences	Personality Lack of experience and knowledge Lack of skills Health of individual	Contribute to increased stress

variability, blood pressure, respiratory rate, or levels of adrenaline, noradrenaline, cortisol, glucose, uric acid, or steroids in samples of urine, blood, sweat, or saliva. It is recommended that a combination of the measures be used.

7.3.2 Coping Strategies

There are different approaches to cope with and manage stress. These approaches can be grouped into two categories by focusing on the work place environment or personal lifestyle. Table 7-8 summarizes the focus on each of these areas (Benson, 1985; Ivancevich and Matteson, 1988; Russell, 1991; Aryee and Tan, 1992).

7.4 JOB ANALYSIS

This section briefly discusses some of the tools that can be used to analyze jobs. The most common, critical, and useful tool for the human factors specialists is the task analysis. Task analysis and critical task analysis are reviewed here. For workload analysis, Chapter 8 discusses mental workload and physical workload is reviewed in Chapter 3.

7.4.1 Task Analysis

Task analysis is a basic tool of human factors specialists in the investigation and design of tasks. It provides the structure for a comparison between the demands the task places on the operator and the capabilities the human operator possesses. Task analysis produces detailed information relevant to the interface of humans and machines or systems. It also can be applied to the development of a new system or the redesign of existing systems. Note that task analysis does not solve problems but points analysts in the direction of potential solutions. This section briefly covers the area of task analysis. For more in-depth information, consult other sources (Drury, 1983; Drury et al., 1987; Kirwan and Ainsworth, 1992).

TABLE 7-8

Stress Coping Strategies

Category	*Approach or Focus on*
Workplace environment	Work overload, underload, and time pressure Lack of control over day-to-day work Lack of training and preparation Uncertainty and ambiguity of tasks and responsibility Individual career and skill development through Self-assessment tools Individual counseling Regular assessment and developmental programs Stress training to modify perceptions or behavior
Personal lifestyle	Physical changes: Regular exercise Yoga and other stretching exercises Breathing exercises Visualization, meditation, imagery Stop smoking Reduce alcohol consumption Improve sleeping habits Emotional or mental stress: Take breaks and vacations Talk and share personal issues with friends and significant others Identify what you can and cannot control Keep real expectations of yourself, others, and events Set priorities Enjoy music and laughter Nutrition: Eliminate high-fat diets Eat healthier fats (i.e., olive oils, fish) Drink plenty of water, about eight 8-oz glasses/day Eat plenty of fruits and vegetables

7.4.1.1 Purpose of Task Analysis

Task analysis techniques provide detailed information related to the prediction and prevention of errors and accidents. Indeed, the use of explicit task analysis approaches should therefore lead to more efficient and effective integration of the human element into system design and operations. Table 7-9 outlines the principle areas where task analysis techniques and approaches are beneficial.

7.4.1.2 When to Use Task Analysis

When task analysis is used in system design, the approach involves the analysis of the human operator tasks in the following order:

1. In an existing system, the type and value of task analysis depends on the relevance and accessibility of the existing system(s) available for analysis, the tasks to be analyzed, and the level of direct correspondence of the tasks between the existing and future systems.
2. During the design process, task analysis is used to evaluate the demands the task places on the operator in the new system.
3. At points during the operational life of the system (the system life cycle), task analysis is used to update an existing system by focusing on specific issues rather than examining the system as a whole. For example, task analysis can be used when an important safety issue is involved, human error is caused by technology or design deficiency, system changes have created levels of uncertainty about system performance and reliability as well as human/system interaction, or there are productivity and availability issues as they relate to human performance.

7.4.1.3 Who Can Perform a Task Analysis?

The users of task analysis methodologies can be grouped into three wide categories:

- System designers, the personnel responsible for the design of the system, equipment layout, displays and controls, and procedure

TABLE 7-9

Principal Areas Where Task Analysis Techniques Are Beneficial

Principal Areas	*Benefits of Task Analysis*
Safety	Identify hazards to the operator in the workplace Achieve a general level of safety through the achievement of good design for human operation Form the basis for the analysis of human error in the system Define what went wrong and help identify remedial measures during an incident or accident investigation
Function allocation	Allocate functions between humans and machines Define the degree of human involvement in the control of the system
Task specification	Define the characteristics and capability requirements of personnel to allow them to carry out the task effectively and safely
Task design	Ensure adequate human/task interface design to carry out the task efficiently and safely in normal and abnormal operations
Skills and knowledge of operators	Identify personnel requirements in terms of skills and knowledge Design training programs Design proper procedures
Job organization	Define the number and variety of personnel Define the organization of team members Define the communications requirements and allocation of responsibilities
Efficiency and productivity	Identify staffing requirements Identify specific training program to ensure efficiency and productivity
Maintenance demands	Identify maintenance demands Define the need for personnel and support tools and systems of work Identify the coordination needed among different areas of the plant

development. These designers, using task analysis techniques, must decide which functions are allocated to the machine and which to the operator.

- Safety personnel, who are involved in risk assessment as well as safety function evaluation and training. Human Factors personnel and ergonomists also fall in this category.
- Supervisors and managers, who are principally involved in the staffing of their departments, training and support system (logistics), and organization of that staff.

7.4.1.4 The Process of Task Analysis

To achieve a meaningful comparison of system demands and operator abilities and capabilities, the process of task analysis involves four phases. These phases follow a top-down structure, where each phase provides an additional, detailed view of the human/system interaction requirements. The four phases are:

1. Plan for a task analysis. Performing a comprehensive, useful task analysis requires the preparation and participation of different types of personnel in addition to the task analysis specialist. The personnel that may contribute to the process are safety engineers and analysts, design engineers (including instrumentation and control engineers), training personnel, procedure writers, operators, and operations supervisors and managers. The input from this cross section of personnel can provide a comprehensive description of the system and the operations necessary to complete the tasks required and produce the output expected.
2. General data collection. This phase involves the collection and documentation of various sources of information about the system. The main goal is to identify all the tasks to provide an overall understanding of the system and its components.
3. Task description. Task description includes a task list for each operating sequence of the system, based on the general data collection phase. This list should include human performance requirements and any critical aspect involved.

4. Specific data collection. The choice of technique for data collection depends on two things: the status of system development and implementation and the experience and availability of the personnel performing the analysis. Several techniques are involved in collecting data and information. These are summarized in Table 7-10. The techniques can be grouped into four categories: documentation review (i.e., analysis of blueprints, plans), questionnaire survey, interview, and observation (i.e., analysis of prototypes, measurements). To obtain complete information, a combination of these techniques is recommended.

7.4.2 Critical-Task Identification and Analysis

As mentioned earlier, task analysis techniques can be applied to existing tasks or tasks that do not yet exist. However, the process described here is concerned mainly with the identification and analysis of tasks associated with a new "project." In this case, a "project" may be the design and construction of a major grassroots facility or the improvement of an existing facility. In new projects, the tasks do not yet exist. Even so, they can be analyzed and to do so, the analyst or the analysis team should have knowledge of the facility in which the operator will work, the equipment the operator will use, the environment in which the operator will perform the task, and the activities the operator will be required to perform. Often, tasks similar to those in the new facility are performed elsewhere and can be used as models for the "nonexistent" tasks analyzed.

For tasks with this gravity described (i.e., they are critical), the identification and their analysis is a responsibility of a project (process and equipment) design team and a prerequisite for the proper human/system interface (HSI) engineering design of the facilities. People who perform these analyses are generally trained in human factors measurement and familiar with the task analysis method. Once identified and analyzed, the tasks can be reviewed routinely throughout the planning and design of the project to ensure that the human factors issues are properly addressed. Stages in which critical tasks are identified most effectively are during

TABLE 7-10

Techniques to Collect Information

Technique	*Benefits*	*Things to Consider*
Documentation review: Books and periodicals; Previous job description; Diary (running record of activities); Prior reports and analyses	Systematically gather larger amounts of data Good for unstructured jobs	Time consuming Does not indicate how tasks are specifically performed
Questionnaire survey:		
Open-ended questions	Large amount of information in short time	Burden to respondent Must be specifically developed Difficult to integrate and analyze information Data tend to be incomplete
Checklist	Easy to administer and tabulate Good for documenting differences between jobs and sites	Requires extensive preliminary work to develop Does not provide integrated picture of job Does not indicate how task is performed
Interview:		
Individual interview (interviewee away from job)	Can use structured interview Combine several interviews Can generate complete detailed information Interviewee is not rushed as in on-the-job interview	Time consuming Depends on recall Depends on interviewee's verbal skills
Group interview (two or more interviewees away from the job)	Interaction between interviewees initiate insight and recall Saves time over individual interviews	All participants may not contribute

continued

Technique	Benefits	Things to Consider
	Can collect complete, detailed information Can use structured format	
Technical conference (two or more experts)	Same group may be able to supply data on several jobs Good for obtaining overview of tasks	No guarantee that experts know the details of the job May rate difficulty on time spent based on their level of skill, rather than less experienced workers Usually lacks very detailed information
Observation:		
Observe but do not talk to operator	First-hand information Can clarify information already collected Can be done in noisy environment Operator can demonstrate task Work situation can act as a cue to recall	Requires prior knowledge for operator to understand tasks May interfere with task performance
Work participation (analyst performs job)	Obtain first-hand knowledge of tasks	Time consuming to train analyst Safety problems May not yield complete information Best for simple, repetitive jobs
Videotaping	Available for review Good for timing tasks Good for manual tasks and timing quick events Good for studying human behavior (i.e., during usability testing)	Not effective for mobile or nonmanual tasks

- Front-end loading, when the major units are being developed [prior to or during design basis memorandum (DBM) or the inherent safety, health, environment (ISHE) review].
- The process design phase, when "process tasks" are defined in detail in the pre-Hazop reviews.
- The equipment design phase, when the equipment and the tasks to be performed by them are specified in the Hazop, nonprocess quality control (NPQC) or model reviews.

The critical-task identification/prioritization and analysis (CTIA) process is divided into three parts: identification of tasks whose analysis is critical, analysis of the critical tasks, and follow-up.

7.4.2.1 Critical-Task Identification Process

Certain safety, health, and environment (SHE) reviews are required, to ensure hazards and risks are appropriately addressed. This process includes hazards and risks attributed to human factors, specifically related to human tasks designed into the project to achieve an HSI design supportive of the human needs to perform a task safe and successful. Therefore, criticality is identified in terms of significant SHE risk potential and significant operations or reliability implications.

7.4.2.2 Critical-Task Analysis

The critical-task analysis (CTA) technique consists of two steps:

1. *Preparation and description of tasks that will be analyzed.* In preparation for the analysis, completion of Form A (Table 7-11) is essential to create the mindset of the team: Identify the people who will perform the task. They may include people remote from the facility but in communication with it.
2. *Conducting the task analysis.* Form B (Table 7-12) allows the analyst to record the results of the analysis. Different forms may be used for unique types of analyses. The one shown in Table 7-12 is recommended for most task analyses. The form consists of five columns:

Form A. Task Analysis Preparation and Description

Date: ______________ **Critical Task Analysis, Form A—Task Description** **Page** ________ **of** ______

Prepared by: **Revision:** ______________

Job Name: ______________________________

Job Objective: ______________________________

Job Function Description

No. of People Involved in Task	**Operating Environment**	**When Is Task Performed?**	**Where Is Task Performed?**	**Personal Protective Equipment in Addition to Normal Issue?** ❑
	Extreme temperatures? High winds? High vibration? Poor lighting? High noise?	Day only 24-hour shift	Indoor Outdoor	Describe:

Communication	**Frequency**		**Vision**
Verbal? Radio? None? With whom? ________	Hourly? Daily? Monthly?	Yearly? As needed?	Need to view controls and/or displays? Yes? No ❑ Describe: ______________

Review Checklist

Review Item
Insert sketch, drawing or model of area (attach drawings if applicable)

TABLE 7-12

Form B. Conducting Task Analysis

Date: ______________ **Critical Task Analysis, Form B—Results Analysis** **Page** ______ **of** ______

Prepared by: **Revision:** ______________

Job Name: ______________________________

Job Objective: ______________________________

Potential Human Factors Issues	**Activity**	**Error Possibility and Consequences**	**Ergonomic and Human Factors Concerns**	**Potential Solution or Mitigation**
Physical activities ***Manual materials handling*** Hose handling Barrels, boxes Valves Height Orientation Force ***Musculoskeletal*** Hands/wrists Upper extremity (head, neck, shoulder) Lower extremity (lower back, leg) Access, walkways, platforms Routes: exit, entrance, stairs **Receiving information** Lighting Noise interference Display design Labeling Signage Color coding **Processing information** Too much information? Menu, checklist present? Short-term memory Long-term memory				

Column 1. Potential human factors issues. This column is designed to help remind people of the human factors issues that should be considered for each activity identified in the task (column 2).

Column 2. Activity. This column is used to identify the sequential tasks performed in the task. Each activity should begin with an action verb to describe the process or action of the personnel, such as installing temporary pressure gauge or filling standby filter.

Column 3. Possible error and consequences. This is evaluated for each activity in column 3. What potential errors can result from performing the activity. When people perform tasks, basically three modes of (mental) information processing are involved: skill, rule, knowledge-based (SRK) information processing (Rasmussen, 1983, 1986). These are defined as follows: Skill-based processing is automated routine requiring little conscious attention; rule-based processing involves the prepackaged units of behavior released when the appropriate rule is applied; and knowledge-based is the improvisation in unfamiliar environments, where no routine or rules are available for handling tasks.

Column 4. Ergonomic and human factors concerns. For each activity, the potential human factors concerns are identified using the job aid in column 1. Once the potential issues are identified (brainstorm everything that could happen), those likely to occur are listed.

Column 5. Potential solutions and mitigations. A solution (or solutions) for each likely issue is developed.

7.4.2.3 Follow-up Documentation

After the CTA is complete, a final report should be prepared that documents the results. The report should be issued to the person who sponsored the CTA from the project. The report documentation should include

- Cover letter that identifies who stewards the recommendations.
- A summary of all the recommendations or findings.
- Documentation of CTA team members and team leader.

- A copy of the completed forms used.
- Any areas of the review that are incomplete.

7.5 TEAM-BASED APPROACH

The team-based approach tools have been successfully used to help improve business, plant teams, and safety performance. The goal in this section is to provide a brief overview of four of these tools (see website addresses). Other tools, not cited here, may exist on the market.

- Kirton Adaptive-Innovative (www.kaicentre.com).
- ACUMEN (www.acumen.com).
- NTL Institute (www.ntl.org).
- Type Resources (www.type-resources.com).
- Temperament Research Institute (www.tri-network.com).
- Creative Problem Solving, CPSB (www.cpsb.com).
- Human Factors International (www.human-factors.com).

7.5.1 Cognitive Problem-Solving Style (KAI)

The Kirton Adaption-Innovation (KAI) inventory or instrument is a measure of cognitive (problem-solving approach) style of individuals as well as for the entire team. The central construct of the Kirton theory is a cognitive-style continuum (AI), all levels of which express creativity but vary qualitatively in terms of how that creativity expresses itself, such as adaptively or innovatively (Kirton, 1994). The product of this instrument is a private, comprehensive hard-copy analysis of the individual's position on the AI scale. In addition, a general analysis is provided to show the makeup of the team in terms of the AI scale.

Adaptors show relatively greater within-paradigm (structure) consistency, while innovators show relatively greater consistency across paradigms. It is very important to emphasize two points:

1. The operational measure of AI yield scores that distribute normally and the interpretations of these scores are relative to other scores. For example, a score of 85 (considered on the moderate adaptive side) can be seen as adaptive to someone with a score of 115 (considered on the innovative side). However, a score of 85 can be seen as more on the innovative side to someone with a score of 65 (considered on the high adaptive side).
2. The AI dimension is independent of creative level and the "technique" aspect of creativity as well as social desirability and general competence, gender, age, culture, and the like.

Normally, most teams have individuals exhibiting a wide range of styles. Differences in style often lead to personal clashes. Teams become less productive and take more time and effort, sometimes requiring outside intervention, in achieving a desired objective. Enlightenment of styles usually results in team members recognizing and accepting individual differences or preferences in approach. Recognition and acceptance help the team collaborate, which greatly increases the team's effectiveness.

7.5.2 Drexler-Sibbet High-Performance Team Model

The Drexler-Sibbet (DS) model describes the team process of achieving high performance. The DS technique measures team progress at each of seven stages. These seven stages are a kind of road map guiding the team through to achieve high performance. These stages include orientation, trust, goals, commitment, implementation, high performance, and renewal. The measurement can be used at any time in a team's life. It quickly pinpoints what the team is doing well and problems need intervention. Thus, the facilitator can work with the team specifically on only those issues impeding the team in reaching its objectives. The recommended use of this technology is at the team's beginning, and then again at midpoint of a team's life.

7.5.3 ACUMEN

The ACUMEN team building technology is utilized for special teams that have a "must do," high-return objective. Therefore, the faster the team can gel and deliver the goods, the faster the profit or technical gain is realized. Trust issues usually get in the way of a team gelling. A facilitated ACUMEN session is aimed at resolving major trust issues and moving the team forward. Therefore, the most optimum time for employing the ACUMEN technology is during the early stages of a team's life. ACUMEN is a software system that privately measures "how" a person thinks and therefore is most likely to act. The analysis is comprehensive and includes a self-perception as well as a private feedback loop requiring the input of four or more associates. The reports are hard copies and belong to the individual for self-improvement. The ACUMEN software also combines individual profiles (through coding for confidentiality) into a team profile. An analysis and a hard-copy report pinpoint the team's major strengths and potential soft spots. This technology has been successfully utilized for major plant and pilot plant start-up teams, operations, research, and technology teams.

7.5.4 Systematic Multilevel Observation of Groups (SYMLOG)

SYMLOG is a theory of personality and group dynamics. The deliverables is a set of practical methods for measuring and changing behavior and values. It is designed for specific groups in their natural environment. It is applicable to many kinds of groups, from small teams to an entire organization, with multiple situations. Specifically, the principle purpose is to understand the group better in order to improve productivity, quality, and safety. The theory and measurements indicate specific ways in which leaders and members can most effectively encourage desirable changes in group performance. Applications include assessment of teamwork and leadership, group composition, and team leader training.

The methods include instruments for measuring values as well as interpersonal behavior. The value measurements are adapted especially for use in survey research, program development and evaluation, cultural climate studies, organizational development, conflict resolution, improvement across interfaces, and the development of organizational policy.

7.6 BEHAVIOR-BASED SAFETY

The objective of all behavior-based safety (BBS) programs is to reduce the number and severity of injuries. In general, safe behavior acts as a barrier to prevent incidents, while unsafe behavior acts as the hazard that initiates the incident. Even though our traits (attitudes, personalities, and values) can affect our behavior, what we observe, however, is only what individuals do or say and not the traits. Behavior-based safety programs are based on field observations, providing feedback to reinforce safe behavior and correct unsafe behavior.

The goal of this section is *not* to rewrite the history and application of BBS programs. Detailed information is available from the following list of the programs and websites. The objective of this section is to present the lessons learned from a summary of a survey, performed by the authors of this book, of different BBS initiatives. In addition, a list of the recommended core and ancillary elements of BBS programs are provided.

- ABC Aubrey Daniels (www.aubreydaniels.com).
- Behavioral Sciences Technology—Thomas Krause (www.bstsolutions.com).
- Dupont Take Two Behavior Modification (www.dupont.com/safety).
- Loss Prevention System—James Bennett (www.Lpscenter.com).
- Safety Performance Solution—E. Scott Geller (www.safetyperformance.com).

7.6.1 Lessons Learned

These lessons learned are based on a survey of BBS programs (Deeb, Danz-Reece, and Smolar, 2000).

7.6.1.1 Implementation

- Review existing programs, study differences in cost and culture, choose one, and adapt it to fit your site.
- Get commitment (buy-in) up front from the top down and bottom up.
- Ensure management is aware of time commitment and provides continuous support.
- Everyone, including management, should be actively involved.
- Expect resistance.
- Be determined and do not rush. The process takes time and hard work.
- Choose a full-time coordinator or facilitator who is credible with management and workers.
- Set up behavior-based safety as a norm or expectation just like any other business function. "This is the way in which we conduct our business."
- Integrate behavior programs within the operation integrity system framework.

7.6.1.2 During Training

- Take time to explain objectives and principles to all employees.
- Do not take shortcuts in initial training.
- Experts should train the trainers, and the trainers should train the remaining employees.
- Provide more refresher training.
- Train supervisors in interpersonal skills.

7.6.1.3 Observations

- Process should not be voluntary; that is, people should have no choice whether to be observed or not.

- Observations should be nonpunitive.
- Establish a strong culture for feedback.
- Set targets for quality not quantity.
- Provide more employee-to-employee observations.

7.6.1.4 Measure

- Ensure regular meetings and feedback.
- Listen and respond to data.
- Ensure systems are in place for feedback.
- Conduct evaluations to measure goals and identify room for improvement.
- Clearly define and communicate the expectations of the process; components should be monitored and measurable.

7.6.1.5 Positive Outcomes

- Increased interface between salaried and wage personnel.
- Top management leadership and support.
- Groups are more cohesive and actively caring because of increased interface.
- Significant improvement in quality and quantity of observations and near-miss reports.
- Dramatically improved safety record.
- Improved morale.
- Increased employee awareness.
- General acceptance at worker level.

7.6.2 Recommended Core and Ancillary Elements of BBS Programs

At a minimum behavior-based safety programs should

- Focus attention on work behavior.
- Define expected behavior.
- Observe if behavior is demonstrated during work.
- Provide feedback to individuals if their behavior meets expectations.

7.6.2.1 Recommended Program Elements

The recommended core elements for a behavior-based safety program are summarized next.

- Implementation:
  The program has features that support continuous improvement.
  The program includes a plan for regular stewardship.
- Buy-in:
  The program expects employee participation.
  The program has visible participation by site and line management.
- Training: The program provides training for employees on
  Basic principles of behavioral reinforcement and correction such as the following:
  Observation skills and improvement, if needed.
  How to deliver feedback and reinforcement.
  Consideration of human factors or ergonomics, both in terms of physical activities and potential for human error.
- Observation:
  Persons who are observed receive immediate feedback or reinforcement on safety behavior issues observed.
  Observers use a checklist or reminder list while observing work activities.
- Tools:
  The program provides tools for consistent observations and analyses.

The program provides plans and tools to effectively define and communicate solutions or learning throughout the organizational levels and shifts and work groups.

- Measurement: Implementation includes initial survey evaluation to ensure that the design of the program meets the needs of the business. Measures and performance targets emphasize results and key activities and are used to set priorities on interventions.
- Contractors: The program extends to appropriate contractors.

For sites with existing programs, the most common questions are these: How complete is my program? How can we improve? The checklist mentioned in Section 7.6.2.2 can also be used to assess if existing programs are complete and, if not, which elements are lacking. To answer the question, How can we improve? one approach is to conduct an assessment of the existing program. In an assessment approach, the perceptions by managers and workers about the elements of the program are surveyed and analyzed. Areas in which perceptions differ significantly are also identified as opportunities for improvement, as are any gaps in the core and ancillary program elements.

7.6.2.2 Practical Considerations for Implementation

In addition to the core and ancillary elements, sites with experience in implementing a program suggest that the following checklist items are important to consider:

- Has the organization evaluated its needs and determined what objectives the program is intended to help achieve?
- How mature and effective are the organization's other safety system elements?
- Are you looking for a program that focuses only on safe work behaviors or something that fills other gaps as well?
- Does the program meet or can it be customized to meet the organization's needs?

- Can it reasonably be integrated with existing safety systems and tools or a mature incident investigation process?
- Does it help diagnose areas of concern and then set priorities or focus efforts on the higher-value opportunities?
- Can training and rollout be tailored to the audience (e.g., management, implementers, field personnel) or is it just a canned package for use with everyone?
- Are you looking for a program that is compatible with the organization's existing culture or one that will help achieve change?
- Do you want a program that focuses on outcomes and results or one that focuses on leading indicators? (The value of BBS is the focus on leading indicators.)
- Is the scientific basis on which the program is designed accepted by or acceptable to the organization?
- Does the program fit with the organization's size and sophistication?
- Will advisors be readily accessible to guide the effort and help with analysis and corrective action, or will small groups be trained and left largely on their own? Larger groups of people allow shared support, greater data collection and analysis, and if desired, more program elements.
- Are implementation requirements consistent with available resources (cost, time, and money)?
- Is the objective to implement a process quickly (off the shelf) or is there time to design or customize a product for the organization?
- How much money is presently available for the project? While the cost of in-house development or use of a third-party product can be nearly the same over the long term, generally a third-party approach has higher up-front costs.

7.7 CASE STUDY

This case study illustrates the use of the task analysis technique.

7.7.1 Introduction

An ergonomic evaluation of a loading rack was performed and consisted of interviews with Mr. X, the full-time technician; observation of work activities performed in this area; and a videotape recording of Mr. X performing his tasks. The on-site data were collected and analyzed (off site) using the following methods: task analysis and biomechanical analysis.

7.7.2 Task Analysis

The major tasks performed by the rack technician were

- Hooking up the railcars.
- Sampling.
- Gauging.
- Unhooking the railcars.

After observing Mr. X perform these tasks, it was concluded that the sampling and gauging activities did not pose significant ergonomic risk to the technician. Therefore, these activities are not addressed here. It was also noted that hooking the railcars essentially involved the same critical tasks as unhooking the railcars. Therefore, this task analysis focuses on the hooking task, with the understanding that any recommendations or concerns also address the corresponding activity as it takes place during the unhooking process. A short version of the task analysis is presented here for demonstration purposes (Table 7-13).

TABLE 7-13

Task Analysis and Recommendations of Loading Rack Operation

Subtasks	*Description*	*Human Factors Concerns*	*Recommendations*
Check cars to verify that they are in the correct spot.	Technician walks up and down the line of railcars to check that each car is in the correct spot prior to loading.		
Chock cars.	Technician throws two chocks per car: one goes in front of the wheel and one goes behind the wheel to prevent the cars from moving while stationed at the rack.	1. The chock is unbalanced—the total weight at the handle is 12.5 lb and the weight at the far end of the chock is 10 lb. A force is placed on the wrist to counteract the forward/downward pull of the chock. 2. It is difficult to place the chock in front of the rail car wheel because obstructions on the rail car make it necessary to twist the neck for the technician to see what he is doing.	1. Redesign the chock in such a way that, when lifting it, the weight at the handle is balanced. 2. Redesign the chocks such that there are two: one that is shorter and can be placed behind the wheel and one that is longer and can be placed in front of the wheel, where obstructions currently dictate that the technician twists the neck when putting the chock in place.
Check car for test dates, certifications.	Technician walks up and down the line of railcars to verify the test dates and		

	certifications for each railcar before authorizing loading.		
Throw derailers and flags.	Technician puts two derailers on the tracks (one on each track) to prevent another train from entering the rack while cars are being loaded; one derailer is made of cast iron steel and the other is made of aluminum.	1. The force required to lift the cast iron steel derailer is 188 lb; biomechanical analysis for a 95th percentile man shows that the back compression upper limit is exceeded, and many lack sufficient hip, torso, or ankle strength to safely perform the task. 2. The derailer is located at ground level, which requires that the technician bend to access it, increasing the risk for lower back strain. The lever arm of the derailer is short, which makes the derailer seem heavier when lifting and makes it more difficult for the technician to get the hand under the derailer when lifting.	1. Both derailers should be made of a lighter material, such as aluminum. The force required to lift the aluminum derailer is much less (80 lb versus 188 lb). However, after performing a biomechanical analysis for a 95th percentile man lifting an aluminum derailer, it was noted that the compression force on the back still exceeded the design limit. So, if possible, the weight of the derailer should be reduced further. 2. Squatting rather than bending at the waist while throwing the derailer would help reduce the compression forces on the low back. Alternatively, adding a handle to the outside of the derailer and giving the technician a hook to grab onto the derailer will minimize the amount of bending the technician must do, thus decreasing the force on the back. This solution also solves the limited hand space/short lever arm problem.

continued

TABLE 7-13

Task Analysis and Recommendations of Loading Rack Operation *Continued*

Subtasks	*Description*	*Human Factors Concerns*	*Recommendations*
Throw loading arm out to car.	Technician pushes loading arm out from the rack to the car.		
Release ramp.	Some of the ramps are hydraulic, while others must be raised and lowered manually, which is accomplished using one hand.	The force required to raise/lower the lift manually is 146 lb; biomechanical analysis for a 95th percentile man shows that the strength upper limit is exceeded for the shoulder and ankle, and many lack the elbow strength to safely perform the task.	Install all air powered lifts. Implement and maintain a maintenance schedule for the ramps to keep them safe and operating efficiently.
Pull plug on dome cap.	Technician must pull the pin that keeps the dome cap closed.		
Open dome cap.	Technician must lift and open the dome cap off the railcar.	The force required to open the dome cap is approximately 90 lb; this exceeds the back compression design limit for a 95th percentile man and many lack sufficient torso, hip, or ankle strength required to safely perform the task.	Use spring assists in all dome caps. The force required to lift a spring-assisted dome cap was measured while on site. A biomechanical analysis was performed for a 95th percentile man and showed a significant decrease in back compression force. Also, implement a quality control program to ensure that the springs used are safe and effective.

7.7.3 Biomechanical Analyses

Biomechanical analyses, using the three-dimensional static strength prediction program (University of Michigan Center for Ergonomics) were performed on the ramp manipulation, derailer, and dome cap lifting tasks. All analyses were performed using a 95th percentile man as the subject, to be consistent with Mr. X's stature.

REVIEW QUESTIONS

1. What factors can affect sleep?
2. Define insomnia.
3. Fatigue is a state caused by a host of factors. Name the factors.
4. What individual differences may have impact on the shift worker?
5. Name and give examples of strategies for coping with shift work.
6. What organizational strategies can contribute to the worker's shift tolerance?
7. Name the sources of stress
8. How is stress measured?
9. What is the purpose of task analysis?
10. What are some benefits of a team-based approach?
11. How does a behavior-based safety program improve safety performance?

REFERENCES

Aanonsen, A. (1964) *Shift Work on Health.* Norwegian Monographs on Medical Sciences. Oslo: Universitets Forlaget. Cited in E. Grandjean, *Fitting the Task to the Man*. Fourth Ed. London: Taylor and Francis, 1988, pp. 217–230.

Aryee, S., and Tan, K. (1992) "Antecedents and Outcomes of Career Commitment." *Journal of Vocational Behavior* 40, pp. 288–305.

Barnes, R. M. (1936) "An Investigation of Some Hand Motions Used in Factory Work." Bulletin 6. Iowa City: University of Iowa, Studies in Engineering. Cited in E. Grandjean, *Fitting the Task to the Man*. Fourth Ed. London: Taylor and Francis, 1988, pp. 217–230.

Barnes, R. M. (1949) *Motion and Time Study*. Third Ed. New York: John Wiley. Cited in E. Grandjean, *Fitting the Task to the Man*. Fourth Ed. London: Taylor and Francis, pp. 217–230.

Benson, H. (1985) "The Relaxation Response." In A. Monat and R. S. Lazarus (eds.), *Stress and Coping: An Anthology*. Second Ed. New York: Columbia University Press, pp. 315–321.

Bhatia, N., and Murrell, K. F. H. (1969) "An Industrial Experiment in Organized Rest Pauses." *Human Factors* 11, pp. 167–174.

Buunk, B. P., and Janssen, P. P. M. (1992) "Relative Deprivation, Career Issues, and Mental Health among Men in Mid-Life." *Journal of Vocational Behavior* 40, pp. 338–350.

Deeb, J. M., Danz-Reece, M. E., and Smolar, T. J. (2000) "Industry Experience with Behavior-Based Safety Programs: A Survey." Proceedings of the International Ergonomics Association 2000 and Human Factors and Ergonomics Society 2000 Congress, San Diego, CA, vol. 2, pp. 213–215.

Drury, C. G. (1983) "Task Analysis Methods in Industry." *Applied Ergonomics* 14, no. 1, pp. 19–28.

Drury, C. G., Paramore, B., Van Cott, H. P., Grey, S. M., and Corlett, E. N. (1987) "Task Analysis." In G. Salvendy (ed.), *Handbook of Human Factors*. New York: John Wiley & Sons, pp. 370–401.

Folkand, S., and Monk, T. H. (1979) "Shift Work and Performance." *Human Factors* 21, pp. 483–492.

Grandjean, E. (1988) *Fitting the Task to the Man*. Fourth Ed. London: Taylor and Francis.

Haworth, N., and Heffernan, C. (1977) "Information for Development of an Educational Program to Reduce Fatigue-Related Truck Accidents." Report 4. Victoria, Australia: Monash University Accident Research Center.

Hildebrandt, G., Rohmert, W., and Rutenfranz, J. (1974) "Twelve and 24-Hour Rhythms in Error Frequency of Locomotive Drivers and the Influence of Tiredness." *International Journal of Chronobiology* 2, pp. 97–110.

Horne, J. A., Anderson, N. R., and Wilkinson, R. T. (1983) "Effects of Sleep Deprivation on Signal Detection Measures of Vigilance: Implications for Sleep Function." *Sleep* 6, pp. 347–358.

Ivancevich, J. M., and Matteson, M. T. (1988) "Promoting the Individual's Health and Well-Being." In C. L. Cooper and R. Payne (eds.), *Causes, Coping and Consequences of Stress at Work*. Chichester, UK: John Wiley & Sons, pp. 267–299.

Karasek, R.A. (1979) "Job Demands, Job Decision Latitude, and Mental Strain: Implications for Job Redesign." *Administrative Science Quarterly* 24, pp. 285–306.

Kelly, R. J., and Schneider, M. F. (1982) "The 12-Hour Shift Revisited: Recent Trends in the Electric Power Industry." *Journal of Human Ergology* 11 (supplement), pp. 369–384.

Kirton, M. (1994) *Adaptors and Innovators: Style of Creativity and Problem Solving.* London, UK: Routledge Publisher.

Kirwan, B., and Ainsworth, L. K. (1992) *A Guide to Task Analysis.* London: Taylor and Francis Publishers.

Lindstrom, K. (1991) "Well-Being and Computer-Mediated Work of Various Occupational Groups in Banking and Insurance." *International Journal of Human-Computer Interaction* 3, no. 4, pp. 339–361.

Matsumoto, K., Matsui, T., Kawamori, M., and Kogi, K. (1982) "Effects of Night Time Naps on Sleep Pattern of Shift Workers." *Journal of Human Ergology* 11 (supplement), pp. 279–289.

Monk, T. H., and Folkard, S. (1992) *Making Shift Work Tolerable.* London: Taylor and Francis.

National Transportation Safety Board (NTSB) (1995) *Factors That Affect Fatigue in Heavy Truck Accidents,* vol. 2. *Case Summaries.* NTSB number: ss-95/02. NTIS number:PB95–917002. Washington, DC.

Presser, H. B., and Cain, V. S. (1983) "Shift Work among Dual-Earner Couples with Children." *Science* 219, pp. 876–878.

Rasmussen, J. (1983) "Skill, Rules, Knowledge: Signals, Signs, and Symbols and Other Distinctions in Human Performance Models." *IEEE Transactions on Systems, Man, and Cybernetics* SMC-13, no. 3, pp. 257–267.

Rasmussen, J. (1986) *Information Processing and Human-Machine Interaction: An Approach to Cognitive Engineering.* New York: Elsevier.

Russell, J. E. A. (1991) "Career Development Interventions in Organizations." *Journal of Vocational Behavior* 38, pp. 237–287.

Rutenfranz, J., and Knauth, P. (1976) *Rhythmusphysiologie und Schichtarbeit in Schicht- und Nachtarbeit.* Vienna: Institut fur Gesellschafts Politik, Sensenverlag. Cited in E. Grandjean, *Fitting the Task to the Man.* Fourth Ed. London: Taylor and Francis, 1988, pp. 217–230.

Sergean, R. (1971) *Managing Shiftwork.* London: Gower Press, Industrial Society.

Smith, M. J., and Carayon, P. (1995) "New Technology, Automation and Work Organization: Stress Problems and Improved Technology Implementation Strategies." *International Journal of Human Factors in Manufacturing* 5, pp. 99–116.

Summala, H., and Mikkolo, L. (1994) "Fatal Accidents among Car and Truck Drivers: Effects of Fatigue, Age, and Alcohol Consumption." *Human Factors* 36, pp. 315–326.

Tepas, D. I. (1985) "Flexitime, Compressed Work Weeks, and Other Alternative Work Schedules." In S. Folkand and T. H. Monk (eds.), *Hours of Work.* New York: John Wiley & Sons.

Tepas, D. I. (1993) "Educational Program for Shift Workers, Their Families, and Prospective Shift Workers." *Ergonomics* 36, pp. 199–209.

Tepas, D. I., and Monk, T. H. (1987) "Work Schedules." In G. Salvendy (ed.), *Handbook of Human Factors.* New York: John Wiley & Sons, pp. 819–843.

Tepas, D. I., Paley, M. J., and Popkin, S. M. (1997) "Work Schedules and Sustained Performance." In G. Salvendy (ed.), *Handbook of Human Factors and Ergonomics.* Second Ed. New York: John Wiley & Sons, pp. 1021–1058.

Thiis-Evenson, E. (1958) "Shift Work and Health." *Industrial Medicine* 27, pp. 493–497.

Weddeburn, A. A. I. (1981) "How Important Are the Social Effects of Shift Work." In L. C. Johnson, D. I. Tepas, W. P. Colquhoun, and N. J. Colligan (eds.), *Biological Rhythms, Sleep, and Shiftwork.* New York: S.P. Medical and Scientific Books, pp. 257–269.

Yang, C. L., and Carayon, P. (1995) "Effects of Job Demands and Social Support on Worker Stress: A Study of VDT Users." *Behavior and Information Technology* 14(1), pp. 32–40.

8

Information Processing

8.1 HUMAN ERROR

8.1.1 Introduction

This chapter reviews the common classifications of human error and the typical types of errors operators make in the petrochemical industry, excerpted from a previous paper by Attwood (1998). Plant signs, procedures, and training are critical sources of information for personnel and should be designed and implemented to reduce the potential for human error. Design of displays, another key aspect of information processing, is covered in Chapter 5.

On February 15, 1982, 84 men drowned when the Ocean Ranger, a semisubmersible oil rig, sank during a storm in the Atlantic Ocean 150 miles east of St. John's, Newfoundland. It was the worst peacetime marine disaster in Canada's history. The learning gained from this incident demonstrates how vitally important it is to provide the information that the operators need to identify trouble, solve problems, and make critical decisions (Heising and Grenzebach, 1989).

A subsequent inquiry by the Canadian Royal Commission on the Ocean Ranger Marine Disaster (1984) found that the rig capsized and sank after water flooded the ballast control room through a porthole and short-circuited the control panel causing seawater to flood the forward ballast tanks and chain lockers. The porthole collapsed because it was not properly sealed. The porthole was used to read draft marks on the columns of the rig. The crew, according to reports, habitually left the "deadlights," which protect the portholes, unbolted. The portholes were not strong enough to withstand the conditions in the North Atlantic.

It was determined that several subsequent errors compounded the problem and led to the sinking. Faulty engineering, inadequate training, and poor procedures reportedly caused the errors. However, the porthole example makes the point that the capsizing would not have begun if the crew had not made an error by not bolting the deadlights, and of course, if the portholes could have been designed in the first place to allow the operators to view the draft marks without unbolting the deadlight. It is generally agreed that people do not intentionally make errors. With few exceptions, the design of the system makes people make errors. So, by designing systems that take into account the capabilities and limitations of humans, the potential for errors should be greatly reduced.

8.1.2 Why Humans Make Errors

When compared to machines, humans appear to be vastly inferior. We cannot process information as fast as computers. We cannot work as fast or as long as machines without getting tired. We cannot exert as much force as machines or exert it as accurately or as long. We have an advantage over machines with our abilities to reason and learn but not with our physical abilities to work quickly, to work for long hours without rest, or to lift heavy objects. Even so, research in human factors founded in experimental and applied psychology provides us the rationale for why even our exceptional abilities break down and we make errors. By understanding the limitations and capabilities of humans, we can design systems to take advantage of their strengths and minimize the impact of their weaknesses. Table 8-1 is organized into three columns. The first column postulates why humans make errors; the reasons are well grounded in experimentation and observation. The second column provides an example of when this limitation might appear in practice. The third column suggests how that system can be designed to eliminate or, at least, reduce the potential of the error being made.

8.1.3 Mental Errors

Table 8-1 illustrates several ways in which higher mental processes can lead to errors. Inattention, for example, can be due to the operator being challenged by conflicting demands for his or her limited attention. The

TABLE 8-1

Why Humans Make Errors and How to Minimize Error Potential

Why Humans Make Errors	*Conditions That Could Contribute to Potential Errors*	*Designing to Minimize Human Error*
Mental errors		
Inattention	Striking a limb on a sharp object or touching a steam line while performing maintenance on a piece of plant equipment	Minimize conflicting demands
	Forgetting to close, bleed the air, or pull the fuses, to disable a remote-operated valve (ROV)	Initiate redundant checks by personnel in critical situations
Forgetting information when distracted	Providing directions to an operator over a radio rather than in writing, then requiring him or her to remember the directions while distracted by other duties	Minimize distractions, interruptions Provide written checklists
Unable to retain enough information	Tags and labels consisting of long strings of random letters and numbers	Chunk information into groups; for example, the number 4162910474 has more meaning when displayed as (416) 291-0474 Develop logical associations between the information and the action
Unable to retain information long enough	Requiring operator to remember large quantities of information without rehearsal or a way of noting it for later reference	Design systems to retain information (visual or auditory) until no longer needed by the operator
System demands exceed human information processing ability	Control system requires operator to perform mathematical calculations to modify the process Operators are poor at consistently making error-free calculations	Design calculation into the process control system
Expectations conflict with reality	Inconsistency in the meaning of red and green between process and electrical systems causing opposite responses to the same color	Site standards should be consistent Provide another method of coding in addition to color
	Imposing North American design standards on cultures where the standards are different	Design to take advantage of human expectations, and culture

Important information is masked by nonimportant details	Many warning messages posted in clusters, critical information buried in the midst of nonimportant information	Highlight critical information Ensure critical information is obvious
Unable to recall information from long-term memory	Computer systems that require the operator to remember complex commands and command sequences Systems that require complex procedures to overcome poor design	Provide checklists, pull-down menus, help screens, set points Refresher training (using procedures) Design to eliminate the need for procedures or the use of long-term memory
Perform activities out of sequence or too slow	Procedures where the sequence of activities is critical to the success of the operation; restarting a compressor, for example, require the operator to perform a series of sequential activities. Missing one activity or performing it incorrectly could require starting the process from the beginning	Design so there is a logical or visible relationship between one activity and the next; for example, numbering valves in the order in which they are operated
Confused by information presentation	Operations that require the operator to perform some intermediate translation to the units required	Ensure that the equipment operates in the units that the operator understands
Environmental factors		
Fatigue	Work schedules that permit the buildup of sleep debt can cause the operator to respond more slowly and with less accuracy in both normal and upset conditions	Design proper shift schedules Minimize overtime, especially when working 12-hour shifts
High temperatures	High-energy activities in hot environments without the proper breaks for rest and water can reduce the ability to detect faults and reduce the accuracy of response, and increase potential for illness or injury.	Provide rest breaks and fluids
Systems factors		
Poor training or instruction (procedural errors)	Control room operators asked to work the board before they have the knowledge or experience required	Proper prerequisites for the job Formal and on-the-job training as appropriate Measurement systems that adequately test operator performance

continued

TABLE 8-1

Why Humans Make Errors and How to Minimize Error Potential *Continued*

Why Humans Make Errors	*Conditions That Could Contribute to Potential Errors*	*Designing to Minimize Human Error*
Information presentation		
Speed versus accuracy	Abnormal operations when the plant is starting up, shutting down, or upset	Design system for the constraints of the task. If speed is required, design the system to assist the operator, to provide better summary information, and to be more forgiving. If accuracy is required, allow sufficient time to perform the task
Inconsistent information presentation	Inconsistent symbols or coding can slow down decision making or cause an operator to mistake one symbol for another	Display similar information in a consistent manner. This includes symbols, color, and navigation
Poor lighting, contrast, or use of color	Signs and labels improperly designed for reading: Contrast is poor, characters too small, or lighting inadequate	Ensure the information contrasts with the background Ensure sufficient background luminance for the task Ensure characters large enough for the reading conditions
	Color used as the only form of coding rather than a redundant coding method	Use color only as a redundant coding method
Displays are confusing or hard to understand	Messages conveyed by signs lead to errors because of confusion or illegibility Directional signs confusing because of their location or design Control activation difficult to detect and interpret, such as the activation of electrical breakers might not be obvious visually The messages that accompany symbolic displays might not be clear or consistent Symbolic displays (e.g., pump designations on process control screens) might not be consistent from one application to the next	Ensure the message is consistent Ensure the information meets the objectives of the task Test the message for understanding before implementing Keep the message simple

	Displays may not convey the information that the operator requires to perform the job	
Displays grouped poorly	Displays may not be properly associated with their respective controls causing errors or delays in operation	Group controls and displays by function
	Navigation of the control panel or process control screen can be slowed or poor navigation lead to errors	Separate functional groups
	Poorly grouped displays cause inefficient operations and slow response in emergencies	Assign high-priority locations to controls and displays used frequently or for emergency
Control/display design not compatible human with movement capabilities	Pull-down menus are not designed efficiently	Assign display size and spacing information based on human capabilities for speed and accuracy
	Clutter on process control screens where commands are activated with pointers promote selection errors	Avoid screen clutter since it promotes targeting errors Ensure hierarchical menus are arranged so the most important information is the easiest to point to
Behavior		
Exhibiting inappropriate learned behavior	Driving too fast on plant roadways	Design a behavior intervention that • Fosters and reinforces the advantages of safe behavior • Eliminates the disadvantages of safe behavior • Influences the perception of "expected" consequences • Eliminates the advantages of unsafe behavior In this example, interventions that meet these principles include speed bumps, which eliminates the advantage of speeding, or a speed feedback system, which congratulates the driver for driving safely

operator may be bombarded with so much information that all of it may be filtered out, including the important item that should demand attention. In these situations, the error is caused when the operator does not attend to the task at hand. The solution may be to provide redundant reminders. For critical situations, such as ensuring that valves are closed or fuses pulled, use duplicate checks by fellow workers, and/or prioritize information that needs immediate attention, such as, restricting alarms to conditions that require operator action.

Failure in the operation of short- and long-term memory can also lead to error. Short-term memory suffers from two inadequacies: lack of storage capacity and sensitivity to distraction. Operators may therefore forget verbal communications or remember them incorrectly. In situations where the operator is required to remember items for later action, the items should be written and not verbal. Long-term memory, while exhibiting an incredible ability to remember items from years past, may not recall them accurately. As a result, operators make errors of omission, commission, or sequence. The solution is to provide operators with memory assistance in the form of checklists or pull-down or pop-up menus.

A final problem with higher mental processes is that the demands of the system can exceed the information processing capacity of operators. In these situations, the operator's responses to information slow down, or in an attempt to maintain speed of response, the operator makes errors. Either way, the system is compromised. The solutions include offloading the information processing requirements from the operator to the system or simplifying the way in which the information is presented to the operator.

8.1.4 Display Errors

The way we present information to the operator can determine how he or she acts on it. Coding that is inconsistent from one system to another can confuse the operator into slowing his or her responses or mistaking one symbol for another. This is especially the case in process control systems, where symbology is common in screen design. Poor lighting or contrast can affect the legibility of alphanumeric characters or symbols, thus increasing the potential for reading errors. Preventing poor legibility requires ensuring that the combination of background luminance, contrast, and character size are designed for the particular task the operator performs.

Some displays may be confusing to the operator or hard to understand. This is especially true of signs whose message has not been designed according to commonly accepted principles (beginning the message with an active verb, keeping the message simple, and using bullets if the message is long). Control and display labeling is another system that can be improperly designed, causing the operator to respond incorrectly. Grouping controls and displays is an effective method of ensuring that the operator properly associates his or her actions with the appropriate system response. Improper grouping is a common cause of operator error and one of the easiest to eliminate during design. Finally, clutter on display panels and process control screens can cause confusion and promote operator selection errors.

8.1.5 Environmental Causes

Higher mental processes are affected by extremes in most environmental variables. Extremes in temperature and relative humidity, for example, can reduce operators' attention to information and affect the speed and accuracy of their responses. Potential interventions include local air conditioning, such as in crane cabs or vehicles, or reducing the need for operator attention and decision making in extreme environments.

8.1.6 System Factors That Lead to Error

Fatigue affects the operator's ability to make decisions in much the same way as extreme temperatures. Fatigue is often the result of inadequate sleep caused by poorly designed work schedules. We know, for example, that sleep debt can build from excessive numbers of back-to-back night shifts or when day shifts start too early in the morning. (See

Chapters 4 and 7 for more detail on how environmental factors and shift work influence human performance.)

In some cases, operators are required to make decisions before they have the training or experience necessary to do so. This condition could be the result of improperly analyzing the tasks included in the training for their decision-making requirements or not considering the full extent of the tasks in the design of the training. In addition, the trainees may not be properly evaluated to ensure that they have achieved the prerequisite learning and skills.

The majority of human error in the oil and gas industry is caused by deficiencies in the design of the equipment, management systems, or work processes—so-called system-induced error (Center for Chemical Process Safety, 1994). Figure 8-1 shows a model, modified from Reason's (1990) accident causation model, for representing how multiple deficiencies can lead to an incident. In the modified model, management systems, work practices, and equipment design are represented as "shutters." Inadequacies in the systems, practices, and design are conditions that adversely affect human performance, represented in the model as holes in the shutters. The figure shows that a combination of inadequacies may coincide (represented by the holes in the row of shutters) and, if triggered by an event, lead to an incident (represented by the trajectory of the triggering event as it passes through the holes of the shutters).

Figure 8-1 illustrates this model using the example of the condensate explosion on the offshore rig, the Piper Alpha (Thomas et al., 2000; Cullen, 1990). When problems were encountered with the main condensate pumps (the triggering event), the operators switched back to the backup pumps. However, the work permit system failed to help identify concurrent mechanical and electrical work in progress on the backup pumps (a hole in work permit shutter). This inadequacy, along with gaps in procedure knowledge (holes in training and procedure shutters), and a communication breakdown between the shifts regarding the backup pump status (hole in communications shutter) were present simultaneously. The backup pumps were started with a relief valve missing and no protection against condensate ignition, leading to a catastrophic incident.

Identifying the situations where error can occur and eliminating them by implementing appropriate interventions can reduce the potential for human error. The error identification process can take many forms. Causes

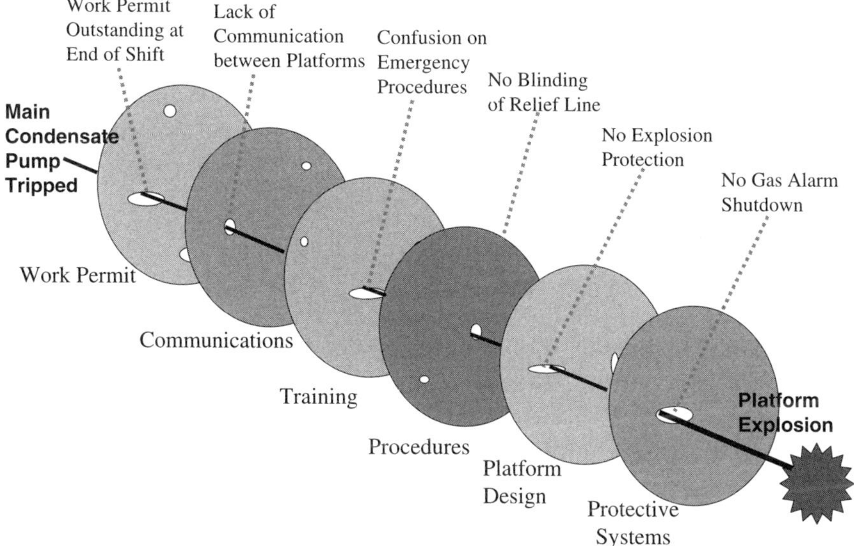

Figure 8-1 Example of the incident "shutter" model applied to a platform explosion. (Modified from *Human Error* by James Reason, 1990.)

of human errors can be identified through incident investigations, the use of human factors analysis tools, or conducting training and awareness programs that include an issue identification component. One example of a human factors analysis tool is a self-help tool used at operating sites (Pennycook and Danz-Reece, 1994; Noerager, Danz-Reece, and Pennycook, 1999; Thomas et al., 2000) to identify potential factors that could influence human performance. Through consensus, the site team rates statements about conditions at their site in 14 general categories for improvement:

1. New installation design.
2. Control room design.
3. Field workplaces.
4. Computer-based process control.
5. Computer-based information.

6. Job design and work planning.
7. Procedures.
8. Training.
9. Work group support.
10. Shift schedules.
11. Communication structure.
12. Employee involvement and feedback.
13. Safe work practices.
14. Reporting near misses.

The tool provides input on opportunities for improvement, based on the team priority ratings of aspects of the 14 categories. This tool was also used to help complete the human factors checklist for the safety case revision required every 3 years for U.K. offshore operations (U.K. Health and Safety Executive, 1999; Exxon Mobil, 2001).

Interventions are as diverse as the problems identified. They can include the redesign of operating equipment, software, work processes, and procedures or the implementation of a behavior-based safety program. It is essential that the introduction of a human error program be coordinated with other programs within the organization. The kernel of any human factors program is the company's management systems responsible for safety, health, productivity, and loss control. Each must be updated to reflect the importance of reducing human error, and each must be visibly supported at the highest levels of the organization.

This chapter includes the following information that is important to consider for preventing human error: plant signs and labels, operations, maintenance and emergency procedures, and skills and knowledge training.

8.2 PLANT SIGNS AND LABELS

Labels and signs are integral to operations in that they remind operators of the identity of equipment and key information, such as procedures or directions. Kletz (1998) refers to equipment labels as a type of personal protective equipment (PPE), meaning that they are an important feature for safety. To help prevent errors, it is critical that plant labels and signs be accurate, complete, consistent, meaningful, and legible. A *label* identifies plant equipment, components, or areas. A *sign* contains a message, such as simple instructions, procedural reminders, personal protective equipment requirements, or hazard information.

8.2.1 Equipment Labeling Program

A labeling program can ensure that labels and signs minimize the potential for error in processing information during operations and maintenance. To implement an equipment labeling program, follow these steps (U.S. Department of Energy, 1993):

1. Identify all equipment, components, and piping that require labeling for example:
   Cabinets
   Circuit Breakers
   Electrical Busses
   Electrical Distribution and Lighting Panels
   Emergency Equipment
      Eyewash Stations
      Fire Alarm Stations
   Equipment Components
      Motors
      Valves
      Pumps
      Etc.
   Equipment Systems and Subsystems (emergency generator, heating and air conditioning, etc)
   Exits, Evacuation Routes
   Facility Location Codes (buildings, units, floor/elevation, columns/rows)
   First Aid Equipment
   Floor Drains
   Fluid and Gas Lines (piping, ventilation ducts, etc.)

Fire Protection Systems and Equipment
Fuse Blocks or Fuse Locations
Gauges, Meters, and Indicators
Motor Control Centers
Protective Equipment
Room Doors (list major equipment inside hazardous or radiation areas)
Safety Devices
Storage Containers/Rooms/Spaces
Test Equipment, Special Tools

2. Check that the labels meet regulatory requirements.
3. Review procedures and check that the content of labels (e.g., names, abbreviations, and identification codes of equipment, components, plant areas, and systems) is consistent with other labels in the plant and those cross-referenced in procedures or control system displays.
4. Ensure that labels are in good condition. Ensure that the label material and the means of attachment are compatible with the components and environment where they are used.
5. Review that labels are properly placed and oriented to enhance readability and component.
6. Verify that labels are accurate.
7. Establish procedures to replace lost or damaged labels and acquire new labels when needed.
8. Create lists of abbreviations for the labeling program to ensure that abbreviations used on labels for systems and components are consistent with other site documentation, such as procedures or computer displays.

8.2.2 Designing Signs and Labels

The design of the sign or label influences how a person processes the information it contains. When designing the sign or label, consider the content of the message, the layout of the words or letters on the surface, the physical appearance of the label, and the location of the message. The recommendations for the content of a sign or label are similar to those for developing written procedures. More detail about the following guidelines can be found in a manual on how to write procedures by Wieringa, Moore, and Barnes (1993).

8.2.2.1 Content of the Message

Follow these guidelines for the content of the message.

1. Use short and positive phrases or messages. For example, instead of "Seal flush system must be blocked in after transfers are completed"
   Use
   "Block in seal flush system after completing transfers."
2. Avoid abbreviations. If they must be used, make sure they can be understood by the most-inexperienced user. Use only standardized abbreviations, and use them consistently. Any new abbreviations should be tested with inexperienced users to ensure that they can identify the full name associated with the abbreviation. Examples of system abbreviations are shown in Table 8-2.
3. For labels or signs that display procedures, break the procedures into short, sequential steps, each possessing a single "action" verb. Each step should be on a separate line.

Acceptable	*Unacceptable*
• Lift receiver • Press button on receiver • Wait for control room to answer • Speak clearly • Replace receiver when finished	To use phone lift receiver and press button. Wait for control room to answer before speaking. Be sure to speak clearly and replace receiver when finished.

4. Messages should be positive and active. Instead of this (16 words),
   "Do not close both inlet and outlet valves unless you open the drain or bypass."
   use this (12 words),

TABLE 8-2

Abbreviations for Plant Systems (DOE, 1993)

Arranged by System Name *System Name*	*Code*	*Code*	*Arranged by Code* *System Name*
Air, compressed	CA	CA	Air, compressed
Air, instrument	IA	CW	Water, cooling
Electrical, low voltage (<240 V)	ELLV	ELLV	Electrical, low voltage (<240 V)
Electrical, medium-voltage (>240 V, <4.16 kV)	ELMV	ELMV	Electrical, medium voltage (>240 V, <4.16 kV)
Heating, ventilation, and air conditioning	HVAC	HPS	Steam, high pressure
Steam, high pressure	HPS	HVAC	Heating, ventilation, and air conditioning
Waste, sanitary	SWR	IA	Air, instrument
Water, cooling	CW	SWR	Waste, sanitary

Source: U.S. Department of Energy (1993).

"Open drain or small bypass before closing both inlet and outlet valves."

5. Use symbols that users expect to see. Symbol signs should include a written message, if possible. Use familiar or standard symbols, if they already exist. New symbols should be tested with the operators to ensure that they understand them and the meaning of the symbol is intuitive to the viewer of the message. For a well designed symbol, at least 80% of those who see the symbol should be able to quickly, confidently, and correctly describe what the symbol means; that is, the symbol has the meaning they expect it to have. Symbols and pictographs are especially recommended for sites where multiple languages are spoken.

8.2.2.2 Laying out the Message

Many messages are often displayed on surfaces of limited size, dictated by the size of small equipment and components (e.g., width of instrument lines and small piping, buttons, and levers within an instrument case). There may be a temptation to squeeze the message onto the surface of label by rotating the text or writing the text around corners of the label. This makes the text difficult to read and could cause an error. All text should be oriented so that it can be viewed horizontally. Longer messages, such as instructions with multiple lines should be displayed on numbered lines and left justified. Shorter messages should be centered on the label or sign.

8.2.2.3 Appearance of the Characters

The style and size of the individual characters influence how legible the message is to the reader. If the message is more than three words, then the characters should be portrayed in mixed upper- and lowercase, i.e., capital and small letters. The characters should be written in simple and clear typeface, and the use of the italic typeface style should be minimized because it is difficult to read under suboptimal environmental conditions. The use of underlining for emphasis should also be minimized, if used at all.

8.2.2.4 Placement of the Sign or Label

To benefit from the content of a message on a label or sign, people must see the message. Therefore, the label or sign must be in the line of sight from where people stand if they work in the area. If a label or sign is placed out of the field of view, then it is likely that the label or sign would not be seen during a visual scan of the area and critical information contained on the sign may be missed. The sign or label should be clearly visible and unobstructed by equipment or shadows. The sign or label must be visible to all personnel; those that are very tall as well as those that are very short in height. It should be obvious to what equipment or area the sign or label refers. If needed, duplicate labels can be used, such as labeling both the equipment switches and the removable panel that covers them or when areas can be approached by personnel from more than one direction or path. Table 8-3 gives examples of typical features that decrease the visibility of characters on a posted sign or label.

TABLE 8-3

Features That Decrease the Visibility of Signs and Labels

Feature	*Example*
Character height too small	The appropriate height of a character is determined by the distance from which it will be viewed. The appropriate minimum character height for a sign that must viewed from 10 ft (3 m) is a minimum of 0.6 in. (1.6 cm). The rule of thumb is to divide the viewing distance by 200 to calculate the character height. However, the character may have to be larger to account for poor viewing conditions, such as dim light; if so, then divide the viewing distance by 120 to calculate the character height.
Inappropriate character style	Use only simple character styles. Italics should never be used in labels and signs posted around the site because italicized characters are difficult to read.
Poor color contrast	Select color combinations that have good contrast and correspond to the site color code (e.g., red background or borders for warnings). Recommended combinations follow, according to the legibility in white light (Eastman Kodak Company, 1987): Very good: Black on white; White on black Good: Yellow on black; Dark blue on white Fair: Red on white; Green on white; Red on yellow; Black on yellow
Incorrect use of upper- and lowercase	Write instructional or informational signs that contain more than three lines of text in mixed upper- and lowercase. Use uppercase (all capital letters) for labels such as equipment identifiers.
Poorly lit or in shadow	Viewing displays under dim light reduces the color contrast, resulting in decreased visibility of the characters. Colors (and therefore the characters) are washed out by glare or lighting that is too bright. Lighting should be improved so that displays designed with appropriate character sizes and colors are clearly visible under day and night viewing conditions. For example, ensure exit routes clearly identifiable in both day and night operations.

8.2.3 Guidelines for Specific Types of Signs and Labels

The following guidelines apply to specific types of signs and labels found in a petrochemical plant.

8.2.3.1 Pipe Labeling (U.S. Department of Energy, 1993)

- Content should be identified with human-readable labels, such as "Acid Gas" not "MTG-601."
- Orient text horizontally so it is read from left to right. If text must be placed vertically, orient it to be read from top to bottom.
- Indicate flow direction with arrows that point away from the text identifying the contents.
- Color must be supplemented with text labels (such as text label with red colored text indicating flammable substance).
- When two or more pipes lay side by side, position labels at same point.
- Locations to label a pipe: intervals along the pipe, at a pipe rack intersection, where the pipe passes through a wall, and at road crossings.

8.2.3.2 Electrical Wire and Cables Labeling

- Use durable materials.
- Labels should be written with ink that is waterproof, smear resistant, and impervious to solvents.

8.2.3.3 Equipment Labels

- Provide only necessary information on the label.
- Ensure labels are consistent and understandable.
- Use durable materials.
- Locate for visibility.

- Provide two pieces of information: functional description and equipment identification (tag) number (corresponding to the process and instrument diagram numbers). The functional description should be specific to the component and "human readable"; that is, not only a series of alphanumeric characters. The functional description should be placed above the equipment identification number. For example,

Glycol Reboiler Steam Isolation N350-05

- If sets of equipment are identical but on separate process trains, emphasize the difference in the code and on the label. For example,

Train **1** Glycol Reboiler Steam Isolation N350-**1-05**

Train 2 Glycol Reboiler Steam Isolation S350-**2-06**

8.2.3.4 Equipment Signs

- Provide information for safety and performance of critical equipment.
- Keep information short and simple.
- Use illustrations or schematics to clarify procedures.
- Utilize procedure design principles to create the signs. Outline the procedure with simple sentences and number each sentence. Left justify the sentences. For nonprocedural information, use line spaces or indentation to break up or separate the information.
- Mount the sign to enable the operator to view it from the position where the work is performed. The sign should be in a position so that the person looks directly at it while using the equipment or while controlling or monitoring the equipment.

8.2.3.5 *Sampling Points*

- Label sample bottle connections.
- If the sampling sequence is complex, provide a graphic schematic that explains the steps with labels that match posted labels in the plant.

8.2.3.6 *Information Signs*

Information signs are those that give emergency or critical information, such as regulatory information.

- For critical information, display the message redundantly; that is, use two types of coding to convey the importance of the message. For example, use combinations of text, symbols, and color to convey the message.
- Information signs usually have signal words, such as *First Aid*, *Lifeboats*, *How to shut down this mixer*. The signal words should be centered and in capital letters. The phrases that follow the signal words should be left-justified and typed in mixed-case letters. Again, use procedures design recommendations for longer instructions, and if symbols are used, use only symbols that are readily recognizable by the plant personnel.
- The number of signs needed and location for each sign is determined by the position of the worker in the plant area and the approaches that the worker would take in getting to the area. Provide as many duplicate signs as there are directions of approach. Mount signs at between 51 and 63 in. (0.13 and 0.16 m) above the floor.
- Mount directional signs in well-traveled aisles and areas.

8.2.3.7 *Hazard Signs*

Hazard signs must include a signal word, such as *DANGER* or *CAUTION*, and a brief description of the hazard, such as *hot surface* or *makes wide right turns*. Although the signal words *Danger*, *Caution*, and *Warning* may intend to mean three levels of hazard, studies show that many people do not distinguish Caution from Warning (Rogers, Lamson, and Rousseau, 2000). Hazard sign format and content may be regulated, as by the U.S. OSHA, and if so these regulations supercede the following recommendations:

- Hazard signs may also include a description on how to prevent the hazard from occurring, such as "don't follow at less than 40 feet," "or hearing protection required."
- Hazard signs may also describe the consequences of the hazard, "may cause explosion" or "may cause skin burns." Text messages should be structured into short, concise statements using active verbs.
- Locate the signal words in the uppermost portion of the sign. A hazard sign should contain the signal words, a description of the hazard, a description of the possible consequences that could happen if the hazard is not heeded, and instructions on how to avoid the hazard (Sanders and McCormick, 1993). The recommended layout for a hazard sign is

SIGNAL WORDS
Hazard Description
Consequences, Preventative Measures

- The text on a hazard sign should be a bold sans serif font. The signal words should be in capital letters twice the height of the remaining text. The signal words should be readable in suboptimal conditions from at least 5 ft (1.5 m). All text should be human readable. Codes should not be used in place of text.
- Ensure symbol use is consistent and understandable. Signs that may be read by multilingual or illiterate persons should use symbols in addition to written text. DANGER and CAUTION signs should also use symbols as well as written text.
- Colors must be used consistently. For example in the United States OSHA requires that the word *DANGER* be capitalized in white on a red oval background, and the remaining nonsignal words be black on a white background. Caution signs should have the signal word in

yellow letters on a black rectangular background, with the remaining nonsignal text in black letters on a yellow background.

- Hazard signs should be made of durable materials and mounted for visibility.

8.3 PROCEDURES

Procedures are a core part of every process operation. There are three types of procedures: operating, emergency, and maintenance. Operating procedures are written, step-by-step instructions and associated information (cautions, warnings, notes, etc.) for safely performing a task within operating limits. Emergency operating procedures are written instructions that address actions to place a process in a safe, stable mode following a system upset. Maintenance procedures are written instructions that address practices needed to ensure system operability and integrity as well as maintenance, testing, and inspection frequency.

Procedures are important because they provide rules to be followed and standardized records of safe and approved operations and maintenance practices. They provide consistent information across the plant and help minimize guesswork, leading to more efficient and safe operations. Certain aspects of procedures may be locally regulated. For example, the U.S. OSHA 29CFR 1910.119 regulation on the Process Safety Management of Highly Hazardous Chemicals requires that all safety critical work must have procedures that specify

- Steps for each operating phase including initial start-up, normal operations, temporary operations, emergency shutdown, normal shutdown and start-up following a turnaround or after an emergency shutdown;
- Consensus deviation from operating limits, and steps required to address deviations; and
- Safety and health considerations of: chemical hazards, precautions to prevent exposure, quality control of raw materials, and hazardous chemical inventory controls.

In addition, the 1910.119 regulation requires that the procedures be clearly written and accurate, include process safety information, be readily accessible to workers, and be kept current. This regulation also requires safe work practices for lockout/tagout, opening process equipment, and control of entrance into a facility.

Effective procedures are usually the single most important tool for improving safety, reducing environmental impact, and ensuring product quality. However, studies of major accidents in the nuclear industry show that 60% of incidents related to human performance were due to ineffective, incorrect, or lack of an appropriate procedure (Reason, 1998). Discussions with petrochemical plant operators about why procedures may not be followed indicate a variety of reasons implicating the nature of the procedures: the procedures were not available or readily accessible, they conflict with actual operations, they are outdated, or they are confusing. There are also reasons that implicate the operators' behaviors; for example, the operators are in a hurry, they are used to doing the job the "old" way, they feel they are too experienced to refer to a procedure, they feel it is acceptable to deviate from a procedure, or they are unaware that a procedure exists. These are barriers to achieving use of procedures and must be identified and overcome to achieve compliance with procedures.

The design and access to electronic documents are as critical as the design and access to paper procedures. Electronic documentation can ensure that the most current version of a procedure is available. Maintaining hard copies of the latest procedures in binders is difficult and may result in older procedures being used inadvertently. The guiding principle behind each procedure and the ultimate goal of all procedures is to increase human performance and reliability while reducing human error. Opportunities to improve procedures to reduce human error associated with procedures include determining when a procedure is needed, developing the procedure, determining why procedures may not be used, and evaluating the procedure.

8.3.1 Guidelines for Determining When a Procedure Is Needed

Procedures may be regulated (e.g., as described (previously) and if so, regulations supersede the following recommendations. A clearly defined process that identifies the tasks that need procedures should be in place. Haas (1999) recommends that the need for procedures is best determined by (1) complexity of the tasks, (2) the frequency that the tasks are executed, and (3) the consequences of possible errors made while executing the job tasks. Tasks with high complexity have nine or more procedural

TABLE 8-4

Guidelines for When a Procedure Is Needed

Task Characteristic	*Procedure, Checklist, or Sign-off Steps Needed*	*Procedure Available for Reference or Review*	*Procedure Not Needed, Learned in Training*
Severity of consequence if error is made	Moderate to high (e.g., injury, process delay, equipment damage)	Moderate (some impact on process or safety)	Low (no impact on process or safety)
Complexity	Moderate to high (more than 9 procedure steps, quick decisions)	Moderate (5 to 9 procedure steps)	Simple (less than 5 steps)
Frequency	Infrequent (less than once per month) to frequent (weekly)	Infrequent to frequent	Very frequent (multiple times per week)

steps and require mental calculations and quick decisions. Tasks with low complexity are simple and require fewer than five steps. Frequent tasks are performed at least once per week, and infrequent tasks are performed less than once per month.

Procedures are needed if the task is undertaken infrequently, is relatively complex, and poses at least a moderate consequence of error. For critical tasks, the procedure may include a checklist or "check-off" of key procedural steps. Reference procedures should be available if the task is relatively infrequent, of moderate complexity, and the consequences are moderate. In this case, the operator should have the option of reviewing the reference documents or referring to the written procedure before performing the procedure. If the procedure is performed frequently, is not complex, and errors are not of significant consequence, then written procedures may not be needed. These guidelines are summarized in Table 8-4, based on the framework by Haas (1999).

8.3.2 Developing Procedures

The method of developing the procedures should ensure that the procedure reflects how the work is done. The users of the procedures should participate in developing them. To effectively develop procedures, the developers need training on how to develop procedures, and the effectiveness of the training should be verified by assessing the procedures that are developed.

Users should be able to readily identify the procedure for the job, and the procedures should be available at convenient locations. A change process should be implemented to track procedures that need modification and updates.

8.3.3 Formatting Written Procedures

A good format of the procedures supports the operator using them and helps prevent errors. One source of detailed information is a guide on writing procedures for nuclear power plant operations (Wieringa et al., 1993). The site should have a style guide that specifies the format of the procedures so that they are styled consistently. Important aspects of procedure format to address in the style guide include

- The format of the procedure should emphasize the steps required to execute the procedure. A narrative style format (consisting of full sentences in prose style similar to how this book is written) is inappropriate for written procedures. A columnar format that clearly shows distinct procedural steps, or groups of steps related to a subtask, is recommended. An example of this format is shown in the end of this chapter case study on how to write a procedure to change a tire.
- A flowchart or decision tree diagram format is useful for procedures that guide troubleshooting.
- Each step in a procedure should be an action described concisely.
- Warnings should be clearly distinguished in the text and must be written before the step with which they are associated.
- Notes for supplemental information should be clearly distinguishable from the procedural action steps.
- Simpler is usually better: Use sans serif fonts and, for emphasis, use either bold or underline style but not italics. Italics can be difficult to read in situations such as low light or shadowy areas.

- Use 12 point font size for procedures in a book or electronic management system. If the procedure is posted as a sign or viewed from a distance, follow the guidelines for character size found in section 8.2 of this chapter to ensure that the characters are readable from a distance.
- Use a simple numbering scheme—two levels are usually sufficient, such as

  (1)
    (1.1)
    (1.2)

  But three levels could be used for complex procedures:

  (2)
    (2.1)
      (2.1.1)
- Each page should contain the procedure identification; that is, title, number, revision or version, date, and page number.
- Consider including job aids with the procedure, which can be pulled for reference and used by the operator out in the plant; for example, checklists, look-up tables, troubleshooting flowchart, or diagrams.

8.3.4 Determining Why a Procedure Was Not Used

An indicator that procedures are well implemented is that employees routinely follow the procedures as they are written. If procedures are not used, then there should be a process to consider why they are not used. This process is similar to an incident investigation process that searches for a root cause by asking "why" in a series of questions.

The questions to ask to determine why a procedure was not used include

- Do conditions in the workplace ensure that procedures can be carried out?
- Is the equipment necessary to comply with procedures readily available?
- Do users routinely follow written procedures (irrespective of skill and experience)?

- Are workers instructed on when to use procedures and how?
- Is enough time available to perform the task according to the procedures?
- Are the written procedures out of date?
- Does the employee feel that his or her method is better, safer, or faster?
- Does the employee understand the consequences of deviating from the procedure?
- Does the culture support "hurry up" or "cutting corners"?
- Does the culture support taking additional risk at the expense of following procedures?
- Has the example of not following procedures been set by the unit supervisor?
- Was the employee rewarded for deviating from procedure?

If the answer to any of these questions is yes, then the improvements in the procedures management system or other management systems, such as training or safety, may be needed.

8.3.5 Evaluating Written Procedures

Procedures can be evaluated with one or a combination of a few methods: observing operators perform the procedure, evaluating the written procedure with a checklist, or conducting a hazard analysis. One way is to observe multiple operators performing the procedure and compare the way they perform the procedure to the documented procedure to uncover deviations. Then, the operators can be interviewed with the preceding "why" questions to determine the cause of the deviations and help identify if the procedure, operators' behaviors, or management systems require improvements.

In addition to this observation, the written procedure should be evaluated based on the following criteria:

- Success in directing the users to accomplish the objective.
- User-friendliness.
- Accuracy.
- Conformation to appropriate standards.
- Consistency in format and layout so users can quickly find information.
- Appropriate level of detail.
- Reference to equipment labels, identification, or names that accurately correspond with actual hardware and other documentation.

The checklist in the Appendix contains the detailed questions that can be used to evaluate written procedures.

8.4 TRAINING

Training is a primary source of information for employees to improve their job skills. The effectiveness of the training, along with practice, can help reduce errors by building the employee's expertise on how to execute procedures. The objective of every training program should be to maximize skills and knowledge and achieve consistency in how procedures are done, while minimizing the time that the trainees require for learning the skills, knowledge, and procedures. Many good books, manuals, and experts are available on how to develop training and how to manage training systems (e.g., Goldstein, 1993). This section offers some highlights from a human factors perspective on training aspects that can help reduce the potential for human error.

An effective training system employs human factors principles that ensure the training program is relevant to the tasks performed, provides workers with meaningful skills and knowledge, and measures training effectiveness. Without enough practice and skills training, workers are unable to progress from novice to expert performance.

8.4.1 Developing Training

An effective way to develop training is to follow these steps (illustrated in Figure 8-2).

1. Analyze the tasks that need to be learned.
2. Identify the knowledge and skills required to carry out the tasks.
3. Select candidates that need the training.
4. Develop course objectives.
5. Develop the test criteria that must be met to demonstrate the material has been learned.

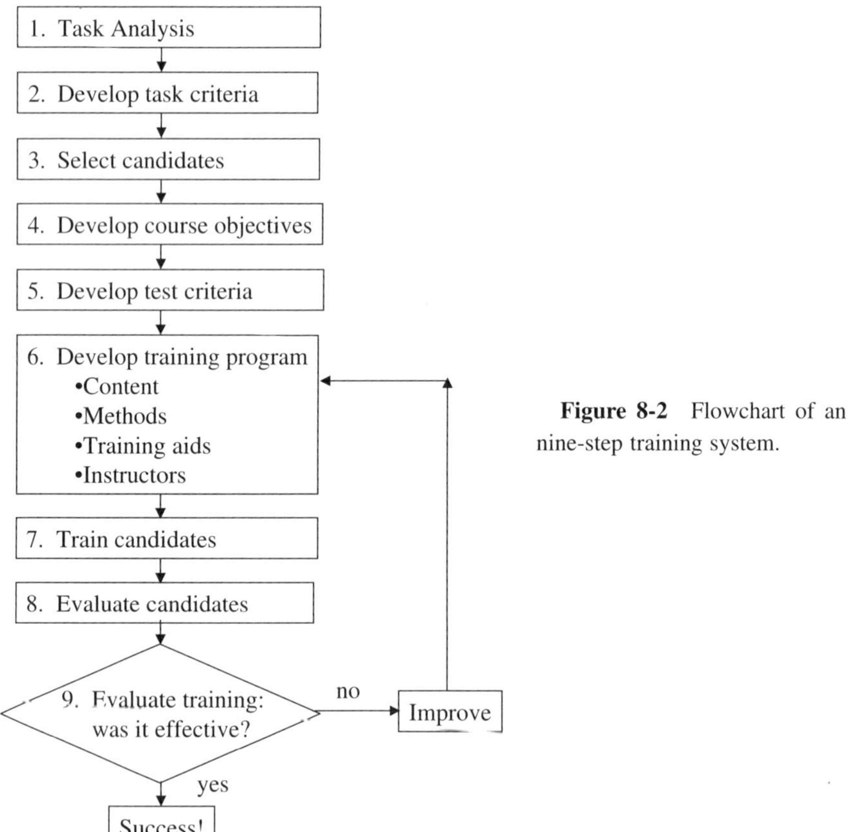

Figure 8-2 Flowchart of an nine-step training system.

6. Develop training resources (content, methods, training aids, instructors).

7. Train candidates.

8. Evaluate candidates. (Give the trainees a test.)

9. Evaluate the training. Ask, Was the training effective? If not, retrain the candidates, improve or redesign the training, or both.

Gordon (1994) describes phases for designing a training program: needs assessment and functional specification, design and development, then evaluation. The needs assessment consists of acquiring information about the needs of the training candidates, such as the types of tasks they perform and the types of knowledge and skills they need to acquire to perform the tasks well. Based on this information, the functional design specification is written. The functional design specification describes the goals, scope, and content of the training program and the constraints, such as resources and timing. This document guides development of the training content and materials. Next, the training package is developed based on the functional specification. The knowledge and skills targeted in the training are identified by analyzing the tasks that the employee must perform. This helps the training program content be directly relevant to the employee's job tasks.

8.4.2 Task Analysis for Training Development

Task analysis is a complete assessment of the tasks (or potential tasks) performed by the employee as part of the job or duties under analysis. Task analyses done with the purpose of training development involve breaking down a job into tasks and subtasks. In addition to providing a basis for the training content, there is also potential that task analyses can provide insight on improvements in equipment, tools, or procedures that could improve employee performance of the task. If a task analysis is used to develop the procedure, then this analysis can be built on for training by specifying the inputs and outputs of task activity and the conditions for performance. Inputs are what the worker needs to know to perform the task step, such as knowledge, skill with instruments, and procedure steps. The outputs are what the worker should achieve when executing the task, such as complete the procedure to calibrate a recorder or remove and replace a pump. Examples of conditions of performance are the time in which the person can do the task, number of errors, or accuracy in executing the procedure. More detail on how to do a task analysis is in Chapter 7.

Gordon (1994) recommends that a technique called *cognitive task analysis* be used to develop the content of the training for jobs where workers must make complex decisions and solve difficult problems, such as the duties of a control room operator. Cognitive task analysis is used to define the knowledge and concepts the employee must understand to perform a job (comprising multiple tasks and procedures). For example, control room operators should understand how one process variable influences another and be trained to identify potential mistakes, consequences of mistakes, and recovery points in the process.

8.4.3 Content of a Training Package

"Training" can consist of one or a combination of different tools that facilitate the learning process. There are basically two types of tools, those that support the trainer in teaching subject matter (e.g., courses, lectures, procedures documentation, manuals, videos, look-up tables) and those that provide hands-on experience (paper and pencil exercises, group problem-solving exercises, use of mock-up equipment, and task simulations) (Chapanis, 1996). Trainees need opportunities to apprentice; that is, apply classroom training to actual operations. The training and procedures systems should be closely linked. Training not only provides knowledge about procedures and when to use them but a knowledge foundation on which to base decisions and discretion, since it is not possible to cover every situation with procedures, instructions, and rules (Kletz, 1991). Training programs should include the specific procedures used on the job and be cross-referenced to the current procedures as a basis for teaching the methods and techniques to be used at the site. Training pace is important because information should be given in stages to avoid overloading the trainees with more information than they can assimilate and apply. Prototype training materials, such as job aids that support written procedures, should be tested with a group of trainees to get their feedback and gauge the effect of the training on their skills and knowledge before widely rolling out the course or program (Gordon, 1994). In some cases, several alternative design concepts (e.g., computerized self-paced instruction versus classroom lecture versus progressive workbook exercises) might be generated for preliminary testing and evaluation. Testing the prototype program on the users is the best way to obtain feedback, because the developer is probably too close to the program and cannot see the training from the learner's perspective. Even if the training is relatively simple, such as training on how to use a checklist, the training should be tested for effectiveness.

8.4.4 Training for Trainers

Trainers should be taught how to train effectively and efficiently. They should have relevant experience in the topic areas for which they are providing the training. Trainers should be aware of cultural differences and how to accomodate them, such as variations in willingness to ask questions publicly. On-the-job or hands-on training should be conducted by personnel with relevant operational experience and a mentor or supervisor should observe the personnel doing the tasks to check their proficiency. The training should be consistent, even when different trainers deliver the training.

8.4.5 When to Provide Training

Providing training in a timely fashion and providing workers the opportunity to expand on their skill sets may help reduce the probability of a human error related to poorly timed training programs. The need for training and the timing of training is critical when

- New employees lack the requisite knowledge and skills to perform a job.
- Workers are failing to perform some aspect their job at a satisfactory level.

- A new job (or job task) has been created and no workers already know how to accomplish the job (or job task).
- New equipment or new procedures are introduced on site.

8.4.6 Evaluating Training

Many trainers solicit feedback from the trainees after a training session by asking if the training was useful, met its objectives, or if some areas need improvement. Although this feedback is important, the effectiveness of training should also be objectively measured to identify aspects that need improvement (National Institute of Occupational Safety and Health, 1988).

Aspects of the training system that should be evaluated are

- Relevance of the training to the work tasks performed.
- Effectiveness of the training materials.
- Measurement to determine if the training system meets its objectives—use of testing to confirm levels of competence defined by the training program or assessment of on-the-job use of skills and knowledge gained through training.

These techniques ensure that training programs are effective in teaching workers the knowledge and skills they need on the job and help reduce error.

8.5 VIGILANCE

Human performance is optimal when we are neither too busy nor too bored. Although the potential for work overload is a common concern, work underload can result in loss of attention and alertness, potentially damaging performance. Jobs with tasks that consist of monotonous, boring work are not only nonmotivating but may be associated with unreliable performance. In these cases, the workload could be increased by supplementing or replacing monotonous tasks with other meaningful (purposeful) work tasks with appropriate skills requirements. Where possible, different tasks can be assigned to prevent boredom and create a change of pace in activity to prevent fatigue due repetition. The goal is to maximize a worker's involvement in the task where appropriate while preventing worker overload and underload.

Human perception allows us to cope with the wide range of inputs that impinge on our senses and discriminate among events. Our attention focuses on those stimuli that

- Are the largest in terms of intensity, size, and time, such as an alerting noise.
- Differ from background "noise," or are novel, such as white letters among a screen full of green letters.
- Involve movement, such as flashing lights.
- Fulfill a need, such as preventing an injury.
- Are of particular interest to the individual, such as hearing one's name in a conversation.
- Conform to expectancy.

Perceptual performance is determined by our ability to focus and maintain attention on the incoming events. The ability to sustain attention is what we call *vigilance*. In vigilance tasks, attention is directed to one or more sources of information over long unbroken periods of time to detect small changes in the information being presented (Davies and Parasuraman, 1982). In the petrochemical industry, the types of tasks that require sustained attention occur in transportation systems, control room operations, mining operations, and driving performance.

In some operations, sustained attention is critical. The difference between critical and important varies with the likelihood of an incident occurring and of the consequence, once it happens. Few would argue that the consequences of reduced attention are critical in chemical and refining process operations, where reactions can quickly get out of hand and the consequences can often bed catastrophic. Accidents during the transport of people and materials can also be life threatening and result in substantial losses. In some cases, accidents can have severe third-party liability, such as a gasoline truck rupture and explosion.

8.5.1 Transportation Systems

In transportation systems, the ability to sustain attention relies on many internal factors. Two of the most important determinants of vigilant behavior are the "circadian state" of the operator and the degree of sleep deprivation (both discussed in Chapter 7).

Many researchers studied the effects of circadian rhythm and the lack of sleep on a person's ability to perform in a vigilance-like situation. D'Amico, Kaufman, and Saxe (1986) employed a full-sized ship simulator to assess the ability of ship's watchkeepers to detect other ships on a collision course with their own. The watches were conducted around the clock to examine the effects of time of day on performance. In addition, half the watchkeepers were deprived of sleep before going on watch. Results indicated that the time required to detect another ship on radar increased with the time on watch. In addition, it was observed that the vigilance decrement was exaggerated among those watchkeepers who were deprived of sleep.

In a similar study, Donderi (1994) observed the performance of watchkeepers during a visual detection task at sea. Watchkeepers on search and rescue vessels were required to detect life rafts tethered to buoys at fixed locations within a 375 square nautical mile area of the coast of Nova Scotia. The search task was performed both day and night. Results showed that visual detection was better during the day than at night, as would be expected. The results also showed that performance on both day and night watches exhibited the classic vigilance decrement by deteriorating after the first half-hour on watch.

8.5.2 Control Room Operations

As the complexity of automated systems increases, the function of the operator changes from an active controller to a passive monitor who detects malfunctions or emergency conditions and executes decisions.

Failures of vigilance among process control operators may constitute an increasingly serious problem as control systems become more automated.

On the other hand, Wickens (1984) argues that the process control task is not a typical vigilance task and identifies two characteristics of the process control operation that distinguish it from typical laboratory vigilance paradigms:

1. In process control, the operator does not wait passively to detect process control system failures but is intermittently engaged in control adjustment activities and the like. These activities maintain at least a moderate level of arousal (alertness).

2. When a failure occurs, it is normally indicated by a visual or auditory alarm sufficiently intense to call attention to itself.

Failure alerting devices, such as alarms and annunciators, are designed to be tolerant to system variations and not to trigger false alarms. So, the operator is expected to monitor trends on process variables and possibly take action before the alarm occurs. This, then, is the vigilance task facing the operator. Process control displays need to be designed to assist the trend detection requirements of operators. One way to improve trend detection is to make the displays more unique and attention getting. Moray (1981) suggests changing the units of displays from absolute physical units to those scaled in terms of their probability of normal value. When all process variables are "normal," each display looks the same. However, when a process variable trends stray out of limits, the display becomes different from the others and draws the operator's attention. In summary, one of the vigilance challenges in the control room is to design displays that attract operator attention to impending out-of-limit situations, without overloading the operator with information. The degree of automation and the trust engendered by the operator can determine the level of vigilance. A system that produces excessive false alarms can cause a voluntary vigilance decrement, whereby the operator pays less attention to what are felt to be unreliable signals. And, this reduces the probability of detecting real system faults. On the other hand, operators may place too much trust in the capability of the automated systems. This false sense of security can lull the operator into complacency, so he or she does not make the needed checks. Danaher (1980) recounts an aviation incident where an air traffic controller noticed the potential intersection of two jetliners but elected to trust the ATC system to decide whether they would collide. As it turned out, collision was avoided when another controller decided that collision was imminent and alerted the pilots.

8.5.3 Mining Operations

Inattention is a contributing factor in off-highway driving incidents. Hulbert, Dompe, and Eirls (1982) report that driver inattention is a major cause of surface mining haulage truck accidents. They define *inattention* as a "state of mind in which the driver is unable to respond properly to an unexpected situation or is unable to make steering corrections in time to maintain his truck on the correct pathway."

Surface haulage driving is not the only mining task that involves vigilance. Other tasks or occupations that required mine workers to maintain vigilance are tasks that are prolonged and continuous, lasting for 30 minutes or more; tasks that tend to be boring and monotonous; and tasks that are seldom disrupted.

Hudack and Duchon (1989) found that over one-third of the surface mining jobs require extreme to high levels of vigilance to adequately perform the job. These include operating bulldozers, dragline and power shovels, forklifts, front-end loaders, haulage trucks, mill machines, and scrapers. These occupations employed about 51% of the approximately 213,000 people who worked in U.S. surface mines in 1986.

Accident statistics reported for mining in 1986 showed that those employed in high-vigilance occupations had twice the accident severity level as those in low-vigilance occupations. When the statistics were corrected by omitting those occurring during low-vigilance activities, the occupations most involved with high-severity accidents are operators of haulage trucks, front-end loaders, and scrapers. Several reasons may contribute to this result, including unfamiliarity with repair and maintenance routines and overexposure to environmental stressors, such as noise and vibration.

8.5.4 Driving Performance

Driving tasks range from maneuvering forklift trucks within warehouses to guiding 250 ton surface haulers at mine sites. To some extent, each of these driving tasks requires sustained attention to some condition that, without constant monitoring, can cause an incident.

In the United States, lapses in driver attention have been identified as a significant contributing factor in as many as 90% of traffic accidents (Bishop et al., 1985). Inattention in these cases has been broadly defined as an attention state where the driver fails to respond to a critical condition. In 38% of all accidents, drivers failed to take any action. In many of these cases, the driver may not have taken action because of drowsiness, physical fatigue, drugs, or the like. But, when all the obvious causes are removed, the data indicate that about 3% were truly due to inattention.

8.5.6 Factors Contributing to Vigilance Decrement

Several factors can contribute to the vigilance decrement: alcohol, drugs, stress, and environmental factors. Driving and industrial tasks require a number of essential skills, including sensory skills, motor ability, perception, and cognition. Research over the past half-century has shown that each of these skills is affected to some extent by alcohol intoxication.

Marijuana affects the functions of the human central nervous system that are important for performance in complex systems (Moskowitz, 1985). Time estimation can be impaired as well as speed of reactions in situations involving complex information processing. Marijuana has also been shown to affect the performance of drivers and machinery operators. In practical terms, the research suggests that, under the effects of marijuana, operators of complex systems would likely miss critical signals entirely or their responses would be delayed for an unacceptable period of time.

Koelaga (1993) reviews 54 studies that investigated the effects of four stimulant drugs—amphetamine, methylphenidate, caffeine, and nicotine—on vigilance performance. In general, caffeine and nicotine showed no effects on performance. Consult the original paper for more details on prescription drugs.

Koelega and Brinkman (1986) conducted a detailed review of the vigilance literature over the past 50 years, looking for results that would demonstrate an effect of moderate levels of noise on vigilance performance. They concluded, based on the results from almost 100 studies, that the effects of noise were equivocal. Noise has been shown to have detrimental, beneficial, and no effects on vigilance performance. While moderate levels of infrequent, irregular noise can benefit signal detection performance, high levels of noise may actually be detrimental (Landstrum, 1990).

Ray (1991) suggests that exercise generally improves monitoring performance and recommends interjecting exercise breaks for personnel engaged in vigilance-like tasks, such as process control or inspection.

Based on a study of the effects of temperature on vigilance, Norin and Wyon (1992), recommend that exercise may improve monitoring performance (i.e., delay the onset of the vigilance decrement), and climate control systems in vehicles that travel as little as a half-hour without interruption should be considered a safety requirement.

8.5.7 Operator Workload Analysis

Several factors contribute to operator workload: design of the displays and alarms, dynamic process characteristics, fitness for duty, environmental factors, control room design, operator interaction with personnel, and operator training and experience (Connelly, 1995). Because of the many contributing factors, using the number of control loops alone is not a good indicator for determining the span of control and, therefore, also is a poor basis for allocating operator workload. Studies show that measures of mental workload (for example, subjective measures, such as operator perception of difficulty, and performance measures, such as number of control moves or alarms per hour) correlate poorly with the number of control loops (Connelly, 1995).

Operator workload can be analyzed with many methods, for example, those that involve the subjective ratings by the workers or measurements of physiological indicators such as heart rate. One simple-to-use method for evaluating workload based on subjective ratings by the workers is the NASA Task Loading Index (TLX). Using this method, the operators rate

TABLE 8-5

Workload Dimensions Analyzed by the NASA TLX Method

Workload Dimension	*Description*
Physical demand	Degree of physical activity (e.g., pushing, pulling, controlling, activating) required to perform the task (e.g., easy or demanding, slow or brisk, slack or strenuous, restful or laborious)
Mental demand	Degree of mental and perceptual activity (e.g., thinking, deciding, calculating, remembering, looking, searching) required (e.g., easy or demanding, simple or complex, exacting or forgiving)
Performance	Perception of success in accomplishing the goals of the task
Temporal demand	Amount of time pressure while performing the task (e.g., slow pace and leisurely, rapid and frantic)
Frustration level	Type of feelings (e.g., insecure, irritated, stressed, annoyed versus secure, gratified, content, relaxed, and complacent) while performing the task
Effort	Degree of mental and physical work required to accomplish the task

their work tasks on the six workload dimensions described in Table 8-5 (Hart and Staveland, 1988; Weimer, 1995). The operators also estimate the percentage of the shift they perform the tasks, and the ergonomist can also measure the percentage of the shift that the tasks are performed by work sampling or direct observation.

The analysis indicates the operators' perceptions of their workload for each task they rate and for each of the six workload dimensions over all of the tasks. Operator tasks for Units A and B in a plant are listed in Table 8-6. An example of the overall TLX workload ratings for Unit A and Unit B operator tasks is shown in Figure 8-3.

This analysis indicates that operators of both units do not perceive their tasks as physically demanding. In general, the tasks performed by Unit A operators have higher mental demands, higher time pressure, and higher frustration levels than Unit B tasks.

One practical use of this type of a TLX analysis is for helping to evaluate if, and how, operator tasks could be combined. The analysis of the

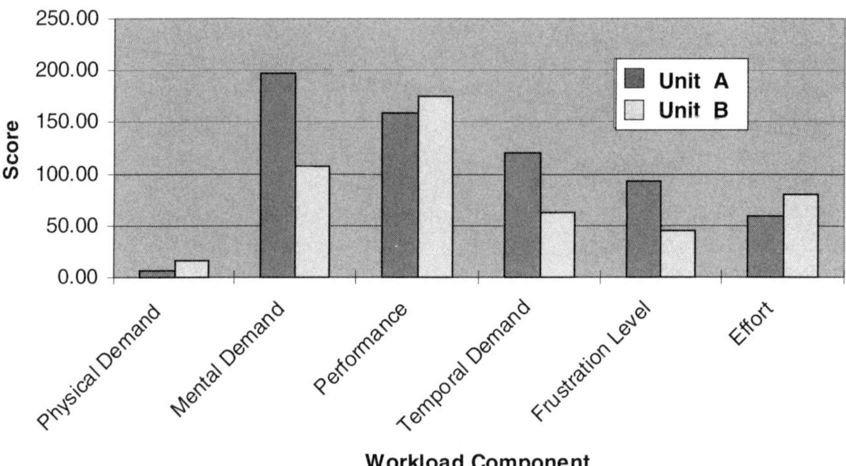

Figure 8-3 Overall ratings of the six workload dimensions for all Unit A and Unit B operator tasks (average for all inside operator tasks).

TABLE 8-6

Operator Tasks for Units A and B

Tasks by Unit A Operator	*Tasks by Unit B Operator*
Coordinate movement of fuel to and from tanks	Approve work permits
Respond to water treatment alarms	Communicate with mechanics
Answer phone calls, radio	Correct deviations in operations
Respond to equipment alarms	Respond to alarms
Coordinate personnel at docks and off site	Manage tank inventory
Communicate with tank gauger	Respond to product scheduling requirements
Monitor lab requests	Communicate tankage changes
Communicate with scheduling dept	Set up blends
Set up and blend fuel A	Monitor lab results
Set up and blend fuel B	
Set up and blend fuel C	

six workload dimensions for each task gives insight on which high-workload tasks should not be performed in parallel by the same operator, unless improvements are made. For example, in this case, there would be potential for a very high workload situation if some Unit A and Unit B tasks were to be performed simultaneously with the tools and methods currently in place.

Detailed analysis of each of the Unit A and Unit B operator tasks indicate that the highest contribution of frustration to workload is not from operating tasks but from telephone interruptions and the coordinating of personnel. Several additional Unit A tasks are perceived as having a high mental demand, with a few of these also having a high time pressure to perform the task. In comparison, Unit B operators reported lower workload requirements for most of the tasks and high mental demands for just a few tasks.

To determine if the high mental or time pressure demands are a potential problem, the tasks should be evaluated from a risk perspective. Each task can be categorized according to high, medium, or low risk, related to the risk if the integrity of the task were jeopardized, such as not completed or completed incorrectly. The results of the risk categorization can be combined with the workload analysis to identify potential workload issues for combined operations.

Table 8-7 lists the Unit A and Unit B operating tasks with the highest workloads. A risk assessment indicated that the tasks marked by an * in

TABLE 8-7

Unit A and Unit B Tasks with the Highest Workload Ratings

Unit A	*Unit B*
Communicate with scheduling department	*Monitor unit alarms, production, and quality control indicators
Blending	Troubleshoot unit
Loading tanks	Rearrange process to achieve product scheduling requirements
*Monitor discharge	
*Monitor loading	

Table 8-7 were also "high" risk tasks. So, the possible pairs of tasks shown in the table should not be performed simultaneously by the operators if they were to operate both Units A and B, unless the current monitoring systems and procedures are enhanced.

8.6 CASE STUDY

A narrative on how to change a tire is transformed into the written procedure shown in Table 8-8, following the guidelines presented earlier in the chapter and summarized in the procedures checklist in the Appendix.

8.6.1 Narrative: How to Change a Tire

Pull off the road, apply the parking brake and place the transmission in park if the car is an automatic. If the car has a standard transmission, place the shift in first gear or reverse. Activate road flares or signal reflectors to warn approaching vehicles of your presence on the roadside. Retrieve the spare tire, jack, and lug wrench. Pry off the flat tire's hubcap using the sharp end of the lug wrench or a screwdriver. To loosen each lug nut, turn the wrench counterclockwise about one turn while the tire is still on the ground.

Place the jack under the reinforced section of the car's body. Most vehicles have a picture with or near the jack, showing lift points and instructions on how to use the jack. Use the jack to raise the car until the flat tire is several inches off the ground. Continue loosening the lug nuts and remove the wheel. Keep track of the lug nuts as you remove them. Lift the spare tire onto the axle hub and align the holes. Replace the lug nuts and tighten each one lightly with your fingers. Use the jack to lower the car to the ground and remove the jack. Then firmly tighten each lug nut by turning the wrench clockwise. As soon as possible, take the car into a service station to have the flat tire repaired and the lug nuts tightened with a torque wrench to specifications.

Table 8-8 illustrates how this narrative should be transformed into a proper written procedure.

TABLE 8-8

Written Procedure on How to Change an Automobile Tire

Procedure: How to change a tire

Procedure No. 01 | Issue date: Jan. 1, 2001 | Page 1 of 2

When to use	Use this procedure to change an automobile tire
Users	Automobile driver or passenger
Reference materials	Automobile owner's manual
Personal hazards	Be aware of the following personal hazards and note the precautions for them:

Hazard	*Precautions*
Traffic	Pull well off the road Use flares or reflectors to deter traffic from your area
Overexertion—while lifting tires, turning lug wrench	Use proper lifting procedures

Equipment: Road flares, jack, lug wrench, spare tire.

Steps:

1. Pull off the road.
2. Apply the parking brake.
3.

If	**Then**
The car is automatic	Place the transmission in park
or	
The car is standard transmission	Place the shift in first gear or reverse

4. Set up road flares or signal reflectors.
5. Retrieve the spare tire, jack, and lug wrench.
6. Pry off the flat tire's hubcap, using the sharp end of the lug wrench.
7. While the tire is still on the ground, loosen each lug nut by turning the wrench counterclockwise about one turn.
8. Place the jack under the reinforced section of the car's body.
   Note: Most vehicles have a picture with or near the jack, showing lift points and instruction on how to use the jack.

continued

TABLE 8-8

Written Procedure on How to Change an Automobile Tire *Continued*

9. **WARNING**: Beware of finger pinch points while using jack.
   WARNING: Do not get into, or under, the car while it is raised on the jack.
   Use the jack to raise the car until the flat tire is several inches off the ground.
10. Continue to loosen the lug nuts and remove the wheel.
    Note: Keep track of the lug nuts as you remove them.
11. Lift off the wheel with the flat tire.
12. Lift the spare tire onto the axle hub and align the holes.
13. Replace the lug nuts and tighten each one with your fingers.
14. Use the jack to lower the car to the ground and remove the jack.
15. Firmly tighten each lug nut by turning the wrench clockwise.
    Note: As soon as possible, take the car into a service station to have the flat tire repaired and the lug nuts tightened with a torque wrench to specifications.

REVIEW QUESTIONS

Test your understanding of the material in this chapter.

1. What is a system-induced error?
2. What should be the content and appearance of a label that describes the hazardous material contained within a pipe?
3. For an investigation of an incident with a potential root cause of procedures were not used, what questions would help determine why the procedures may not have been used?
4. What type of training should be provided to ensure that technicians can maintain and calibrate new sensors that will be installed in a plant?

APPENDIX 8-1. Procedures Evaluation Checklist

Procedures Checklist (based on Center for Chemical Process Safety, 1996, and NUS Training Corporation, 1995, Center for Chemical Process Safety, 1994; and Wieringa *et al.* 1993).

Procedure Title: ________________ **Plant site:** ________________

Date of evaluation: ________________ **Evaluator:** ________________

Checklist Item	*Confirmed* (✓)
Presentation and Usefulness	
1. Does the procedure appear to be concise and easy to use?	
No details that do not contribute to work performance, safety or quality	
Only "need to know"	
No "nice to know"	
2. Is the procedure written so that the detail is appropriate to the range of experience of the users and their capabilities?	
Technical terms familiar to the reader	
No jargon	
All acronyms are familiar	
3. Is the procedure written so that the detail is appropriate to the complexity of the job, including	
Criticality?	
Potential hazards?	
Ease of performing?	
4. Are conditional instructions easy to understand? If an action must meet more than two requirements, are the requirements listed?	
5. Are calculations clear and understandable? For complicated or critical calculations, is a formula or table included or referenced?	
6. Can graphs, charts, and tables be easily and accurately extracted and interpreted?	
7. Are steps written in short, concise statements?	
8. Are the same terms used consistently for the same components or operations?	
Format, Layout, and Design	
9. In general, the procedures	
Contain plenty of white space?	
Contain tab markers to help locate them quickly, such as black	

continued

Checklist Item	Confirmed (✓)
tabs along the right edge of the page to help when searching through and locating major sections?	
Are not cluttered or busy?	
Use lines or white space to separate groups of related items?	
Are written in Times Roman?	
Use a consistent font size throughout and at least 12 point?	
Use capital letters for major titles?	
Use mixed upper- and lowercase text throughout?	
Left justify text?	
Identify steps by their own unique number?	
List each step in sequential order as it should be performed?	
Begin each step with an action verb?	
10. Are cautions and warnings placed immediately before the step to which they apply?	
11. Do cautions and warnings stand out from the proceduress steps?	
12. Are P&IDs or flowcharts placed ahead of the relevant steps or included with them?	
13. If conditions or criteria are used to help the user make a decision or recognize a situation, do they precede the action?	
14. Is the procedure written for the lowest education level allowed among user (plant or contractor) personnel; for example,	
Few syllables per word	
Complexity count = 1 [complexity count is calculated by dividing the number of verbs in action steps by total number of steps in the procedure]	
Content	
15. Does the title accurately describe the nature of the activity?	
16. If the procedure is over 5 pages, is it equipped with a Table of Contents on the first page?	
17. Does the first page of the procedure have the following information?	
Procedure title	
Objective (purpose): Clearly defines the goal of the procedure	
Background: Tells the user how this procedure fits into the "big picture" of the process and why it is important	
References: Documents that support the use of the procedure	
Special equipment list: Any special equipment that must be ready before starting the procedure	

continued

Checklist Item	*Confirmed (✓)*
Precautions (hazard summary): Conditions, practices or procedures that must be observed to avoid potential hazards	
Prerequisites: Any initial conditions that the worker must satisfy or actions that need to be performed before starting the procedure	
Contact person or author	
Authorized signature	
18. Is the necessary procedure control information included on each page, such as	
Facility or unit name or identifier	
Procedure title	
Procedure number	
Date of issue, approval date, required review date, and effective date	
Revision number	
Page number and total pages	
19. Is the last page of the procedure clearly identified?	
20. Are temporary procedures clearly identified?	
21. Does every procedure have a unique and permanent identifier?	
22. For duplicate processes, are the procedures complete and accurate for each process?	
23. Is all information necessary for performing the procedure included or referenced in the procedure?	
24. Does the procedure include all steps required to complete a task? For example, are any steps missing? (That is, reference other procedures to describe the step)	
25. Does the procedure match the way the task is done in practice? For example, any steps out of sequence?	
26. Are all items referenced in the procedure listed in the References section?	
27. Are items listed in the References section of the procedure correctly and completely identified?	
28. Do the References contain a list of supporting documents and locations?	
29. If more than one person is required to perform the procedure, is the person responsible for performing each step identified?	
30. Are steps that can be done simultaneously noted?	
31. Is a sign-off line provided for verifying critical steps of a procedure? (Optional)	

continued

Checklist Item	*Confirmed* (✓)
32. If the procedure requires coordination with others, does it contain a checklist, sign-off, or other method for indicating the steps or actions have been performed or completed?	
33. If a step contains more than two items, are they listed rather than buried in the text?	
34. If two actions are included in a single step, can the actions actually be performed simultaneously or as a single action?	
35. Are steps that must be performed in a fixed sequence identified as such?	
36. Are operating or maintenance limits or specifications written in quantitative terms?	
37. Does the procedure provide instructions for all reasonable contingencies?	
38. If the contingency instructions are used, does the contingency statement precede the action statement?	
39. Do procedures that specify alignment, such as valve positions, pipe and spool configurations, or hose station hookups:	
Specify each item?	
Identify each item with a unique number or designator?	
Specify the position in which the item is to be placed?	
Indicate where the user records the position if applicable?	
40. Do emergency operating procedures contain provisions for verifying	
Conditions associated with an emergency? (Initiating conditions)	
Automatic actions associated with an emergency?	
Performance of critical actions?	
41. Do maintenance procedures include required follow-up actions or tests and tell the user who must be notified?	
42. If a procedure must be performed by someone with a special qualification, are the required technical skill levels identified?	

REFERENCES

Attwood, D. A. (1998) "Understanding and Dealing with Human Error in the Oil, Gas and Chemical Industries." Presentation for the Center for Human Process Safety, San Antonio, TX.

Bishop, H., Madnick, B., Walter, R., and Sussman, E. D. (1985) *Potential for Driver Attention Monitoring System Development.* Report DOT-TSC-NHTSA-85-1. Cambridge, MA: U.S. Department of Transportation.

Canadian Royal Commission on the Ocean Ranger Marine Disaster. (1984) *Report One: The Loss of the Semisubmersible Drill Rig Ocean Ranger and Its Crew.* Publication no. Z1-1982/1-1E. St. John's, Newfoundland: Canadian Government Publishing Centre.

Center for Chemical Process Safety. (1994) *Guidelines for Preventing Human Error in Process Safety.* New York: American Institute of Chemical Engineers.

Center for Chemical Process Safety. (1996) *Writing Effective Operating and Maintenance Procedures.* New York: American Institute of Chemical Engineers.

Chapanis, A. (1996) *Human Factors in Systems Engineering.* New York: John Wiley & Sons.

Connelly, C. S. (1995) "Toward an Understanding of DCS Control Operator Workload." *Instrument Society of America Transactions* 34, pp. 175–184.

Cullen, Lord. (1990) *Report on Piper Alpha Accident.* London: Her Majesty's Stationers Office.

D'Amico, A. D., Kaufman, E., and Saxe, C. (1986) *A Simulation Study of the Effects of Sleep Deprivation, Time of Watch and Length of Time on Watch on Watchstanding Effectiveness.* Tech. Rep. CAORF-16-8122-02, U.S.D.O.T. Washington, DC: National Maritime Research Center.

Danaher, J. W. (1980) "Human Error in Air Traffic Control (ATC) System Operations." *Human Factors* 22, pp. 535–545.

Davies, D. R., and Parasuraman, R. (1982) *The Psychology of Vigilance.* New York: Academic Press.

Donderi, D. C. (1994) Visual Acuity, Color Vision and Visual Search Performance at Sea. International Ergonomics Association 1994 Congress, Vol.4, pp. 224, Toronto, August 1994.

Exxon Mobil. (2001) *Corporate Safety, Health and Environment Progress Report.* Available at ExxonMobil.com.

Goldstein, I. (1993) *Training in Organizations.* Belmont, CA: Wadsworth.

Gordon, S. E. (1994) *Systematic Training Program Design.* Englewood Cliffs, NJ: Prentice-Hall.

Haas, P. M. (1999) "Human Performance Engineering: A Practical Approach to Application of Human Factors." Paper no. NSC-99-108. 1999 National Petrochemical Refiners Association (NRPA) National Safety Conference, Dallas, TX.

Hart, S. G., and Staveland, L. E. (1988) "Development of NASA-TLX (Task Load Index): Results of Experimental and Theoretical Research." In P. A. Hancock and N. Meshkati (eds.), *Human Mental Workload.* Amsterdam: North Holland Press, pp. 129–183.

Heising, C. D., and Grenzebach, W. S. (1989) "The Ocean Ranger Oil Rig Disaster: A Risk Analysis." *Risk Analysis* 9, no. 1, pp. 55–62.

Hudack, S. D., and Duchon, J. C. "A Safety Risk Evaluation of Vigilance Tasks in the U.S. Surface Mining Industry." Proceedings of the International Symposium on Off Highway Haulage in Surface Mines. May 1989. Edmonton, Alberta I. S. Golosinski and V. Srajer (eds.) Balkema Publishers.

Hulbert, S. F., Dompe, R. J., and Eirls, J. L. (1982) *Driver Alertness Monitoring System for Large Haulage Vehicles.* Bureau of Mines OFR 119-83 Contract H0282006, p. 120, Tracor MBA, NTIS PB 83-220640. Springfield, VA.

Kletz, T. (1991) *An Engineer's View of Human Error.* Rugby, Warwickshire, UK: Institute of Chemical Engineers.

Kletz, T. (1998) *What Went Wrong? Case Histories of Process Plant Disasters,* Fourth Ed. Houston, TX: Gulf.

Koelega, H. S. (1993) "Stimulant Drugs and Vigilance Performance: A Review." *Psychopharmacology* 111, pp. 1–16.

Koelega, H. S., and Brinkman, J.-A. (1986) "Noise and Vigilance: An Evaluative Review." *Human Factors* 28(4), pp. 465–481.

Landstrom, U. (1990) "Noise and Fatigue in Working Environment." *Environment International* 16, pp. 471–476.

Moray, N. (1981) "The Role of Attention in the Detection of Errors and the Diagnosis if Failures in Man-Machine Systems." In J. Rassmussen and W. B. Rouuse (eds.), *Human Detection and Diagnoses of System Failures.* New York: Plenum Press.

Moskowitz, H. (1985) "Marijuana and Driving." *Accident Analysis and Prevention,* 17(4), pp. 323–342.

National Institute of Occupational Safety and Health. (1988) *Training Requirements in OSHA Standards and Training Guidelines.* OSHA Publication No. 2254. Washington, DC: U.S. Department of Labor.

Noerager, J. A., Danz-Reece, M. E., and Pennycook, W. A. (1999) "HFAST: A Tool to Assess Human Error Potential in Operations." Offshore Technology Conference, Houston TX.

Norin, F., and Wyon, D. P. (1992) *Driver Vigilance—The Effects of Compartment Temperature.* Report No. 920168. Society of Automotive Engineers at SAE World Congress, Warrendale, PA: SAE.

NUS Training Corporation. (1995) *Procedure Writing Workshop Manual.* Seventh Ed. Gaithersburg, MD: NUS Training Corporation.

Pennycook, W. A., and Danz-Reece, M. E. (1994) "Practical Examples of Human Error Analysis in Operations." Paper no. SPE 27262. Society of Petroleum Engineers International Conference on Health, Safety and Environment, Jakarta, Indonesia.

Ray, P. S. (1991) "Effects of Fatigue and Heat Stress on Vigilance of Workers in Protective Clothing." Proceedings, 35th Annual Meeting of the Human Factors Society, pp. 885–889.

Reason, J. (1990) *Human Error.* Cambridge, UK: Cambridge University Press.

Rogers, W. A., Lamson, N., and Rousseau, G. (2000) "Warning Research: An Integrative Perspective." *Human Factors and Ergonomics* 42, no. 1, pp. 102–139.

Sanders, M. S., and McCormick, E. J. (1993) *Human Factors Engineering and Design,* Seventh Ed. New York: McGraw-Hill.

Thomas, J. J., Noerager, J. A., Danz-Reece, M. E., Pennycook, W. A. (2000) "Systematic Assessment of Human Error Potential in Operations." Paper no. SPE 60999. Society of Petroleum Engineers International Conference on Health, Safety and Environment, Stavanger, Norway.

U.K. Health and Safety Executive. (1999) *Reducing Error and Influencing Behavior.* HSG48. London, U.K.

U.S. Department of Energy. (1993) *Guide to Good Practices for Equipment and Piping Labeling.* DOE-STD-1044-93. Washington, DC: U.S. Department of Energy.

U.S. OSHA 29CFR 1910.119. Regulation on the Process Safety Management of Highly Hazardous Chemicals.

Weimer, J. (1995) *Research Techniques in Human Engineering.* Englewood Cliffs, NJ: Prentice-Hall.

Wickens, C. D. (1984) *Engineering Psychology and Human Performance.* Glenview, IL: Scott, Foresman and Company.

Wieringa, D., Moore, C., and Barnes, V. (1993) *Procedure Writing.* Columbus, OH: Battelle Press.

9

The Use of Human Factors in Project Planning, Design, and Execution

9.1 INTRODUCTION

Previous chapters demonstrated the benefits of examining human factors issues in terms of the potential to reduce injuries, errors, and costs. Chapter 1 introduces a model that demonstrates how to identify issues, conduct ergonomic assessments, and develop solutions. The necessary tools to conduct assessments and design solutions are also provided. So, at this point, you know what to do, how to do it, and why. This chapter suggests when the initiative should be taken to include human factors in a plant modification or new project.

The inclusion of a human factors perspective in capital projects has been shown to be good business (Robertson, 1999). Human factors, if considered early enough in the project, can be extremely cost effective. In other words, a project costs no more to design and build while considering the human during the design. This chapter explains how this is done.

In most process plants, human factors is typically built into projects to improve equipment and facilities for one of three reasons:

1. To reduce the potential for injury.
2. To prevent potential damage to existing equipment and facilities.
3. When the benefits of making the modifications exceed the costs (Attwood and Fennell, 2001).

Experience has shown that, except for these reasons, facilities and equipment are unlikely be improved on strictly human factors grounds.

But, as already said, a human factors perspective, when properly applied, does *not* increase the cost of projects (even those conducted for reasons other than human factors). Projects are conducted every day in all process plants. They are typically not major, capital projects but smaller projects, often local or base projects or just plant modifications. The process of changing, repairing, or maintaining a major piece of equipment is a project. So, there is now a fourth reason for considering human factors in existing plants:

4. To ensure that local projects consider the capabilities and limitations of humans.

This chapter is about the inclusion of human factors in base and major projects. It provides a model that shows how to manage human factors in projects; when to apply human factors tools during the planning, design, and implementation of the project; and describes the tools available. The distinction between the way major capital projects and base projects are conducted profoundly influences the way that human factors are considered in each and the design of the tools used. It is important to remember that not all "projects" are created equal.

9.2 PROJECT MANAGEMENT

Contrary to popular belief, the role of project management is more than just "on time and under budget." Project management should also

- Integrate human factors into the project execution process.
- Identify human factors issues throughout the execution of each phase of the project.
- Address each human factors issue during detailed design.
- Ensure that a human factors approach is applied during construction.
- Review the equipment and facilities for human factors completion prior to start-up.
- Conduct a postproject review to ensure that high learning value issues are identified and incorporated into future projects to continuously improve the application of the human factors approach.

9.2.1 Management of Major Projects

The responsibilities of the various members of the project team throughout the life of a major project are sketched in Figure 9-1. As this example illustrates, the primary responsibility for project execution shifts from one project team member to another as a project progresses through various phases of execution.

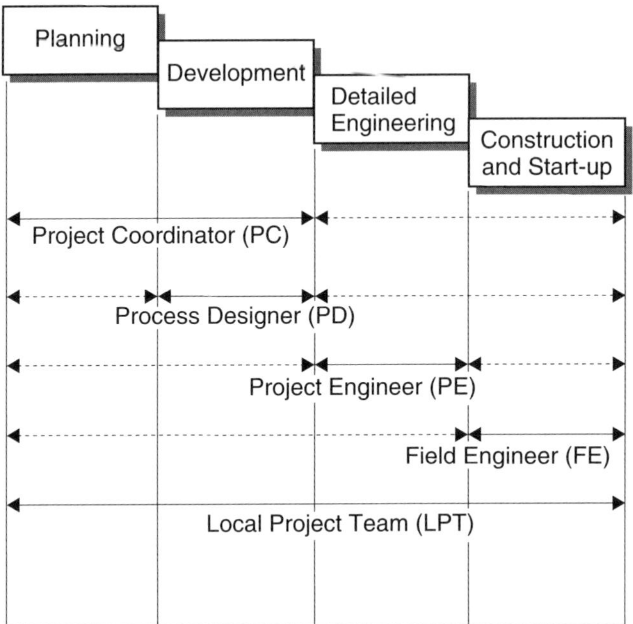

Figure 9-1 Project staff responsibilities during each project phase.

In the example, the project coordinator (PC) is responsible for initiating the project and has primary responsibility for defining the project basis and job specification by involving all stakeholders, a process also called *front-end loading* (FEL). Once this is achieved, the PC is responsible for the execution of the project through to the beginning of detailed design phase. The process designer (PD) joins the project at the process specification phase, following up on the FEL requirements and ensuring that the chemical process is sufficiently defined and specified. The project engineer (PE), who assumes primary responsibility for the project at the detailed design phase, is responsible for designing and specifying the equipment involved in the chemical process. It is essential at this juncture that the equipment purchased is what was specified. Quality Assurance/Quality Control (QA/QC) checks are required to ensure that this happens. The field engineer assumes primary responsibility for construction and equipment turnover to the operator.

The engineering contractors are specified early in the project and often involved early in the planning and FEL phase. But, they are primarily responsible for detailed engineering and, in most cases (i.e., the major projects), procurement and construction as well.

9.2.2 Management of Base Projects

While local or base projects, such as capacity expansions, plant modification, and improvement follow the same project model, the base project is typically managed by a local project team (LPT) comprising unit operating and maintenance personnel and engineering contract staff. The responsibility for human factors in base projects may reside with unit first-line supervisors or plant operating staff who have been trained as specialists in human factors (HF). In the latter case, local projects from all units would be reviewed by these HF specialists.

In addition, because of their comparatively short duration, base projects have fewer established reviews throughout planning, design, and execution as major projects. Some other differences for base projects include

- Contractors for base projects could be on a long-term plant contract. So, they are likely working on a number of small projects at the same time. This differs from major projects, where the staff of the engineering contractor is usually dedicated to that project.
- Staff turnover within the contractors' organizations follows the area and local project demand cycles and is high at times.
- Project reviews are fewer and generally less rigorous. They are often ongoing and may be unscheduled.
- Drawings are often two-dimensional CAD. Three-dimensional computer models are usually not available.
- Safety and health reviews are conducted only if the product being produced or those involved in the production meet a "trigger" level of concern.
- Full hazard and operability (HAZOPs) reviews are not always required; review and approval of the project specifications by the plant safety committee is normally sufficient.

9.3 HUMAN FACTORS TOOLS FOR PROJECT MANAGEMENT

Figure 9-2 illustrates the human factors tools designed for use in project development and design and reviewed in this chapter. The figure also indicates the stage in the project where the tools are intended for use.

The following paragraphs describe each of the human factors tools introduced in Figure 9-2 and outline the recommended use of each tool by project phase and provider. Some tools stand alone, such as LINK analysis; others are integrated into existing key project tools, such as HAZOPs.

9.3.1 Human Factors Tracking Database

The human factors tracking database (HFTDB) consists of a standard spreadsheet that captures the human factors design issues as they are iden-

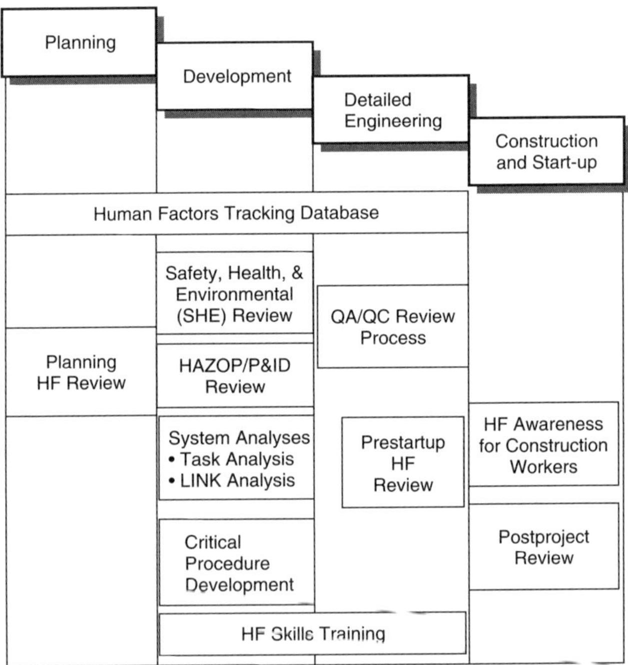

Figure 9-2 Human factors tools for use during project planning, development, design, and construction.

tified. The HFTDB passes through various members of the project team as follows:

- It is initiated by the project coordinator using a generic human factors checklist during the business planning cycle.
- It is used by the development team to specify equipment, including package unit designs.
- It is handed off to the project engineer during the design phase and to the field engineer during the construction phase.

In our model, the project coordinator has overall responsibility to ensure that all issues are addressed during each phase of the project. The database carries each issue through to resolution at the appropriate phase of the project.

It is easy to envisage the HF tracking database being an integral part of the tracking tools usually required for all follow-up items in capital or base projects. This database is valuable also as part of the project "management of change" (MOC) process.

9.3.2 HF Review: Planning Phase

As soon as a project is conceived, the people involved know something about what the new plant or equipment will look like and how the product will be produced. Their knowledge may be sketchy, but it is sufficient to allow them to decide whether the engineering intentions of the new process unit might involve complex operations or work tasks that could exceed the limitations and capabilities of the workforce.

The human factors tool designed for use during this phase is a list of screening questions intended to identify basic human factors information about the proposed plant. The questions are listed in Table 9-1.

If the answers to any of these questions are yes or don't know, the issue is added to the human factors tracking database for review at some appropriate time later in the project, as more information and process plant definition become available.

TABLE 9-1

Human Factors Screening Questions to Consider during Project Planning and Development

Will equipment or facility design require operations or maintenance personnel to

1. Work in abnormal environmental conditions requiring PPE or other temporary measures due to
   Ambient temperatures (hot or cold)?
   Air quality?
   Lighting?
   Noise or vibration?
2. Work in small, confined spaces (restricting movement, cannot stand up, walk around) for periods of time?
3. Perform tasks requiring periods of manual materials handling, repetitive motion, or unusual body position?
4. Perform complex tasks (especially if required during night shift), such as those
   Having a high number of sequential steps?
   With a high information processing load (e.g., operator would have to consider many process readings, alarms, etc. and then decide on the right action)?
5. Perform tasks that, if not done correctly, have significant safety, health, environmental, or production consequences?
6. Operate a process or work on equipment that is new to the plant (little or no experience) or for which past experience demonstrates that incidents occur more frequently?
7. Work in situations that are atypical for the site, such as batch processes at a site that historically has managed continuous processes only.

9.3.3 Safety, Health, and Environmental Review

Early in the project, a review is conducted to identify and address the key safety, health, and environmental (SHE) hazards that could harm people or damage property or the environment. The objective of the SHE review is to identify the hazards and to find ways to eliminate or control them: that is, prevent a hazard from occurring, alert the workforce and public if it does occur, and mitigate the effects of the occurrence. If hazards cannot be prevented or controlled to achieve an acceptable level of risk, the project may not go forward.

While most of the SHE review is based on toxicological or engineering hazard identification and control, there is an opportunity to examine the physical hazards that have a human factors focus. The tool for conducting the human factors evaluation is the list of screening questions from Table 9-1. At this stage of the project, more information is likely to be available to answer the screening questions than during the planning phase. In addition to the questions in Table 9-1, other questions can be asked at this stage. These are listed in Table 9-2. Again, a yes or don't know answer is sufficient to add the issue to the human factors tracking database.

Completing the SHE process is typically the responsibility of the project coordinator but is usually conducted by a project person with responsibility for safety, health, and environment for the project, such as the process safety engineer.

For base projects, the SHE review is conducted when the process triggers are reached.

9.3.4 Human Factors Training for the Project Team

Training modules should provide the project teams with a basic understanding of human factors in all areas of facilities and equipment design.

TABLE 9-2

Additional Human Factors Screening Questions to Be Asked at the SHE Review Stage of the Project

Are tasks associated with the venture that

1. Have the potential for high SHE consequences?
   Planned SHE-critical tasks, such as to respond to a critical alarm
   Production-critical tasks (abnormal situations) with SHE implications, such as response to loss of power
2. Involve interaction with several other personnel (high-demand situations)?
3. Require PPE beyond the basic?
4. Be difficult to recover from a mistake?
5. Unplanned, time-critical, or complex tasks?

In addition, the training should allow the participants to identify the issues associated with the project. The content of the course should be tailored to meet the needs of the attendees. The following are considered to be minimal requirements of project team training.

9.3.4.1 Course Objectives

- Familiarize participants with human capabilities and limitations.
- Generate awareness of the influence of human factors on process design.
- Train participants so they can identify human factors issues as they arise.
- Provide ideas through case studies and demonstrations on how to approach issues from a human factors perspective.
- Familiarize participants with the processes and tools for building a human factors approach into projects.
- Develop follow-up action plans.

9.3.4.2 Course Content

The modules for human factors training should include

- Introduction to human factors in process and equipment design in base and capital projects.
- Influence of human capabilities and limitations on facilities and equipment design.
- Individual and group differences.
- Body size and strength.
- Physical activity factors.
- Work physiology.
- Information processing and human error.

- Control room operator issues.
- Access to facilities and equipment.

Equipment topics should include

- Displays, signs, and labels.
- Electrical buildings (equipment and instruments).
- Instruments.
- Field control panels.
- Emergency shutdown equipment.
- Operability and maintainability of equipment.
- Construction of facilities.
- Package units.
- Control room design.
- Pumps and compressors.
- Process control system (interface design).
- Procedures.
- Sample points.
- Loading racks.
- Alarm management and the hierarchy of alarms.
- HF tools for process and equipment designers.
- Issues review and an action plan for human factors implementation in projects.
- HF in project management systems.

For capital projects, training should be offered to the project team twice during the project. The first is during the design specification phase, well before the start of detailed design. At this point, the unit's major tasks are defined and its procedures beginning to be developed. Critical operator tasks can be identified and analyzed for major human factors issues. The key training modules to be given at this time are

- Human information processing, which addresses the cognitive limitations of the uscr population.
- Alarm management, which addresses the alarm philosophy and designation.
- Control room design.
- Operability and maintainability, which addresses issues such as valve identification and location of major equipment.
- Procedures design.

The second set of training should be conducted at the beginning of the detailed engineering phase. For this training, all modules are applicable, but they should be selected based on the needs of the project. The target audience for these modules consists of equipment designers from the site's own staff and engineering contractors.

For base projects, training should be offered at routine intervals (e.g., every 6 months) for ongoing members of the site project community and the contractors who support the site. New employees should attend the first available course.

9.3.5 Human Factors in the Hazard and Operability Reviews

For major projects, a hazard and operability (HAZOP) review analysis is conducted during the development phase, prior to detailed engineering. For base projects, a process and instrumentation drawing (P&ID) review is conducted during the development phase.

9.3.5.1 A Tool to Identify Human Factors Concerns during the HAZOP Process

The objective of this tool is to assist the HAZOP team in the identification of human factors concerns associated with particular pieces of equipment while conducting the HAZOP review. The tool should also quickly target the major items of HF concern, be user-friendly, require minimal training for its use, and not duplicate those design issues covered by any HF design guidelines or standards.

Previous experience with the implementation of human factors in a HAZOP process has pointed out the following issues:

- The current HAZOP protocol is less systematic than it should be to address HF concerns such as human/machine interface (HMI). It relies on the use of a checklist without recommending a process for applying it. Since, human factors concerns are not the prime reason to perform a HAZOP review, the HAZOP leader often elects not to review the checklist during the HAZOP proceedings, especially during a HAZOP review for major capital projects. More likely, the checklist is reviewed at the end of the HAZOP study, resulting in a less-critical, systematic approach.
- The principle objective of the HAZOP review is to identify process hazards, especially those that affect the containment integrity. Some of these hazards are a result of human factors; that is, with either the process equipment or instrumented process control. Other tools are available that focus on HF issues such as critical task analysis.
- The HAZOP review is a sequential process, a line-by-line and equipment-by-equipment review. It requires that each line or process system on the process and instrumentation drawing be followed and checked and that the hazards associated with the equipment located at each node be identified and discussed. The HF issues need to be highlighted at these nodes as an integral part of the HAZOP review and not at the end of the review, when the team has completed each P&ID.

This HAZOP process introduced here has the following characteristics:

- It utilizes a human factors perspective in a targeted manner.
- It determines the difference between HF-relevant and HF-irrelevant systems quickly, so the team can move through the P&ID in a timely fashion.
- It considers only the pieces of equipment or systems that have a human factors interface.
- It focuses attention on the operations that can be hazardous to the process, operators, or maintenance workers.

- It allows the HAZOP team to use the process with little previous knowledge and minimal practice. After a short introduction to HF and a brief exposure to the tool and its use, the team should be able to quickly distinguish between relevant and irrelevant systems and pinpoint the HF concerns with the relevant systems.
- It recognizes that the procedure identifies HF issues beyond those covered by a design specification; that is, issues specific to the particular process and engineering design reviewed by the HAZOP team.

Human factors engineering is seen as an integral part of the HAZOP process. The following outlines the process of implementing HF during HAZOP reviews:

1. A trained HF practitioner is included on the team. For large projects, this would be a HF specialist or a person with extensive HF background.
2. The team members, if not trained in HF, are given a short introduction, which should cover:
   Capabilities and limitations of operators.
   Areas where consideration of HF may improve the safety of the system.
   Familiarization with the simple tools used to distinguish between relevant and irrelevant human-system interfaces.
3. Human factors experts on the HAZOP team, if coming from outside the engineering function, must understand the HAZOP process and be able to read and understand the information on a P&ID. This includes all the symbols (e.g., valve types) and what functions the equipment performs (e.g., a "double block and bleed" piping arrangement).
4. Human factors "screening" lists are available to the HAZOP team and used by designated HAZOP team members.
5. The screening lists are referred to each time a node in the P&ID is reached and considered. The purpose of the screening lists is to remind the team to determine whether the equipment requires a HF

review and follow-up. The type and format of the HF screening list is given later.

6. If a human factors concern is identified, it is recorded on the HAZOP worksheet and in the HFTDB. Sufficient information and knowledge may be available to resolve the issue during the HAZOP review. However, the resolution of the issue most likely takes place outside the HAZOP review.

Each screening list is a simple checklist that identifies the most important HF characteristics of the targeted equipment or system so the team can quickly decide whether a potential problem exists and what action to recommend.

A screening list is designed for each of the following pieces of equipment:

- Valves—sequential, critical, frequent.
- Blind and blanks—weight, lifting, location, frequency, clearing, draining.
- Pumps and Compressors—maintaining, draining, in start-up; in shutdown.
- Field displays—critical, start-up, shutdown.
- Fire fighting and deluge systems—critical, accessing, escaping, directing.
- Field instrumentation—rodding, clearing, critical.
- Sample points—exposure, PPE.
- Reactors and dryers—critical, switching or regenerating, clearing, entering, catalyst dumping, lifting.
- Vessels (including exchangers)—cleaning, exposure, PPE, lifting, entering.
- Furnaces and fired heaters—light-off, decoking, flame controlling, switching, PPE, entering.
- Filters—switching, cleaning, clearing, PPE, weight, lifting, entering.

- Loading and unloading facilities—height, space, exposure, weight, clearing, disconnecting.

An example of a screening list for sample points is given in Table 9-3. Appendix 9-1 provides a screening list for each of the 12 pieces of targeted equipment identified. Appendix 9-2 provides a sample of the additional information that the user needs to use the screening lists. The data are provided for the sample point screening list. Column 1 of the table repeats the screening list as it was given in Table 9-3.

Column 2 of Appendix 9-2 provides example situations where the answer to the screening question might be positive and action should be taken. Column 3 provides a series of potential solutions if action is required. Unfortunately, space constraints prevent listing the information in Appendix 9-2 for each of the remaining 11 pieces of equipment.

The human factors concerns for each piece of targeted equipment are captured on both the HAZOP worksheet and on the human factors tracking database. For each concern, three basic follow-up actions need to be considered:

1. The HAZOP team assesses the HF concern raised and agrees on a sound solution based on the team's HF knowledge. The improvement recommendation is recorded on the HAZOP worksheet and the HFTDB for follow-up.

TABLE 9-3

HAZOP Human Factors Screening List

Sample Points

Check box if the statement is true. HF follow-up is required for this sample point if any of the boxes is checked:

☐ The product being sampled is in a hazard class where defined exposure criteria must be observed.

☐ The sample point is located above grade; that is, sample must be carried to elevated location.

☐ The sample system does (does not) use a closed purge system.

Operator access to the sample point is restricted.

2. The HAZOP team recommends a thorough examination of the human/machine interface design. The team recognizes the need to assess the HF concern outside the HAZOP proceedings to ensure the HF issue is addressed thoroughly using all available resources.
3. The HAZOP team recommends a critical task analysis (see Section 9.3.7.1) to assess the HF problems and their solutions in a rigorous (formal, systematic, critical) approach.

All three actions ensure that appropriate HF design and engineering principles are used in the design and engineering process, so that the chance of a human error and human injury are eliminated or adequately reduced.

9.3.6 Procedures

The purpose of a procedure is to guide a process worker, step-by-step, through the task. Not every task requires a procedure. If the task has few steps, if it is performed frequently, and if the consequence of error is minor, it likely does not require a procedure. However, if the task is complex, if it is performed infrequently, or if the consequence of making an error is high, a procedure likely is required. The conditions for requiring procedures are discussed in Chapter 8.

From the viewpoint of a project, each critical task an operations or maintenance worker performs requires a procedure. The tasks on which the procedure is based on should be specified prior to detailed engineering, since the way in which equipment is used affects its design. For example, the layout of a field panel that controls the operation of a fired heater or a compressor should be based on the sequence in which the equipment is operated. The placement of equipment manifolds should be dictated by the sequence in which the valves are manipulated. And, the design of filters can affect the methods used to change cartridges. The design and placement of each of these pieces of equipment is affected by or affects the procedures used to operate or maintain them.

Surprisingly, procedures are addressed at this early stage of design. But, every design uses people to execute numerous operations and maintenance tasks. Hence, in the early phases of a project, it is very clear which activities are conducted by people and which are performed automatically by the equipment. The earlier that the human tasks are addressed and the hazards identified, the better the human/machine interface can be designed without incurring additional costs.

The sequence of activities and the operations performed in each activity guide the human factors considerations in each task. So, procedures are required to perform critical-task analyses (CTAs) and link analyses. CTAs and link analyses, which are briefly discussed in the following section, are explained in detail in earlier chapters.

9.3.7 Analysis Techniques

Prior to detailed engineering, the human factors issues identified and included on the HF tracking database should be resolved, since each issue could affect the design of the facilities and equipment. Issue resolution often involves analyzing worker tasks to ensure that the design of the facilities or equipment supports how operators work. Two tools that help specify the design and layout of facilities and equipment are task and link analyses.

9.3.7.1 Critical-Task Analysis

The critical-task analysis is discussed in detail in Chapter 7. For projects, the objective of the CTA is to provide a structured, systematic approach for process and equipment designers to identify and analyze the tasks performed on a unit that are critical to the safe operation of the process. The tasks might be performed during abnormal or high-cognitive-demand situations, such as emergencies or when the plant is put into "standby" mode. The tasks might also be selected for analysis because they have the potential for serious consequences if not performed correctly.

People are an essential and integral part of a process plant. The design of a process plant must take into account the capabilities and limitations of the people who operate and maintain it. The safe operation of a process relies on the successful completion of tasks performed by people, some of whom may be critical to the safety, environmental integrity, or reliability of the operation. These critical tasks take precedence during the design of the unit. Task analysis techniques can be applied to existing tasks, as demonstrated in Chapter 7, or to tasks that do not yet exist. The technique is the same in each application. The difference between the two applications is the amount of information available for the analysis.

The process described here is concerned mainly with the identification and analysis of tasks associated with a new project. In this case, a project may be the design and construction of a major grassroots facility or the improvement of an existing facility. In new projects, the tasks do not yet exist. Even so, they can be analyzed, and to do so, the analyst or the analysis team should have knowledge of

- The facility in which the operator will work.
- The equipment the operator will use.
- The environment in which the operator will perform.
- The activities the operator will be required to perform, typically captured by a procedure as discussed in the previous section.

Often tasks that are similar to those on the new facility are performed elsewhere can be used as models for the nonexistent tasks to be analyzed.

For projects, the critical-task analysis process is divided into two parts: the identification of critical tasks for analysis and the analysis of those tasks.

The identification of critical tasks begins at the planning phase of the project and continues through to the start of detailed engineering. Each should be identified on the HFTDB. Critical tasks are those that

- Require people to work in abnormal environmental conditions for excessive periods.
- Are performed in confined spaces for excessive periods.
- Require substantial periods of manual materials handling.
- Have the potential for significant safety, health, or environmental incidents.
- Require workers to perform critical procedures [e.g., under process hazard analysis (PHA) regulation definitions].

- Have the potential for serious consequences if not performed correctly.

Case Study: Change of Boron Tri-Fluoride (BF_3) Spheres

The case study that follows demonstrates the CTA process conducted for the design of a batch process plant. Table 9-4 illustrates the tasks chosen by the design team as the most critical in terms of the risk to plant personnel and property. The analysis was conducted using the process, which was introduced in Chapter 7. The data available for the task analysis consisted of the plastic model shown in Figure 9-3, P&ID drawings, and some two-dimensional plans and elevations.

The process was lead by a process designer and included on the process design team an operations representative, a safety engineer, and a human factors specialist. The results of the analysis are shown in Table 9-5. Note that the format of the table is slightly different from the one in Chapter 7.

9.3.7.2 Link Analysis

Link analysis was discussed in detail in Chapter 6. For projects, the objective of the link analysis is to ensure that equipment is laid out to allow the operators and maintenance staff to work efficiently. The layout of a plant depends to a large extent on the way an operator or maintenance technician performs the task. For this reason, it is essential that procedures for the tasks be written prior to detailed engineering. The case study to

TABLE 9-4

Critical Tasks Identified during the Design of a Chemical Plant

Remove pump, drain, and flush
Unload fresh caustic
Change boron tri-fluoride (BF_3) spheres
Preventive maintenance (PM) of leak detection system
Sampling caustic
Unload carbon monoxide (CO)

Figure 9-3 Plastic model used to illustrate the design of a chemical plant.

follow illustrates the importance of having procedures available during the design process.

Case Study: Preparing a Reactor for Start-Up

The case study described here shows how a design can be evaluated using the operations start-up procedures. The objective of this exercise was to determine whether the design of the valve manifolds considers the physical effort required by the operator. Clearly, a bad design makes the operator move unnecessarily around the unit. Conversely, a good design has the operator working in the same general area until the tasks are complete before moving to another area to complete another set of tasks.

Figure 9-4 is a plan view of the reactor deck. The figure shows four distinct manifold areas. Using the procedures, the sequential links were identified between valves. Each link is drawn and numbered in sequence

TABLE 9-5

Task Analysis: Changing a BF_3 Sphere (removing the connection from the sphere)

Activity	*Human Factors Issues*	*Potential Mitigation, Comments*
Access platform outside of the building next to the sphere	Height of the platform is too low	Set height according to the stature of the population of users
Reach through glove box	Reach to the connection is too far Balance as operator reaches in	Reach based on diameter of the sphere and arm reach of users; revise design to move operator closer to valve
Pick up wrench beside the sphere to release the connector	Grasp of wrench through gloves Location of wrench Orientation of wrench Manipulation of wrench Vision of connector through the window	Gloves prevent good feel of tools; suggest tool be designed as part of the connector with a good orientation, adequate grip, sufficient leverage
Disconnect the valve	Vision of connector Ability to determine when the connector is fully disconnected	Adequate vision through the window requires unobstructed view of the connector and excellent lighting in day and night conditions; measure luminance
Close valve on the sphere	Access to valve, about 12 in. from the disconnect	Adequate reach to valve
Check alarm to ensure no BF_3 in building	Ability to sense alarm—sufficiently loud or bright	

continued

TABLE 9-5 *Continued*

Activity	*Human Factors Issues*	*Potential Mitigation, Comments*
Open door to building	Interlock between alarm and door prevents door from opening if interlock fails, where is emergency shower and alarm?	Ensure operator can identify when interlock operates and what it means, by training and signs
Complete disconnect	Reach connector from within the building Orientation to the connector Manipulation of wrench	Simulate before finalizing the design
Remove sphere with forklift	Operation of forklift Vision of sphere from forklift Room to manipulate forklift	Should be simulated using a mockup of the design prior to finalizing the design of the building
Replace with new sphere	Removal of sphere from storage and replacement of spent sphere	
Orient new sphere so the connectors line up		
Connect sphere to line		
Exit building		
Open valve from glove box		

in the drawing. The task starts at activity 1 and progresses sequentially through to activity 17. The analysis demonstrates the following issues:

- In most cases, the process designer located the equipment so the distance the operator travels between activities is minimized.
- Exceptions are apparent: Travel between activities 5 and 6 is excessive; travel is required to the next level (6 m) to perform activity 10.

The stairs designed to travel to the next higher level are at the end of the structure. If a set of stairs is installed between the levels where most of the activity takes place, the physical stress on the operator is reduced substantially. In addition, if valve B_3 were relocated to manifold 63, the

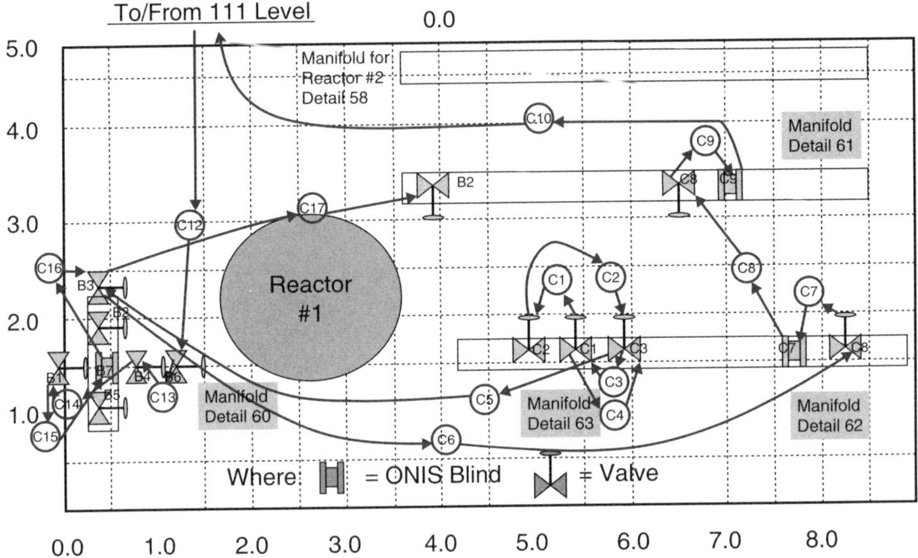

Figure 9-4 Plan view of a reactor deck. Sketch includes the activity links between pieces of equipment.

total travel distance to and from the valve—activities 5–6 and 16–17—are shortened.

9.3.8 Quality Assurance/Quality Control Review Process

9.3.8.1 Introduction

Quality assurance/quality control (QA/QC) is an ongoing process to review the progress of the design that begins at the process specification phase and continues through to the end of detailed engineering design. QA/QC specialists are technical people knowledgeable in both the design of process equipment and facilities and human factors. So, the specialist can come into the project from one of two directions, either as a safety engineer trained in human factors or human factors specialist experienced in plant processes.

It should be emphasized that the QA/QC process is an integral part of any capital investment strategy. So, the activity needs to be planned carefully and communicated widely among project personnel.

The responsibilities of the QA/QC, human factors specialist are to

1. Train the project team, plant personnel, and contractors, in human factors issues. After the training, each member of the team should be aware of how the capabilities and limitations of operations and maintenance personnel are affected by the design of the equipment and facilities for which each is responsible.
2. Ensure that the issues identified on the HFTDB are resolved by the plant and contractor project staff.
3. Evaluate the impact of design changes on the capabilities and limitations of the plant operations and maintenance personnel.
4. Work closely with QA/QC specialists in other disciplines, especially safety, to ensure that human factors matters are adequately reviewed. Other specialists may be responsible for
   Rotating machinery (pumps and compressors).
   Mechanical (including piping, valves, blinds).
   Structural (including stairs, ladders, platforms).
   Vessels (including reactors, exchangers, fired heaters).
   Instruments and electrical equipment.
   Off-site equipment (e.g., storage tanks).

The QA/QC process begins after the process design specification and is completed at the end of detailed engineering.

9.3.8.2 Conducting the QA/QC Process

QA/QC process is conducted by the application of a systematic process that is explained in the following paragraphs.

Visits to the Contractor's Offices

The initial visit of the HF specialist to the contractor's office serves several highly important functions and is significant in establishing the human factors QA/QC control of the job. The first purpose of this initial visit is to review the items on the HFTDB identified throughout the project. Decisions are made on how these items affect design decisions.

In addition to the HF tracking items, areas that require early resolution and review include

- Layout. The HF specialist should be involved in any plot plan changes that result from increased equipment sizes, structural changes affecting access, or overall process changes.
- The type of all emergency block valves (EBVs) and the frequency of operation of all valves to determine their access.
- Function, layout, labeling, and location requirements for all control panels and consoles.
- Labels and safety sign requirements for equipment.
- Access requirements for complex or skid-mounted equipment (e.g., pump units, compressors, lube oil units, and furnace burners).

Finally, any special needs and requirements specific to the individual job, where problems may be anticipated, should be discussed at this time.

Preparation of a QA/QC Checklist

The purpose of the human factors QA/QC checklist is to communicate the scope of work of the HF specialist. While the human factors QA/QC checklists should be as extensive as possible, the topics on the initial checklists are identified in discussions with the project management team and the contractor during the initial visit to the contractor's office. For example, the availability of a three-dimensional computer-aided design (3D CAD) model is an important factor in determining to what extent preliminary and final drawings of the contractor should be reviewed.

Human factors QA/QC checklists vary with the project. Most checklists are one to two pages long, listing the project documents requested for the review and the approximate timing of major topics to be reviewed. Appendix 9-3 contains a comprehensive master checklist covering most human factors involvement (assuming a detailed model is available). The checklist helps improve communication between project engineers and HF specialists during the detailed engineering phase of a project.

HF Training for Contractor's Staff

In many cases, the location of equipment, controls, and display panels determines whether a facility is "designed for human use." The project personnel who most influence the location of equipment are the contract designers tasked with laying out the plant. The most effective way to convince contractors of the importance of HF to good design is to provide them with training that contains concrete examples of facilities with good human factors features. Each designer requires human factors information in his or her specialty, such as mechanical design, not in those of others. So, the QA/QC training should be optimized by presenting selected modules from the full project-training course in specific topics, such as mechanical design, instrument and electrical design, or process control systems design. The subjects of the training modules and their intended audiences are listed in the matrix in Table 9-6. Training costs money. So, to ensure that the training takes place, it must be agreed to ahead of time and included in the services contract.

Human Factors Model Review

The main objective of the human factors model review is to confirm that the critical-task and design issues identified on the human factors tracking database (Section 9.3.1) have been addressed during the engineering design. This tool provides a method to permit the project design team to review, systematically, plastic models, 3D CAD models, or drawings for human factors and ergonomic issues; identify specific human factors problems in the detailed design; and suggest areas or designs that need improvement. This approach allows for review at different stages of the design development, but the review should provide feedback early in the detailed engineering process to avoid costly late changes.

The model review process normally begins at about the 30% completion stage of the engineering design. Most large equipment and major facilities have been laid out and most large piping and valves (>8 cm, 3 in.) are in place.

The model should be capable of

- Providing accurate distances between equipment and vertical distance between equipment and platforms or grade.

TABLE 9-6

Training Matrix for Key Plant Positions Recommended Modules by Position

Key Positions	*Intro to Human Factors*	*Human Factors in Project Design and Implementation*	*Human Factors in Mechanical Design*	*Human Factors in Instrumentation Design*	*Human Factors in Electrical Design*	*Human Factors in Process Control Design*	*Human Factors Project Tools*	*Human Factors in Construction*	*Comments and Estimate Time Commitment (hr)*
Project development committee	✓								
Project coordinators/ engineers	✓	✓	✓	✓	✓	✓	✓	✓	Full course: about 14 hr
Process designers	✓	✓	✓	✓	✓	✓	✓		
Process engineers	✓	✓	✓	✓	✓	✓	✓		
Field engineers	✓	✓	✓	✓	✓	✓	✓	✓	
Contractor engineering staff	✓	✓	✓	✓	✓	✓	✓	✓	
Contractor design and construction lead specialists:	✓	✓	✓				✓	✓	Lead specialist
• Mechanical contractor design and construction lead specialists	✓	✓		✓			✓	✓	Lead specialist training: about 4 hr
• Instrumentation contractor design and construction lead specialists	✓	✓			✓		✓	✓	Lead specialist training: about 4 hr
• Electrical contractor design and construction lead specialists	✓	✓				✓	✓		Lead specialist training: about 4 hr
• Process control systems									
Area project teams	✓	✓	✓	✓	✓	✓	✓	✓	

- Visually comparing vertical and horizontal reaches with the 5th percentile user population (see Chapters 2 and 6).
- Visually comparing vertical and horizontal clearances with the 95th percentile user population (see Chapters 2 and 6).

Two techniques can be used to evaluate the model. The first is to evaluate the design against the results of the task analyses conducted prior to the start of the engineering design. Critical-task analyses are described in Section 9.3.6. The second is to apply review checklists to evaluate the model: one for maintainability and a second for operations.

The maintenance checklist (Appendix 9-4) can be used as each drawing section is completed. This allows for easy and early review of the drawing design. The walkthrough or rounds checklist (Appendix 9-5) requires that all stairs, ladders, and platforms be completed to simulate the movements of an operator through the plant. The walkthrough simulates the movements and actions that an operator would actually perform in the field once the plant is in service. It should catch any deficiencies not previously caught in the maintenance reviews.

The design is then modified by the solutions generated at the 30% design review. The proposed solutions are re-reviewed at the 50% design stage. A final review, at about the 80% design completion stage, can be used to evaluate the design of local control panels and the placement of sample points.

The following case study illustrates the benefits of an early model review. Figure 9-5 is a 3D CAD model of a unit designed to mix an additive with a blender containing existing product. In this design, the additives are contained in cylinders. The cylinders are transported to the structure in groups of four by a forklift and lifted to the top of the structure using a cable winch. Referring to the numbers on the figure, the following issues were discovered in the review:

1. The connection to the hoist is awkward and time consuming with the potential for operator error.
2. The load is permitted to swing freely and turn on the cable as it is lifted to the third floor.

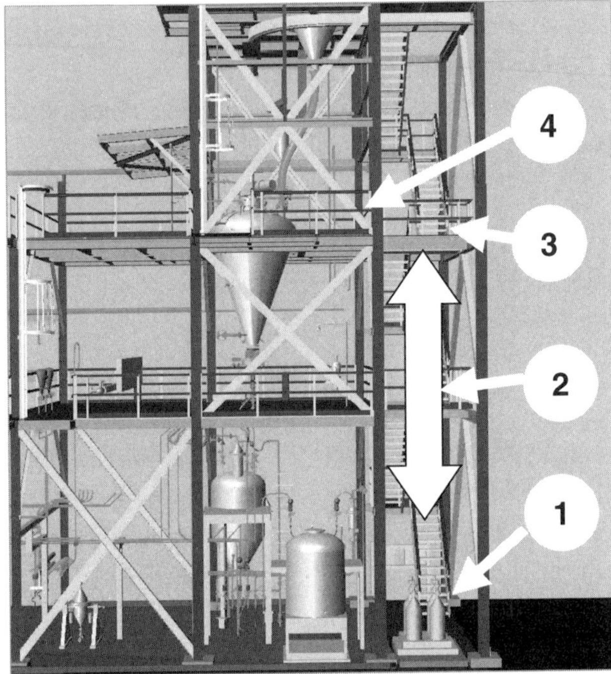

Figure 9-5 Reactor structure showing paths to access stairs to upper deck.

3. At the top of the lift, the operator is expected to steady the load with one hand and operate the lift control with the other.

4. The operator must have two lift-control stations, one at level 1, the other at level 3. The question of control interlocks becomes complicated.

The cylinders were then transported on a rail to a storage position (Figure 9-6). The issues associated with this operation are illustrated by number in Figure 9-6 and explained next:

5. At the top, operator is expected to work under a live load as he or she guides the four-pack on the monorail around to the storage area.

6. The four-pack must clear the screw blender as it is guided to its staging position.

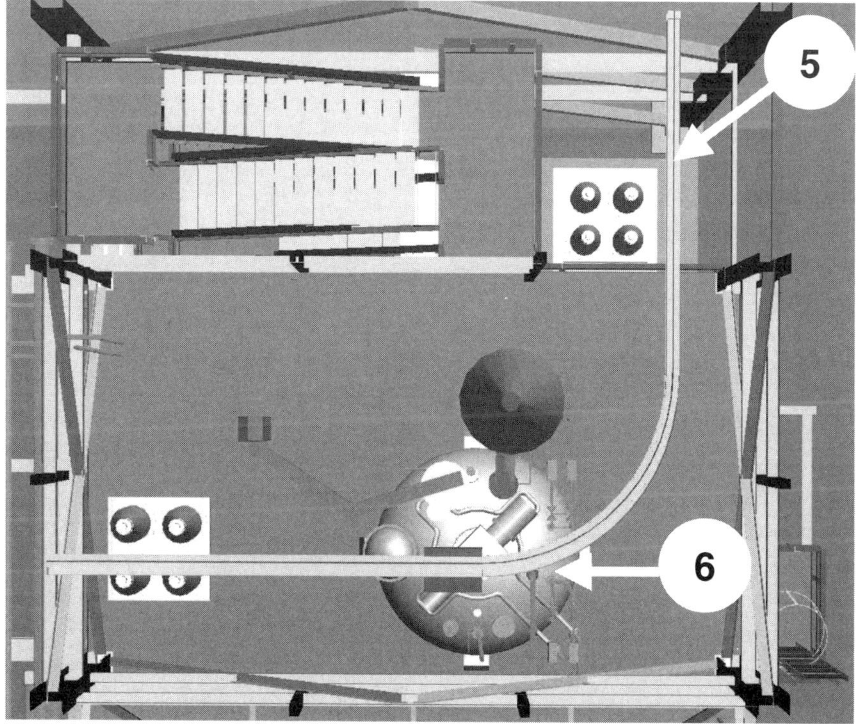

Figure 9-6 Plan view of mixing deck showing transporting rail.

An articulating arm is used to carry the cylinders to the blender (Item 7, Figure 9-7), rotate them upside down, and mate them with the drum. The arm returns the empty cylinders to storage. Referring to the numbers in Figure 9-7, the following issues are identified:

7. Each cylinder is grabbed, rotated upside down, and mated with a connector on top of the blender. The operation requires precise motor movements.

8. Once the cylinders are empty, they are returned to the pallet. How the cylinders in the rear of the pallet are accessed is not considered in the design.

These notes illustrate the problems faced by the operator with this design. As it turns out, the problems cannot be easily resolved and the

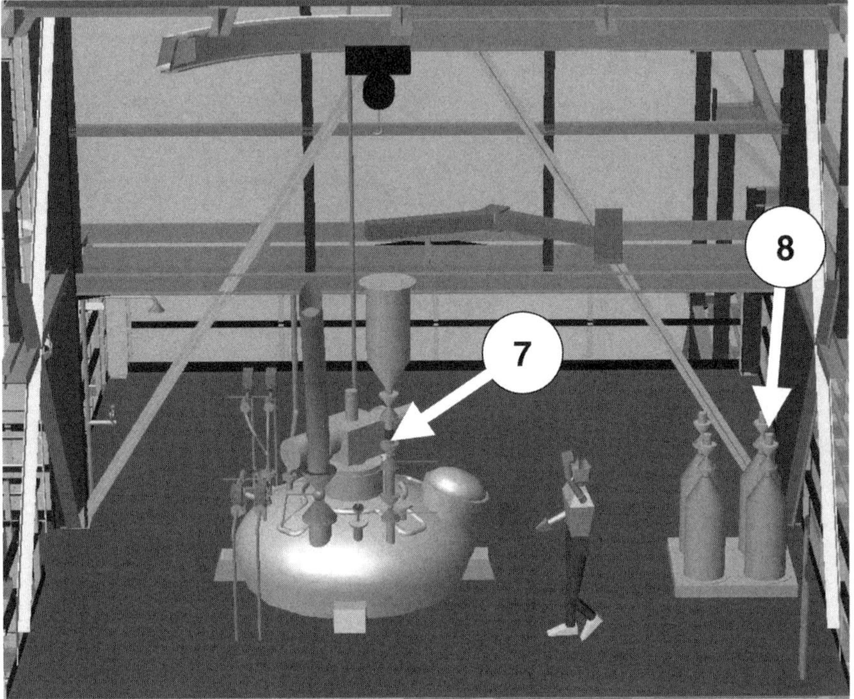

Figure 9-7 Perspective view of chemical mixing deck.

concept is scrapped. A new concept, designed to blow additive from grade to the mixing drum (Figure 9-8), replaces the original concept. The numbers on Figure 9-8 illustrate these considerations:

9. The additive now is delivered in a drum.
10. The additive is blown to a weigh pot.
11. The whole operation is controlled from a panel at grade.

The results of the model review are estimated to have saved the company almost $1 million in terms of redesign, new equipment, and business interruptions—not to mention the safety of the operator.

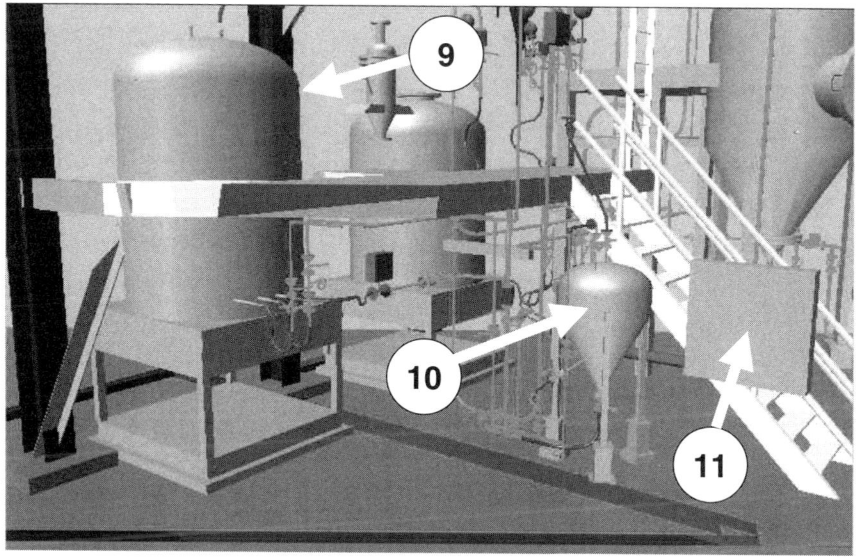

Figure 9-8 Revised design of mixing facility of grade level.

Wrap-up Meeting

A final wrap-up meeting with the contractor should be planned near the conclusion of the human factors QA/QC effort. The purpose of this meeting is to verify the status of the contractor's compliance with corrections identified during previous QA/QC visits. The meeting should be scheduled after each human factors item has been reviewed. The final review is carried out primarily with the project engineer who had been coordinating the safety QA/QC effort. It also includes those contractor representatives involved in areas where significant problems had been identified.

9.3.9 Prestart-up Human Factors Review

The prestart-up human factors review is a plant walkthrough (surveillance) process conducted with a checklist. For capital projects, the prestart-up review is conducted at about 90% of plant construction completion. The same review should also be conducted for all "change" work (base projects and maintenance). The surveillance team should be composed of an HF specialist as well as operations, maintenance, safety, and engineering personnel. Ideally, the review covers all SHE aspects of the project, including human factors. A good practice is to ensure that the team includes members who represent the 5th and 95th percentile statures of the user population. The questions relating to visual performance of the equipment should be conducted at night, to ensure that sufficient lighting has been provided, and during the day, to ensure that the sun's glare is not an issue.

An essential part of the prestart-up review is to check the status of all HF follow-up items from the beginning of the project. The HFTDB is key to this activity. Follow-up activities are developed for those items that fail the review or require further investigation. The HF prestart-up checklist is located in Appendix 9-6.

9.3.10 HF Awareness for Construction Contractors and Company Personnel

During construction, the project should have two major human factors objectives: Build the project without injury to the contract or project work force, and ensure that the facilities are built to be safe and efficient for the owner to operate and maintain. Objective 1 means paying attention to the safety of the workers. Objective 2 means making sure that equipment is installed to minimize rework due to unsafe or inefficient implementation.

For capital projects, brief awareness sessions should be conducted by contractor specialty (e.g., piping, electrical, instrumentation) up through the foreperson level. The sessions should be based on the awareness packages developed for the teams participating in the QA/QC reviews. For base projects, the training could be conducted every 6 months to ensure new field workers and forepersons are trained.

Demonstrations of the proper methods for field-run wiring, routing small bore piping, and the like are conducted using existing plant equipment, sketches, mock-ups, and so forth. Appendix 9-7 provides a summary of recommendations to field workers that could be used in a training session. Each recommendation should be illustrated with a photo to make the training more meaningful and interesting.

HF awareness through training and mock-ups helps raise the awareness of construction personnel, demonstrates how to field-run equipment, and instructs the workers what to do if they encounter a potentially unsafe or inefficient design. The program minimizes the potential for rework to correct human factors issues.

9.3.11 Postproject Review

The first objective of the postproject review is to capture the "learning" gained, positive and negative, that helps improve the project execution process. By so doing, future projects are planned, designed, and executed more efficiently and the quality of the finished product is continually improved.

The format of the postproject review and the level of effort spent on it depend on the size of the project. Clearly, larger, more-complicated projects have more opportunity for mistakes than smaller ones. So, more review effort is typically required for capital than for base projects.

For major projects, the process coordinator is best positioned to "own" the postproject review process. However, the learning gained should be gathered during each project phase and the collection process should be managed by those responsible for the completion of each project phase (Figure 9-1). For base projects, the local project team would gather the items learned.

To achieve the most benefit from the process, learning gained should be captured at the end of each project phase, at the point where the project is handed off to the next person responsible. Memories fade with time and project notes are often lost or misfiled. The most complete information is obtained if gathered as soon as possible after each project phase is completed. Finally, the owners of each project phase should come together within 6 months after the project is complete to share and discuss the learning gained and document such learning in a final project report. This section recommends how to capture the learning gained on the human factors component of the project. The human factors learning process is based on answering the questions in Table 9-7.

The final process in the postproject review is to "benchmark" performance against that achieved by other projects both within the site or

TABLE 9-7

Questions to Capture Human Factors Learning Gained from a Project

Question	*Response or Discussion*
What HF issues were identified during each phase for the project and added to the HF issues tracking list?	
Did we identify the right things?	
Did we identify them at the right time?	
Did we use the proper tools and processes?	
Did we engage the right individuals in the process?	
Did we follow up properly?	
Can we identify the benefits of incorporating a human factors perspective in terms of	
• Potential reduction of injuries or illnesses (to contractor personnel during the construction of the facility? to plant personnel during the operation of the facility?)	
• Potential reduction in the cost to operate and maintain the facility in terms of personnel, equipment, and losses (production or equipment)	
How much did the human factors effort cost in terms of	
• People (time of staff, consultants)	
• Equipment that had to be added solely as a result of human factors improvements	
• Schedule delays solely as a result of human factors changes	
How can we improve the application of human factors approaches during each phase of the project?	

company or outside the company. Benchmarking consists of comparing the answers to the questions posed in Table 9-7 to those obtained from similar projects elsewhere. Most comparisons are based on qualitative information, such as "by developing our operations procedures prior to detailed engineering, we were able to perform a more complete 3D CAD review than in previous projects." However, other answers might be quantitative. For example, from our experience, the cost of the human factors effort in large capital projects is typically in the range of 0.05% of the cost of the project. So, for a $100 million project, the typical cost for plant and contractor effort is $50,000. This cost varies between 0.03 and 0.075%. The percentage cost typically increases as the value of the project decreases. The benchmarking process could compare the HF effort expended in each project against historical data.

For base projects, the benchmarking responsibility rests with the local project owner, who obtains data from the local project team. The responsibility for human factors benchmarking in major projects is typically with the project coordinator (Figure 9-1) with the assistance of the project staff.

REVIEW QUESTIONS

1. Task procedures should be written (choose one)
   a. Prior to the planning stage of the project?
   b. As soon as possible after the QA/QC process has been completed?
   c. Between construction completion and postproject review?
   d. Prior to the start of detailed engineering?

2. Base or local projects differ from capital projects in the following ways (True or False)
   a. Base projects do not require the use of a project model.
   b. Base projects are typically managed by site teams.
   c. HAZOPs are always conducted.

3. Which phrase does not belong in the following list?
   a. Human factors tracking database.
   b. NIOSH lifting assessment.
   c. Hazard and operability review.
   d. Critical-task analysis.

4. Answer yes or no: Human factors training is conducted as part of the QA/QC assessment.

5. Answer true or false: The purpose of the prestart-up human factors review is to capture the learning from the accomplishments and mistakes identified in the project.

6. Answer true or false: The purpose of benchmarking is to ensure that the results of the project are reported to senior management of the company?
7. Fill in the blanks: This chapter is about the use of human factors in project ______________, ______________, and ______________.
8. Answer true or false: Human factors issues are not identified in the project until the detailed engineering phase.

APPENDIX 9-1

HAZOP Human Factors Screening Lists

Valves

Check box if the statement is true.

☐ The valve is used for maintenance or turnaround or is used so infrequently as to have no human error or musculoskeletal significance.

The location and use of the valve does not have to be analyzed. The HF provisions are sufficient.

HF follow-up is required for this valve if any of the following boxes is checked.

☐ Valve is operated manually in an emergency, so the operator must be able to get at it quickly and operate it easily and without mistake.

☐ Valve is operated sequentially and its position and operation must be considered in relation to other valves.

☐ The incorrect operation of the valve exceeds safe operating envelope and no apparent mitigation strategy has been implemented into the design. In this event, a task analysis must be considered in case the valve manipulation is an integral part of a task exceeding 7 steps or requires interaction with other operations.

☐ The valve is operated remotely from a field location, so the design of the control panel is important to consider.

Blinds and Blanks

Check box if the statement is true. HF follow-up is required for this blind if any of the following boxes is checked.

☐ Access to the blind is not defined or is difficult [too high, reach is excessive, obstructed, weight is >23 kg (51 lb)].

☐ The blind has no associated bleed and drain valves or no clearly defined safe drain location.

☐ The blind is an unfamiliar design not used before in the plant (e.g., proprietary blind).

☐ The blind is used in an extreme service (toxic, exposure criteria, pressure >50 barg, temperature >200°C).

Pumps and Compressors

Check box if the statement is true. HF follow-up is required for this pump or compressor if any of the following boxes is checked.

☐ Must be blown to a closed system before maintenance, and product or process, is hazardous.

☐ Local controls and displays are associated with start-up, shutdown, or surveillance.

☐ Start-up, shutdown, or abnormal operations require tasks with more than seven steps and interaction with other operations.

☐ High noise levels are expected.

☐ Auxiliary systems are package units for which design standards may not fully apply.

Field Display Panels

Check box if the statement is true. HF follow-up is required for this panel if any of the following boxes is checked.

☐ Use of the panel requires a procedure with more than seven steps.

☐ Multiple panels that operate similar equipment are used on the unit (design is consistent and layout provided).

☐ Panels are used under all light conditions (lighting sufficient and glare eliminated).

☐ Panels are used under emergency conditions (clarity of layout sufficient).

☐ Panels are used during normal surveillance (located in surveillance path).

Fire Fighting and Deluge Systems

Check box if the statement is true. HF follow-up is required at the model or field review for these systems if any of the following boxes is checked.

☐ Operator of the “fire monitor” does not have a clear line of sight to the target.

☐ Operator’s access to fire monitor or hydrant is obstructed.

☐ Local deluge activation box is not located on an escape route.

☐ Main deluge panels are not located at unit periphery.

☐ Main deluge panels contain controls for three or more geographical coverage areas.

Instrumentation Systems

Check box if the statement is true. HF follow-up is required for these systems if any of the following boxes is checked.

HAZOP Human Factors Screening Lists *Continued*

- ☐ Instrument must be accessed frequently for calibration, clearing, rodding.
- ☐ Instrument is located in an area potentially hazardous to the operator, such as heat, elevation, exposure to process.
- ☐ Clearing or rodding requires more space than provided by design standards.
- ☐ Instrument is itself inherently dangerous (e.g., radioactive).

Sample Points

Check box if the statement is true. HF follow-up is required for this sample point if any of the boxes is checked.

- ☐ The product being sampled is in a hazard class where defined exposure criteria must be observed.
- ☐ The sample point is located above grade.
- ☐ The sample system is not using a closed purge system.
- ☐ Access to the sample point does not consider the operator wearing PPE.

Reactors and Dryers

Check box if the statement is true. HF follow-up is required for this equipment if any of the boxes is checked.

- ☐ Reactor must be entered for cleaning, repair, catalyst filling.
- ☐ Reactor catalyst is toxic, reactive, pyrophoric, and must be handled by personnel.
- ☐ Multiple reactors must be switched frequently.

Vessels: Towers, Drums, and Exchangers

Check box if the statement is true. HF follow-up is required if any of the boxes is checked.

- ☐ Vessel must be entered for cleaning or maintenance.
- ☐ Vessel contents are hazardous, reactive, pyrophoric, or decomposing and entry requires wearing PPE or taking other precautions.

Furnaces and Fired Heaters

Check box if the statement is true. HF follow-up is required if any of the boxes is checked.

- ☐ Furnace is manually lighted.
- ☐ Furnace is automatically lighted from a remote location and no local panel is provided.
- ☐ Inside of the fire box requires frequent visual monitoring.
- ☐ Furnace must be frequently decoked, switched, or visited.

Filters

Check box if the statement is true. HF follow-up is required if any of the following boxes is checked.

☐ Filter must be frequently cleaned, cleared, drained, or switched.

☐ Product being filtered is hazardous or PPE must be worn by operator.

☐ Weight of filter element or filter cover exceeds 23 kg (51 lb).

☐ Height of top of filter is above grade or floor of platform *and* length of filter element exceeds shoulder height of the female user population.

Loading and Unloading Facilities

Check box if the statement is true. HF follow-up is required if any of the following boxes is checked.

☐ The product being loaded or unloaded is in a hazard class where defined exposure criteria must be observed.

☐ The loading or unloading operation requires a task with more than five steps and interaction with others (people, computer) during the operation.

☐ The weight of facilities (arm, hose, platforms, etc.) to be manipulated manually exceeds 23 kg (51 lb).

☐ Top loading or unloading is required.

APPENDIX 9-2

Assistance Using HAZOP Screening Lists (includes examples of poor design and potential solutions)

Screening Question (Sample Points)	*Example Situation*	*Potential Solution*
The product sampled is in a hazard class where defined exposure criteria must be observed	Caustic sample point. Sample valve and outlet are not located close together.	Operator has unobstructed access to safety showers and eyewash stations as per standards
The sample point is located above grade	Sample is hazardous or open and must be carried downstairs.	Sample accessible from a platform If a harmful substance, two escape means provided
The sample system does not use a closed purge system	Sample bomb in closed purge system Open system for caustic samples Strahman valves	If closed purge, the sample loop is long enough to allow operator access to the sample point without obstruction If not closed purge, then operator has unobstructed access to sampling station
Access to the sample point does not consider the operator wearing PPE	Hazardous sample without a closed purge system	Unobstructed access to sample point is assured if operator is wearing PPE. Sample point should be at grade. Consider modifying to a closed loop sample system to eliminate need for PPE.

APPENDIX 9-3

Quality Assurance/Quality Control Checklist

Concerns to Be Appraised	*Project Stage**				*Notes*
	1	*2*	*3*	*4*	
Access and Spacing					
1. Equipment layout					
Normal walkthrough paths	x		x	x	
Emergency operations paths	x		x		
Maintenance access	x		x	x	
Emergency block value (EBV) control stations	x		x		
Utility stations	x		x	x	
Manual sampling points	x		x	x	
Safety showers	x		x	x	
Access for mobile stairs	x		x		
Firefighting facilities	x		x		
2. Structural design					
Interference with access paths at equipment	x		x	x	
Access to battery limit and EBVs	x		x		
Access from platforms to frequently operated equipment (e.g., filters, air-fins, manual valves, instruments, controls)	x				
Access at vessel manways	x		x		
Access to blinds, bleeders	x		x		
Headroom under structures	x		x		
Trolley beams for lifts	x		x		
Connected platforms	x		x		
3. Package unit drawings (e.g., chemical injection, compressor lube units)					
Access to equipment	x	x			
Space around the skid	x		x		
Location of controls	x	x			
4. Fired heater drawings					
Access to burners/pilots	x	x			
5. Compressor/blower drawings					
Access to filters, probes	x	x			
Access to controls	x	x			

Quality Assurance/Quality Control Checklist *Continued*

Concerns to Be Appraised	*Project Stage** *1*	*2*	*3*	*4*	*Notes*
6. Piping design					
Valve reach	x	x	x	x	
Valve spacing	x	x	x	x	
Valve position indication	x			x	
Access to safety valves	x		x		
Access to instruments	x	x	x	x	
Manual sampling stations	x		x	x	
Control interface					
1. Control panel layouts					
Alarm layout/location	x	x			
Controls layout/location	x	x			
Type of emergency shutdown (ESD) switches	x	x			
Display convention	x	x			
Consistency of controls	x	x			
Visibility day/night	x	x			
Labels	x	x			
2. Audible alarm data					
Alarm variability (H_2S, fire)	x				
Alarm locations/audibility	x				
3. Control room (CR) console drawings					
Priority 1 alarm layout/location		x			
Communication module		x			
TV monitors		x			
Storage space for manuals		x			
Electrical interface					
1. Building layout, doors, ramps, breaker lifts	x		x		
2. Transformers	x		x		
3. Switchgear	x	x			
4. Motor control center	x	x			
5. Battery storage	x		x		
6. Labels/tagging	x	x			

Concerns to Be Appraised	*Project Stage** *1*	*2*	*3*	*4*	*Notes*
Work Environment					
1. Lighting layout					
Night illumination	x				
2. Climate control design					
Temperature control	x				
Humidity control	x				
Labels					
1. Label system data					
Equipment/piping	x				
Instruments/controls	x				
Tagging requirements info	x				
Label size/visibility	x				
2. Color coding data					
Emergency valves	x		x		

*Time sequence at stage of execution:

1. Kickoff meeting/training session.
2. Vendor drawings (compressors, furnaces, control panels).
3. Contractors detailed drawings (plot plant, piping layouts).
4. Final Computer assisted drawing (CAD) model review.

APPENDIX 9-4

Maintenance Review Checklist (maintenance accessibility)

Review Item	*Completed*
Check for operator access to valves, review valve heights and orientations	
Check for accessibility and clear view of instrument panels from work locations	
Review location of emergency block valves (EBVs); ensure accessibility and proper location of control panel	
Check that headroom clearance is provided over platforms and walkways (use 95th percentile person for given population)	
Check for adequate tool access to flanges near platforms	
Check for adequate clearance for maintenance around pumps, compressors, and valves	
Check that no grade level tripping hazards exist	
Ensure that clearance is provided to access bleeder valves and drainage hoses and ease of accessibility to operate bleeder valves	
Check that sight glasses are accessible, viewable, and do not obstruct pathways or pose a hazard	
Check that adequate access has been provided for equipment and personnel around pumps and motors	
Check for is proper clearance for the safe operation and use of ladders and moveable platforms	
Check pathways for proper access for mobile equipment	
Check the need for davits, location, and accessibility	
Has an allowance been made for any insulation and thermal growth in this clearance?	
Can the employee maintain both feet on a solid surface when performing the designated task?	
Are the height, angle, and reach to hand wheels or levers of manually operated valves within the anthropometric guidelines?	
Is headroom sufficient to perform inspection and maintenance without crawling?	
Do areas where work is done have a minimum clearance of 4 ft (1.2 m)?	
Is clearance sufficient for removal and replacement of equipment?	
Has space been provided for tools and manipulating the tools (swinging, turning, gripping)?	
Are blinds greater than 70 kg (150 lb) oriented vertically and accessible by lifting devices?	
If personal protective equipment (PPE) must be worn at all times, is clearance sufficient to operate and maintain equipment with PPE?	

APPENDIX 9-5

Walkthrough/Rounds Review Checklist

Review Item	*Checked*
Is body or manual access safe and unrestricted from all required sides?	
Is access sufficient to components, such as valves and bolts?	
Is headroom sufficient for unobstructed access under elevated process equipment and platforms?	
Has an allowance been made for any insulation and thermal growth in this clearance?	
Has clearance been made to protect individuals from dangerous objects (sharp surfaces, sharp or pointed objects, or electrical service)?	
Can the employee maintain both feet on a solid surface when performing the designated task?	
Are the height, angle, and reach to hand wheels or levers of manually operated valves within the anthropometric guidelines?	
Are the height, angle, and reach to remove covers and media of frequently cleaned filters within the anthropometric guidelines?	
Is headroom sufficient that inspection and maintenance can be done without crawling?	
Do areas where work is done have a minimum clearance of 4 ft (1.2 m)?	
Is clearance sufficient for removal and replacement of equipment?	
Is space sufficient for tools and for manipulating the tools (swinging, turning, gripping)?	
Are any blinds greater than 70 kg (150 lb) oriented vertically and accessible by lifting devices?	
Is space sufficient for impaired physical ability (cramped space clearance, arm reach, hand clearance, shape of valve handles)?	
If personal protective equipment (PPE) must be worn at all times, is clearance sufficient to operate and maintain equipment with PPE?	
Are pathways kept clear of obstructions?	
Are pathways free of interference by open doors or covers?	
Have the sampling points been designed for ease of access by employee and sample container?	
Has adjacent equipment been connected by walkways or platforms to minimize repetitive climbing and descending (connecting tanks for sampling and gauging, adjacent towers monitored during walkthroughs)?	
Are the connecting platforms and work landings at a uniform level?	
Are indicating devices for frequently monitored instruments installed high above grade visually accessible from grade?	
Are valves spaced such that there is clear and unobstructed access?	

Walkthrough/Rounds Review Checklist *Continued*

Review Item	*Checked*
Are valve stems and gear operators oriented so that they do not obstruct the access or walkthrough pathway in front of the valve?	
Are adjacent hand wheels and equipment sufficiently spaced to avoid knuckle injury if hands slip [6 in. (15 cm) clearance preferred]?	
Has wrench-assisted operation been considered?	
Are the handles of ball and plug valves oriented so they do not obstruct the access or walkthrough pathway in front of the valve?	
Can operators see the instrument panel from the path during normal walkthrough?	
Can operators monitor valve position from the path during normal walkthrough?	
Is the position and movement of motor-operated values (MOVs) in the operator's line of sight and visible from the MOV control station?	
Are block valves centralized into a minimum number of accessible locations?	
Are sequentially operated valves located close together and access unobstructed?	

APPENDIX 9-6

Prestart-up Human Factors Review Checklist

Item	*Yes (✓)*	*No (✓)*	*N/A (✓)*	*Remarks*
1. General Considerations				
1.1 Have each of the issues from the human factors tracking database been followed up in the design?				
1.2 Have the human factors issues from the HAZOPs been resolved?				
1.3 Have operations procedures been completed?				
1.4 Have operations personnel been trained on the new facility?				
1.5 Have maintenance procedures been completed?				
1.6 Have maintenance personnel been trained on the equipment in the new facility?				
2. Utility Systems				
2.1 Have utility stations been properly identified?				
2.2 Are the hose colors correct for the service?				
2.3 Can you see the colors and the labels at night?				
2.4 Do the hose connections match the connections on the hard pipes?				
2.5 Can the hard connections be accessed easily?				
3. Safety Relief Systems				
3.1 Are the safety valves easy to identify?				
3.2 Can you read the identification at night?				
3.3 Can the safety valves be reached without scaffolding?				
3.4 Are rupture discs correctly tagged and installed facing the proper direction?				
3.5 Are bleeders accessible?				

Prestart-up Human Factors Review Checklist *Continued*

Item	*Yes (✓)*	*No (✓)*	*N/A (✓)*	*Remarks*
4. *Pipes and Valves*				
4.1 Are lines and valves properly identified?				
4.2 Can you see the identification at night?				
4.3 Are valve actuators properly oriented for access?				
4.4 Are valve stems oriented to prevent obstruction?				
4.5 Do valve handles protrude in the walking paths?				
4.6 Are all process-significant valves easily accessible for operation?				
4.7 Can all process-significant valves be easily opened and closed?				
4.8 Are flanges and valves accessible for removal or maintenance?				
4.9 Does insulation interfere with operational or maintenance access to valves or flanges?				
4.10 Are control valves located at grade or accessible from a platform?				
4.11 Are bypass valves accessible?				
4.12 Are bleeders accessible?				
4.13 Are all high-point vents and low-point drains accessible?				
5. *Personnel Protection*				
5.1 Is the lighting adequate for both operation and maintenance?				
5.2 Are escape routes and emergency exits provided from all locations and identified?				
5.3 Can emergency signs be identified at night?				

Item	*Yes (✓)*	*No (✓)*	*N/A (✓)*	*Remarks*
5.4 Are safety showers, eyewash stations, fire stations, and emergency equipment stations properly identified?				
5.5 Can they be seen easily at night?				
5.6 Can safety showers and eyewash stations be accessed in a direct line with no obstructions or turns?				
5.7 Is the water in the shower and eyewash stations temperate in both hot and cold climates?				
6. Fire Protection Facilities				
6.1 Can the location of fire hydrants, extinguisher stations, and fire monitors be easily identified?				
6.2 Can they be easily identified at night?				
6.3 Are hydrants and monitors properly labeled?				
6.4 Are hydrants, monitors, extinguishers, and emergency equipment easily accessible?				
7. Structures				
7.1 Has proper lifting equipment or the provision for outside equipment been installed for all heavy items?				
7.2 Do chain hoists have properly posted load limit signs?				
7.3 Are monorails posted with load limit signs?				
7.4 Are ladders installed with proper toe clearances?				
7.5 Are handrails and ladder rails free of finger catchers, metal burrs, and do they have adequate clearance from nearby pipes and structures?				

Prestart-up Human Factors Review Checklist *Continued*

Item	*Yes (✓)*	*No (✓)*	*N/A (✓)*	*Remarks*
7.6 Are stair heights and treads adequate?				
7.7 Are all stairsteps the same height?				
7.8 Are any overhead obstructions in the surveillance path (especially around stairs and ladders)?				
7.9 Are toe plates installed around platforms?				
7.10 Do toe plates interfere with the operation or maintenance of equipment?				
7.11 Is each structure equipped with the proper number and type of exits?				
8. Buildings				
8.1 Are buildings properly identified?				
8.2 Can the building identifiers be seen at night?				
8.3 Does each building have the proper number and type of exits?				
8.4 Are the proper alarm systems installed and tested?				
8.5 Are the proper emergency instructions posted?				
9. Towers and Drums				
9.1 Are vessels properly identified?				
9.2 Can vessels be identified at night?				
9.3 Can vessels nameplates be seen and read from the surveillance path?				
9.4 Are manways accessible?				
9.5 Are any manways located at the termination of stairs or ladders?				
9.6 Are manway covers equipped with davits or hoisting lugs for removal?				

Item	*Yes (✓)*	*No (✓)*	*N/A (✓)*	*Remarks*
9.7 Are manway covers accessible for removal?				
9.8 Is the power for tools to remove manways accessible nearby?				
9.9 Are rod-out ports accessible?				
9.10 Is it easy to travel between drums and towers related to each other without first traveling to grade?				
9.11 Is access to rod-out ports blocked by the structure?				
10. Exchangers				
10.1 Are all exchangers properly numbered and labeled?				
10.2 Are vessel nameplates visible from the surveillance path?				
10.3 Are vessel nameplates visible at night?				
10.4 Are vessel inspection ports accessible and usable?				
10.5 Are valves accessible from grade or platforms?				
10.6 Can you travel between exchangers without encountering physical obstructions?				
11. Furnaces				
11.1 Are furnaces properly numbered?				
11.2 Can the furnace identifications be seen and read at night?				
11.3 Can the identifications be seen from the normal surveillance path?				
11.4 Are peepholes and access doors accessible by the extremes of the user population?				
11.5 Are dampers and operators accessible?				
11.6 Can critical isolation valves be accessed easily?				
11.7 Are burner controls beneath the furnaces properly labeled and visible (day and night)?				

Prestart-up Human Factors Review Checklist *Continued*

Item	*Yes (✓)*	*No (✓)*	*N/A (✓)*	*Remarks*
11.8 Is adequate headroom available beneath the furnaces?				
11.9 Are draft gauges provided and clearly identified?				
11.10 Can all controls operated together or in close sequence be easily reached from the same location?				
11.11 Are emergency shutdowns properly identified?				
12. Tanks, Spheres, and Off-Site Drums				
12.1 Are vessels properly identified?				
12.2 Are stairways and railings in good condition?				
12.3 Are handrails provided on tank roofs where required?				
12.4 Are vessel nameplates visible and readable from the surveillance path?				
12.5 Have stairs and walkways been provided over large lines?				
12.6 Are the rungs on the ladders on floating roofs properly designed to ensure good footing at all angles?				
12.7 Have proper support facilities (lighting, access, platforms) been provided for gauging?				
12.8 Has access between tanks been provided on the roofs?				
12.9 Can site glasses be read from grade, platform, or ladder?				
13. Machinery				
13.1 Are all pumps, blowers, compressors, fin-fans, and the like properly numbered?				

Item	*Yes (✓)*	*No (✓)*	*N/A (✓)*	*Remarks*
13.2 Can the labels be seen and read at night?				
13.3 Can the labels be read from the surveillance path?				
13.4 Is switchgear associated with machinery properly labeled?				
13.5 Are filters and strainers accessible?				
13.6 Is the hardware required to operate or maintain filters and strainers accessible?				
14. Instruments and Analyzers				
14.1 Are all instrument wires and terminals properly labeled?				
14.2 Can the labels be read day and night?				
14.3 Are control valves properly oriented?				
14.4 Are orifice taps properly oriented?				
14.5 Are all transmitters accessible from grade, platforms, stairs, or ladders?				
14.6 Can transmitters be easily removed or installed from ladders?				
14.7 Is there proper access to instrument leads and transmitters for maintenance?				
14.8 Are pressure, flow, and temperature gauges located where they can be read, day and night?				
14.9 Are all alarms identified by color or location?				
14.10 Has sufficient clearance been provided between instrument cabinets in analyzer buildings?				
14.11 Is proper lighting provided in analyzer buildings?				

Prestart-up Human Factors Review Checklist *Continued*

Item	*Yes (✓)*	*No (✓)*	*N/A (✓)*	*Remarks*
14.12 Are instruments located to be readable?				
14.13 Is the hardware provided to open cabinets accessible by the extremes of the user population?				
15. Electrical Equipment				
15.1 Are all breaker panels, starters, motors, and MOVs properly labeled to identify the equipment served?				
15.2 Are underground cables and duct backs properly labeled to identify the equipment served?				
15.3 Is the lighting of the proper type?				
15.4 Has the lighting been placed in the proper locations for operation and maintenance?				
15.5 Is the emergency lighting properly designed and placed?				
15.6 Are batteries accessible for maintenance?				
15.7 Is building-mounted equipment (e.g., HVAC systems) accessible from grade?				
15.8 Are platforms provided on the building exits and entrances to allow switchgear to be easily installed or removed?				
15.9 Have hoists been provided for moving heavy equipment?				
15.10 Have benches been provided for maintaining electrical equipment?				
15.11 Is the access to electrical equipment safe?				

Item	*Yes (✓)*	*No (✓)*	*N/A (✓)*	*Remarks*
15.12 Are all motor start/stop stations and switches with on/off positions properly labeled?				
16. Field Controls and Displays				
16.1 Do all displays that are monitored during rounds face the surveillance path?				
16.2 Are all displays visible at night?				
16.3 Are all displays visible and readable in full sunlight?				
16.4 Does the display panel face the equipment that it controls?				
16.5 Are panels that control two or more pieces of identical equipment laid out in the same configuration as the equipment they control?				
16.6 Is panel labeling readable?				
16.7 Is the relationship between the label and the control or display ambiguous?				
16.8 For pumps, compressors, and other machinery controlled in the field, can the displays required for start-up or shutdown be seen from the control position?				

APPENDIX 9-7

Summary of Recommendations to Construction Workers Installing Field-Run Equipment

---

Construction Safety

Ensure that you and your fellow workers are working safely at all times

- Ensure that all lifting is conducted safely:
  Never lift more than 50 lb by yourself, if it weighs more than 50 lb, get someone to help or use lifting equipment
  Never twist with a load, move your feet not your back
- Welding can lead to burns to the skin and eyes. Protect yourself and others by always shielding yourself and others from the bright arc light and sparks
- Use the staging equipment and personal protective equipment provided by the company—*do not take shortcuts*
- Tie off your ladders
- Do not climb on pipes and equipment. Use the ladders and stepstools provided
- Ensure that the floor of your platforms have no holes to fall into
- Think about how to assemble equipment. If it is not heavy, it could be assembled at waist height to reduce bending
- Make sure that your vision is unobstructed when using power equipment

Field-Run Facilities Installation Tips

- Conduit, valves, blinds, and piping should not be installed where it interferes with the operation and maintenance of equipment
- Install drain valves and bleeders so they can be operated without pinch points and the outlet can be connected to a temporary hose
- Insulation should not be installed where it will interfere with future maintenance
- In locations where piping and vessels can creep because of extreme changes in temperature, clearance with the structure must be ensured
- Tell your foreperson if you think you are installing equipment that could be dangerous, inefficient, or where the installation might not work
- Route cables and cable trays so they do not interfere with normal surveillance paths or the operation of valves or blinds
- Panels and push buttons should be laid out the same as the equipment they operate
- Keep access ways clear and open. Valves and instruments should not extend into walkways where they can become path obstruction hazards. Cycle the valve actuator to ensure that it does not obstruct personnel in any part of the cycle
- Maintain overhead and side clearances to allow for small crane access to pumps
- Identify opportunities for slips, trips, and falls, so they can be fixed
- Locate and orient field displays to ensure they can be read from the control position
- Install tubing to control valves so it can be removed easily for maintenance
- Install services and utilities so they are *accessible* but not *in the way* when not used

---

REFERENCES

Attwood, D. A., and Fennell, D. J. (2001) "Cost-Effective Human Factors Techniques for Process Safety." CCPS International Conference and Workshop, Toronto, ON, Canada.

Robertson, N. (1999) "Starting Right: Sable Offshore Energy Project's HFE Program." Proceedings, 1999 Offshore Technology Conference, Houston, TX.

Index

F

G

H

I

J

K

L

M

N

O

P

Q

R

S

W